BIRKHÄUSER

# Singularly Perturbed Boundary-Value Problems

Luminiţa Barbu and Gheorghe Moroşanu

Birkhäuser
Basel · Boston · Berlin

Authors:

Luminiţa Barbu
Department of Mathematics and Informatics
Ovidius University
Bd. Mamaia 124
900527 Constanta
Romania
lbarbu@univ-ovidius.ro

Gheorghe Moroşanu
Department of Mathematics and Its Applications
Central European University
Nador u. 9
1051 Budapest
Hungary
Morosanug@ceu.hu

2000 Mathematics Subject Classification: 41A60, 35-XX, 34-XX, 47-XX

Library of Congress Control Number: 2007925493

Bibliographic information published by Die Deutsche Bibliothek
Die Deutsche Bibliothek lists this publication in the Deutsche Nationalbibliografie; detailed bibliographic data is available in the Internet at http://dnb.ddb.de.

ISBN 978-3-7643-8330-5 Birkhäuser Verlag AG, Basel • Boston • Berlin


Basel • Boston • Berlin
P.O. Box 133, CH-4010 Basel, Switzerland
Part of Springer Science+Business Media
Printed on acid-free paper produced from chlorine-free pulp. TCF ∞
Printed in Germany

ISBN 978-3-7643-8330-5
9 8 7 6 5 4 3 2 1

e-ISBN 978-3-7643-8331-2
www.birkhauser.ch

*To George Soros,*
*a generous supporter*
*of Mathematics*

# Contents

# Preface

It is well known that many phenomena in biology, chemistry, engineering, physics can be described by boundary value problems associated with various types of partial differential equations or systems. When we associate a mathematical model with a phenomenon, we generally try to capture what is essential, retaining the important quantities and omitting the negligible ones which involve small parameters. The model that would be obtained by maintaining the small parameters is called the perturbed model, whereas the simplified model (the one that does not include the small parameters) is called unperturbed (or reduced model). Of course, the unperturbed model is to be preferred, because it is simpler. What matters is that it should describe faithfully enough the respective phenomenon, which means that its solution must be "close enough" to the solution of the corresponding perturbed model. This fact holds in the case of regular perturbations (which are defined later). On the other hand, in the case of singular perturbations, things get more complicated. If we refer to an initial-boundary value problem, the solution of the unperturbed problem does not satisfy in general all the original boundary conditions and/or initial conditions (because some of the derivatives may disappear by neglecting the small parameters). Thus, some discrepancy may appear between the solution of the perturbed model and that of the corresponding reduced model. Therefore, to fill in this gap, in the asymptotic expansion of the solution of the perturbed problem with respect to the small parameter (considering, for the sake of simplicity, that we have a single parameter), we must introduce corrections (or boundary layer functions).

More than half a century ago, A.N. Tikhonov [43]–[45] began to systematically study singular perturbations, although there had been some previous attempts in this direction. In 1957, in a fundamental paper [50], M.I. Vishik and L.A. Lyusternik studied linear partial differential equations with singular perturbations, introducing the famous method which is today called the Vishik-Lyusternik method. From that moment on, an entire literature has been devoted to this subject.

> This book offers a detailed asymptotic analysis of some important classes of singularly perturbed boundary value problems which are mathematical models for various phenomena in biology, chemistry, engineering.

We are particularly interested in nonlinear problems, which have hardly been examined so far in the literature dedicated to singular perturbations. This book proposes to fill in this gap, since most applications are described by nonlinear models. Their asymptotic analysis is very interesting, but requires special methods and tools. Our treatment combines some of the most successful results from different parts of mathematics, including functional analysis, singular perturbation theory, partial differential equations, evolution equations. So we are able to offer the reader a complete justification for the replacement of various perturbed models with corresponding reduced models, which are simpler but in general have a different character. From a mathematical point of view, a change of character modifies dramatically the model, so a deep analysis is required.

Although we address specific applications, our methods are applicable to other mathematical models.

We continue with a few words about the structure of the book. The material is divided into four parts. Each part is divided into chapters, which, in turn, are subdivided into sections (see the Contents). The main definitions, theorems, propositions, lemmas, corollaries, remarks are labelled by three digits: the first digit indicates the chapter, the second the corresponding section, and the third the respective item in the chapter.

Now, let us briefly describe the material covered by the book.

The first part, titled *Preliminaries*, has an introductory character. In Chapter 1 we recall the definitions of the regular and singular perturbations and present the Vishik-Lyusternik method. In Chapter 2, some results concerning existence, uniqueness and regularity of the solutions for evolution equations in Hilbert spaces are brought to attention.

In Part II, some nonlinear boundary value problems associated with the telegraph system are investigated. In Chapter 3 (which is the first chapter of Part II) we present the classes of problems we intend to study and indicate the main fields of their applications. In Chapters 4 and 5 we discuss in detail the case of algebraic boundary conditions and that of dynamic boundary conditions, respectively. We determine formally some asymptotic expansions of the solutions of the problems under discussion and find out the corresponding boundary layer functions. Also, we establish results of existence, uniqueness and high regularity for the other terms of our asymptotic expansions. Moreover, we establish estimates for the components of the remainders in the asymptotic expansions previously deducted in a formal way, with respect to the uniform convergence topology, or with respect to some weaker topologies. Thus, the asymptotic expansions are validated.

Part III, titled *Singularly perturbed coupled problems*, is concerned with the coupling of some boundary value problems, considered in two subdomains of a given domain, with transmission conditions at the interface.

In the first chapter of Part III (Chapter 6) we introduce the problems we are going to investigate in the next chapters of this part. They are mathematical models for diffusion-convection-reaction processes in which a small parameter is present. We consider both the stationary case (see Chapter 7) and the evolutionary one (see Chapter 8). We develop an asymptotic analysis which in particular allows us to determine appropriate transmission conditions for the reduced models.

What we do in Part III may also be considered as a first step towards the study of more complex coupled problems in Fluid Mechanics.

While in Parts II and III the possibility to replace singular perturbation problems with the corresponding reduced models is discussed, in Part IV we aim at reversing the process in the sense that we replace given parabolic problems with singularly perturbed, higher order (with respect to $t$) problems, admitting solutions which are more regular and approximate the solutions of the original problems. More precisely, we consider the classical heat equation with homogeneous Dirichlet boundary conditions and initial conditions. We add to the heat equation the term $\pm \varepsilon u_{tt}$, thus obtaining either an elliptic equation or a hyperbolic one. If we associate with each of the resulting equations the original boundary and initial conditions we obtain new problems, which are incomplete, since the new equations are of a higher order with respect to $t$. For each problem we need to add one additional condition to get a complete problem. We prefer to add a condition at $t = T$ for the elliptic equation, either for $u$ or for $u_t$, and an initial condition at $t = 0$ for $u_t$ for the hyperbolic equation. So, depending on the case, we obtain an elliptic or hyperbolic regularization of the original problem. In fact, we have to do with singularly perturbed problems, which can be treated in an abstract setting. In the final chapter of the book (Chapter 11), elliptic and hyperbolic regularizations associated with the nonlinear heat equation are investigated.

Note that, with the exception of Part I, the book includes original material mainly due to the authors, as considerably revised or expanded versions of previous works, including in particular the 2000 authors' Romanian book [6].

The present book is designed for researchers and graduate students and can be used as a two-semester text.

The authors December 2006

# Part I

# Preliminaries

# Chapter 1

# Regular and Singular Perturbations

In this chapter we recall and discuss some general concepts of singular perturbation theory which will be needed later. Our presentation is mainly concerned with singular perturbation problems of the boundary layer type, which are particularly relevant for applications.

In order to start our discussion, we are going to set up an adequate framework. Let $D \subset \mathbb{R}^n$ be a nonempty open bounded set with a smooth boundary $S$. Denote its closure by $\overline{D}$. Consider the following equation, denoted $E_\varepsilon$,

$$L_\varepsilon u = f(x, \varepsilon), \quad x \in D,$$

where $\varepsilon$ is a small parameter, $0 < \varepsilon \ll 1$, $L_\varepsilon$ is a differential operator, and $f$ is a given real-valued smooth function. If we associate with $E_\varepsilon$ some condition(s) for the unknown $u$ on the boundary $S$, we obtain a boundary value problem $P_\varepsilon$. We assume that, for each $\varepsilon$, $P_\varepsilon$ has a unique smooth solution $u = u_\varepsilon(x)$. Our goal is to construct approximations of $u_\varepsilon$ for small values of $\varepsilon$. The usual norm we are going to use for approximations is the sup norm (or max norm), i.e.,

$$\|g\|_{C(\overline{D})} = \sup\{|g(x)|;\ x \in \overline{D}\},$$

for every continuous function $g : \overline{D} \longrightarrow \mathbb{R}$ (in other words, $g \in C(\overline{D})$). We will also use the weaker $L^p$-norm

$$\|g\|_{L^p(D)} = \left( \int_D |g|^p \, dx \right)^{1/p},$$

where $1 \le p < \infty$. For information about $L^p$-spaces, see the next chapter.

In many applications, operator $L_\varepsilon$ is of the form

$$L_\varepsilon = L_0 + \varepsilon L_1,$$

where $L_0$ and $L_1$ are differential operators which do not depend on $\varepsilon$. If $L_0$ does not include some of the highest order derivatives of $L_\varepsilon$, then we should associate with $L_0$ fewer boundary conditions. So, $P_\varepsilon$ becomes

$$L_0 u + \varepsilon L_1 u = f(x, \varepsilon), \quad x \in D,$$

with the corresponding boundary conditions. Let us also consider the equation, denoted $E_0$,

$$L_0 u = f_0, \quad x \in D,$$

where $f_0(x) := f(x, 0)$, with some boundary conditions, which usually come from the original problem $P_\varepsilon$. Let us denote this problem by $P_0$. Some of the original boundary conditions are no longer necessary for $P_0$. Problem $P_\varepsilon$ is said to be a *perturbed problem* (*perturbed model*), while problem $P_0$ is called *unperturbed* (or *reduced model*).

**Definition 1.0.1.** Problem $P_\varepsilon$ is called *regularly perturbed* with respect to some norm $\|\cdot\|$ if there exists a solution $u_0$ of problem $P_0$ such that

$$\|u_\varepsilon - u_0\| \longrightarrow 0 \text{ as } \varepsilon \to 0.$$

Otherwise, $P_\varepsilon$ is said to be *singularly perturbed* with respect to the same norm.

In a more general setting, we may consider time $t$ as an additional independent variable for problem $P_\varepsilon$ as well as initial conditions at $t = 0$ (sometimes $t$ is the only independent variable). Moreover, we may consider systems of differential equations instead of a single equation. Note also that the small parameter may also occur in the conditions associated with the corresponding system of differential equations. For example, we will discuss later some coupled problems in which the small parameter is also present in transmission conditions. Basically, the definition above also applies to these more general cases.

In order to illustrate this definition we are going to consider some examples. Note that the problem of determining $P_0$ will be clarified later. Here, we use just heuristic arguments.

*Example* 1. Consider the following simple Cauchy problem $P_\varepsilon$ :

$$\frac{du}{dt} + \varepsilon u = f_0(t), \quad 0 < t < T; \quad u(0) = \theta,$$

where $T \in (0, +\infty)$, $\theta \in \mathbb{R}$, and $f_0 : \mathbb{R} \longrightarrow \mathbb{R}$ is a given smooth function. The solution of $P_\varepsilon$ is given by

$$u_\varepsilon(t) = e^{-\varepsilon t}\Big(\theta + \int_0^t e^{\varepsilon s} f_0(s)\, ds\Big), \quad 0 \le t \le T.$$

Obviously, $u_\varepsilon$ converges uniformly on $[0, T]$, as $\varepsilon$ tends to 0, to the function

$$u_0(t) = \theta + \int_0^t f_0(s)\, ds,$$

which is the solution of the reduced problem

$$\frac{du}{dt} = f_0(t), \quad 0 < x < T; \quad u(0) = \theta.$$

Therefore, $P_\varepsilon$ is regularly perturbed with respect to the sup norm.

*Example* 2. Let $P_\varepsilon$ be the boundary value problem

$$\varepsilon \frac{d^2u}{dx^2} + \frac{du}{dx} = 2x, \quad 0 < x < 1; \quad u(0) = 0 = u(1).$$

Its solution is

$$u_\varepsilon(x) = x(x - 2\varepsilon) + \frac{2\varepsilon - 1}{1 - e^{-1/\varepsilon}}\left(1 - e^{-x/\varepsilon}\right).$$

Note that

$$u_\varepsilon(x) = (x^2 - 1) + e^{-x/\varepsilon} + r_\varepsilon(x),$$

where $r_\varepsilon(x)$ converges uniformly to the null function, as $\varepsilon$ tends to 0. Therefore, $u_\varepsilon$ converges uniformly to the function $u_0(x) = x^2 - 1$ on every interval $[\delta, 1]$, $0 < \delta < 1$, but not on the whole interval $[0, 1]$. Obviously, $u_0(x) = x^2 - 1$ satisfies the reduced problem

$$\frac{du}{dx} = 2x, \quad 0 < x < 1; \quad u(1) = 0,$$

but $\|u_\varepsilon - u_0\|_{C[0,1]}$ does not approach 0. Therefore, $P_\varepsilon$ is singularly perturbed with respect to the sup norm. For a small $\delta$, $u_0$ is an approximation of $u_\varepsilon$ in $[\delta, 1]$, but it fails to be an approximation of $u_\varepsilon$ in $[0, \delta]$. This small interval $[0, \delta]$ is called a *boundary layer*. Here we notice a fast change of $u_\varepsilon$ from its value $u_\varepsilon(0) = 0$ to values close to $u_0$. This behavior of $u_\varepsilon$ is called a boundary layer phenomenon and $P_\varepsilon$ is said to be a *singular perturbation problem of the boundary layer type*. In this simple example, we can see that a uniform approximation for $u_\varepsilon(x)$ is given by $u_0(x) + e^{-x/\varepsilon}$. The function $e^{-x/\varepsilon}$ is a so-called *boundary layer function* (*correction*). It fills the gap between $u_\varepsilon$ and $u_0$ in the boundary layer $[0, \delta]$.

Let us remark that $P_\varepsilon$ is a regular perturbation problem with respect to the $L^p$-norm for all $1 \leq p < \infty$, since $\|u_\varepsilon - u_0\|_{L^p(0,1)}$ tends to zero. The boundary layer which we have just identified *is not visible in this weaker norm*.

*Example* 3. Let $P_\varepsilon$ be the following Cauchy problem

$$\varepsilon \frac{du}{dt} + ru = f_0(t), \quad 0 < t < T; \quad u(0) = \theta,$$

where $r$ is a positive constant, $\theta \in \mathbb{R}$ and $f_0 : [0,T] \to \mathbb{R}$ is a given Lipschitzian function. The solution of this problem is given by

$$u_\varepsilon(t) = \theta e^{-rt/\varepsilon} + \frac{1}{\varepsilon}\int_0^t f_0(s)e^{-r(t-s)/\varepsilon}ds, \ 0 \le t \le T,$$

which can be written as

$$u_\varepsilon(t) = \frac{1}{r}f_0(t) + \left(\theta - \frac{1}{r}f_0(0)\right) e^{-rt/\varepsilon} + r_\varepsilon(t), \ 0 \le t \le T,$$

where

$$r_\varepsilon(t) = -\frac{1}{r}\int_0^t f_0'(s)e^{-r(t-s)/\varepsilon}ds.$$

We have

$$\mid r_\varepsilon(t) \mid \le \frac{L}{r}\int_0^t e^{-r(t-s)/\varepsilon}ds \le \frac{L}{r^2}\varepsilon,$$

where $L$ is the Lipschitz constant of $f_0$. Therefore, $r_\varepsilon$ converges uniformly to zero on $[0,T]$ as $\varepsilon$ tends to 0. Thus $u_\varepsilon$ converges uniformly to $u_0(t) = (1/r)f_0(t)$ on every interval $[\delta, T]$, $0 < \delta < T$, but not on the whole interval $[0,T]$ if $f_0(0) \neq r\theta$. Note also that $u_0$ is the solution of the (algebraic) equation

$$ru = f_0(t), \quad 0 < t < T,$$

which represents our reduced problem. Therefore, if $f_0(0) \neq r\theta$, this $P_\varepsilon$ is a singular perturbation problem of the boundary layer type with respect to the sup norm. The boundary layer is a small right vicinity of the point $t = 0$. A uniform approximation of $u_\varepsilon(t)$ on $[0,T]$ is the sum $u_0(t) + \left(\theta - \frac{1}{r}f_0(0)\right) e^{-rt/\varepsilon}$. The function $\left(\theta - \frac{1}{r}f_0(0)\right) e^{-rt/\varepsilon}$ is a boundary layer function, which corrects the discrepancy between $u_\varepsilon$ and $u_0$ within the boundary layer.

*Example* 4. Let $P_\varepsilon$ be the following initial-boundary value problem

$$\varepsilon u_t - u_{xx} = t\sin x, \quad 0 < x < \pi, \quad 0 < t < T,$$

$$u(x,0) = \sin x, \quad x \in [0,\pi]; \quad u(0,t) = 0 = u(\pi,t), \quad t \in [0,T],$$

where $T$ is a given positive number. The solution of this problem is

$$u_\varepsilon(x,t) = t\sin x + e^{-t/\varepsilon}\sin x + \varepsilon\big(e^{-t/\varepsilon} - 1\big)\sin x,$$

which converges uniformly, as $\varepsilon$ tends to zero, to the function $u_0(x,t) = t\sin x$, on every rectangle $R_\delta = \{(x,t) : \ 0 \le x \le \pi, \ \delta \le t \le T\}, \ \ 0 < \delta < T$. Note that $u_0$ is the solution of the reduced problem $P_0$,

$$-u_{xx} = t\sin x, \quad u(0,t) = 0 = u(\pi,t).$$

However, $u_0$ fails to be a uniform approximation of $u_\varepsilon$ in the strip $B_\delta = \{(x,t) : 0 \le x \le \pi, \quad 0 \le t \le \delta\}$. Therefore, $P_\varepsilon$ is a singular perturbation problem of the boundary layer type with respect to the sup norm on the rectangle $[0,\pi] \times [0,T]$. The boundary layer is a thin strip $B_\delta$, where $\delta$ is a small positive number. Obviously, a boundary layer function (correction) is given by

$$c(x, t/\varepsilon) = e^{-t/\varepsilon} \sin x,$$

which fills the gap between $u_\varepsilon$ and $u_0$. Indeed, $u_0(x,t) + c(x,t/\varepsilon)$ is a uniform approximation of $u_\varepsilon$.

It is interesting to note that $P_\varepsilon$ is regularly perturbed with respect to the usual norm of the space $C\big([0,\pi];\, L^p(0,T)\big)$ for all $1 \le p < \infty$. The boundary layer phenomenon is not visible in this space, but it is visible in $C([0,\pi] \times [0,T])$, as noticed above. In fact, we can see that $P_\varepsilon$ is singularly perturbed with respect to the weaker norm $\| \cdot \|_{L^1(0,\pi;\, C[0,T])}$.

*Example* 5. Let $D \subset \mathbb{R}^2$ be a bounded domain with smooth boundary $\partial D$. Let $P_\varepsilon$ be the following typical Dirichlet boundary value problem (see, e.g., [48], p. 83):

$$\begin{cases} -\varepsilon \Delta u + u = f(x,y,\varepsilon) \quad \text{in } D, \\ u = 0 \quad \text{on } \partial D, \end{cases}$$

where $\Delta$ is the Laplace operator, i.e., $\Delta u := u_{xx} + u_{yy}$ and $f$ is a given smooth function defined on $\overline{D} \times [0,\varepsilon_0]$, for some $\varepsilon_0 > 0$, such that $f(x,y,0) \ne 0$ for all $(x,y) \in \partial D$. It is well known that problem $P_\varepsilon$ has a unique classical solution $u_\varepsilon(x,y)$. Obviously, $P_0$ is an algebraic equation, for which the boundary condition is no longer necessary. Its solution is

$$u_0(x,y) = f(x,y,0), \quad (x,y) \in \overline{D}.$$

Clearly, in a neighborhood of $\partial D$, $u_\varepsilon$ and $u_0$ are not close enough with respect to the sup norm, since $u_{\varepsilon|\partial D} = 0$, whereas $u_0$ does not satisfy this condition. Therefore, $\|u_\varepsilon - u_0\|_{C(\overline{D})}$ does not converge to 0, as $\varepsilon \to 0$. According to our definition, problem $P_\varepsilon$ is singularly perturbed with respect to $\| \cdot \|_{C(\overline{D})}$. Moreover, this problem is of the boundary layer type. In this example, the boundary layer is a vicinity of the whole boundary $\partial D$. The existence of the boundary layer phenomenon is not as obvious as in the previous examples, since there is no explicit form of $u_\varepsilon$. Following, e.g., [48] we will perform a complete analysis of this issue below. On the other hand, it is worth mentioning that this $P_\varepsilon$ is regularly perturbed with respect to $\| \cdot \|_{L^p(D)}$ for all $1 \le p < \infty$, as explained later.

*Example* 6. In $D_T = \{(x,t);\ 0 < x < 1,\, 0 < t < T\}$ we consider the telegraph system

$$\begin{cases} \varepsilon u_t + v_x + ru = f_1(x,t), \\ v_t + u_x + gv = f_2(x,t), \end{cases} \qquad (S)_\varepsilon$$

with initial conditions

$$u(x,0) = u_0(x), \quad v(x,0) = v_0(x), \quad 0 < x < 1, \tag{$(IC)_\varepsilon$}$$

and boundary conditions of the form

$$\begin{cases} r_0 u(0,t) + v(0,t) = 0, \\ -u(1,t) + f_0(v(1,t)) = 0, \quad 0 < t < T, \end{cases} \tag{$(BC)_\varepsilon$}$$

where $f_1, f_2 : \overline{D}_T \to \mathbb{R}$, $f_0 : \mathbb{R} \to \mathbb{R}$, $u_0, v_0 : [0,1] \to \mathbb{R}$ are given smooth functions, and $r_0, r, g$ are constants, $r_0 > 0,\ r > 0,\ g \geq 0$. If in the model formulated above and denoted by $P_\varepsilon$ we take $\varepsilon = 0$, we obtain the following reduced problem $P_0$:

$$\begin{cases} u = r^{-1}(f_1 - v_x), \\ v_t - r^{-1}v_{xx} + gv = f_2 - r^{-1}f_{1x} \text{ in } D_T, \end{cases} \tag{$(S)_0$}$$

$$v(x,0) = v_0(x), \quad 0 < x < 1, \tag{$(IC)_0$}$$

$$\begin{cases} rv(0,t) - r_0 v_x(0,t) + r_0 f_1(0,t) = 0, \\ rf_0(v(1,t)) + v_x(1,t) - f_1(1,t) = 0, \quad 0 < t < T. \end{cases} \tag{$(BC)_0$}$$

In this case, the reduced system $(S)_0$ consists of an algebraic equation and a differential equation of the parabolic type, whereas system $(S)_\varepsilon$ is of the hyperbolic type. The initial condition for $u$ is no longer necessary. We will derive $P_0$ later in a justified manner.

Let us remark that if the solution of $P_\varepsilon$, say $U_\varepsilon(x,t) = (u_\varepsilon(x,t), v_\varepsilon(x,t))$, would converge uniformly in $\overline{D}_T$ to the solution of $P_0$, then necessarily

$$v_0'(x) + ru_0(x) = f_1(x,0),\ \forall x \in [0,1].$$

If this condition is not satisfied then that uniform convergence is not true and, as we will show later, $U_\varepsilon$ has a boundary layer behavior in a neighborhood of the segment $\{(x,0);\ 0 \leq x \leq 1\}$. Therefore, this $P_\varepsilon$ is a singular perturbation problem of the boundary layer type with respect to the sup norm $\|\cdot\|_{C(\overline{D}_T)^2}$. However, using the form of the boundary layer functions which we are going to determine later, we will see that the boundary layer is not visible in weaker norms, like for instance $\|\cdot\|_{C([0,1];\, L^p(0,T))^2}$, $1 \leq p < \infty$, and $P_\varepsilon$ is regularly perturbed in such norms.

*Example* 7. Let $P_\varepsilon$ be the following simple initial value problem

$$\begin{cases} \varepsilon \frac{du_1}{dx} - u_2 = \varepsilon f_1(x), \\ \varepsilon \frac{du_2}{dx} + u_1 = \varepsilon f_2(x), \quad 0 < x < 1, \\ u_1(0) = 1, \quad u_2(0) = 0, \end{cases}$$

where $f_1,\ f_2 \in C[0,1]$ are given functions. It is easily seen that this $P_\varepsilon$ is singularly perturbed with respect to the sup norm, but not of the boundary layer type. This conclusion is trivial in the case $f_1 = 0,\ f_2 = 0$, when the solution of $P_\varepsilon$ is

$$u_\varepsilon = \big(\cos(x/\varepsilon), -\sin(x/\varepsilon)\big).$$

**Definition 1.0.2.** Let $u_\varepsilon$ be the solution of some perturbed problem $P_\varepsilon$ defined in a domain $D$. Consider a function $U(x,\varepsilon),\ x \in D_1$, where $D_1$ is a subdomain of $D$. The function $U(x,\varepsilon)$ is called an asymptotic approximation in $D_1$ of the solution $u_\varepsilon(x)$ with respect to the sup norm if

$$\sup_{x\in D_1} \|u_\varepsilon(x) - U(x,\varepsilon)\| \to 0 \ \text{ as } \ \varepsilon \to 0.$$

Moreover, if

$$\sup_{x\in D_1} \|u_\varepsilon(x) - U(x,\varepsilon)\| = \mathcal{O}(\varepsilon^k),$$

then we say that $U(x,\varepsilon)$ is an asymptotic approximation of $u_\varepsilon(x)$ in $D_1$ with an accuracy of the order $\varepsilon^k$. We have similar definitions with respect to other norms. In the above definition we have assumed that $U$ and $u_\varepsilon$ take values in $\mathbb{R}^n$, and $\|\cdot\|$ denotes one of the norms of this space.

For a real-valued function $E(\varepsilon)$, the notation $E(\varepsilon) = \mathcal{O}(\varepsilon^k)$ means that $|E(\varepsilon)| \le M\varepsilon^k$ for some positive constant $M$ and for all $\varepsilon$ small enough.

In Example 4 above $u_0$ is an asymptotic approximation of $u_\varepsilon$ with respect to the sup norm in the rectangle $R_\delta$, with an accuracy of the order $\varepsilon$. Function $u_0$ is not an asymptotic approximation of $u_\varepsilon$ in $[0,\pi]\times[0,T]$ with respect to the sup norm, but it has this property with respect to the norm of $C\big([0,\pi];\ L^p(0,T)\big)$, with an accuracy of the order $\varepsilon^{1/p}$, for all $1 \le p < \infty$. Note also that the function $t\sin x + e^{-t/\varepsilon}\sin x$ is an asymptotic approximation in $[0,\pi]\times[0,T]$ of $u_\varepsilon$ with respect to the sup norm, with an accuracy of the order $\varepsilon$.

In the following we are going to discuss the celebrated **Vishik-Lyusternik method** [50] for the construction of asymptotic approximations for the solutions of singular perturbation problems of the boundary layer type. To explain this method we consider the problem used in Example 5 above, where $\varepsilon$ will be replaced by $\varepsilon^2$ for our convenience, i.e.,

$$\begin{cases} -\varepsilon^2\Delta u + u = f(x,y,\varepsilon) \quad \text{in } \ D, \\ u = 0 \quad \text{on } \ \partial D. \end{cases}$$

We will seek the solution of $P_\varepsilon$ in the form

$$u_\varepsilon = u + c, \tag{1.1}$$

where $u$ and $c$ are two series: $u = \sum_{j=0}^{\infty}\varepsilon^j u_j(x,y)$ is the so-called *regular series* and does not in general satisfy the boundary condition; the discrepancy in the boundary condition is removed by the so-called *boundary layer series* c, which will be introduced in the following. Let the equations of the boundary $\partial D$ have the following parametric form:

$$x = \varphi(p), \quad y = \psi(p), \quad 0 \le p \le p_0.$$

More precisely, when $p$ increases from 0 to $p_0$, the point $(\varphi(p), \psi(p))$ moves on $\partial D$ in such a way that $D$ remains to the left. Consider an internal $\delta$-vicinity of $\partial D$, $\delta > 0$ small, which turns out to be our boundary layer. Any point $(x, y)$ of the boundary layer is uniquely determined by a pair $(\rho, p) \in [0, \delta] \times [0, p_0]$. Indeed, let $p \in [0, p_0]$ be the value of the parameter for which the normal at $(\varphi(p), \psi(p))$ to $\partial D$ contains the point $(x, y)$. Then $\rho$ is defined as the distance from $(x, y)$ to $(\varphi(p), \psi(p))$. It is obvious that $(x, y)$ and $(\rho, p)$ are connected by the following equations

$$x = \varphi(p) - \rho\psi'(p)/\left(\varphi'(p)^2 + \psi'(p)^2\right)^{1/2},$$

$$y = \psi(p) + \rho\varphi'(p)/\left(\varphi'(p)^2 + \psi'(p)^2\right)^{1/2}.$$

We have the following expression for the operator $L_\varepsilon u = -\varepsilon^2 \Delta u + u$ with respect to the new coordinates $(\rho, p)$

$$L_\varepsilon u = -\varepsilon^2 \Big( u_{\rho\rho} + (p_x^2 + p_y^2) u_{pp} + (\rho_{xx} + \rho_{yy}) u_\rho + (p_{xx} + p_{yy}) u_p \Big) + u.$$

We stretch the variable $\rho$ by the transformation $\tau = \rho/\varepsilon$. The new variable $\tau$, called *fast*indexfast variable or *rapid variable*, helps us to describe the behavior of the solution $u_\varepsilon$ inside the boundary layer. The construction of the fast variable depends on the problem $P_\varepsilon$ under investigation (see, e.g., [18] and [29]). It turns out that for the present problem $\tau = \rho/\varepsilon$ is the right fast variable. If we expand the coefficients of $L_\varepsilon$ in power series in $\varepsilon$, we get the following expression for $L_\varepsilon$ with respect to $(\tau, p)$

$$L_\varepsilon u = \big( -u_{\tau\tau} + u \big) + \sum_{j=1}^{\infty} \varepsilon^j L_j u,$$

where $L_j$ are differential operators containing the partial derivatives $u_\tau, u_p$ and $u_{pp}$. We will seek the solution of problem $P_\varepsilon$ in the form of the following expansion, which is called *asymptotic expansion*,

$$u_\varepsilon(x, y) = u + c = \sum_{j=0}^{\infty} \varepsilon^j \Big( u_j(x, y) + c_j(\tau, p) \Big). \tag{1.2}$$

Now, expanding $f(x, y, \varepsilon)$ into a power series in $\varepsilon$ and substituting (1.2) in $P_\varepsilon$, we get

$$\sum_{j=0}^{\infty}\varepsilon^j\Big(-\varepsilon^2\Delta u_j(x,y)+u_j(x,y)\Big)+\sum_{j=0}^{\infty}\varepsilon^j\Big(-c_{j\tau\tau}(\tau,p)+c_j(\tau,p)\Big)$$
$$+\sum_{j=0}^{\infty}\varepsilon^j\Big(\sum_{i=1}^{\infty}\varepsilon^i L_i c_j(\tau,p)\Big)=\sum_{j=0}^{\infty}\varepsilon^j f_j(x,y), \tag{1.3}$$

$$\sum_{j=0}^{\infty}\varepsilon^j\Big(u_j(\varphi(p),\psi(p))+c_j(0,p)\Big)=0. \tag{1.4}$$

We are going to equate coefficients of the like powers of $\varepsilon$ in the above equations, separately for terms depending on $(x,y)$ and $(\tau,p)$. This distinction can be explained as follows: the boundary layer part is sizeable within the boundary layer and negligible outside this layer, so in the interior of the domain we have to take into account only regular terms, thus deriving the equations satisfied by $u_j(x,y)$; then we continue with boundary layer terms. For our present example we obtain

$$u_j(x,y)=f_j(x,y),\quad j=0,\,1,$$
$$u_j(x,y)=f_j(x,y)+\Delta u_{j-2}(x,y),\quad j=2,3,\ldots.$$

For the boundary layer functions we obtain the following ordinary differential equations in $\tau$

$$c_{j\tau\tau}(\tau,p)-c_j(\tau,p)=g_j(\tau,p),\quad \tau\geq 0, \tag{1.5}$$

where $g_0(\tau,p)=0,\quad g_j(\tau,p)=\sum_{i=1}^{j}L_ic_{j-i}(\tau,p)$ for all $j=1,2,\ldots,$ together with the conditions

$$c_j(0,p)=-u_j(\varphi(p),\psi(p)). \tag{1.6}$$

In addition, having in mind that the boundary layer functions should be negligible outside the boundary layer, we require that

$$\lim_{\tau\to\infty}c_j(\tau,p)=0,\quad \forall p\in[0,p_0]. \tag{1.7}$$

We can solve successively the above problems and find

$$c_0(\tau,p)=-u_0(\varphi(p),\psi(p))e^{-\tau},$$

while the other $c_j$'s are products of some polynomials (in $\tau$) and $e^{-\tau}$. Therefore,

$$|c_j(\tau,p)|\leq K_je^{-\tau/2},\quad j=0,1,\ldots,$$

where $K_j$ are some positive constants. In fact, these corrections $c_j$ should act only inside the boundary layer, i.e., for $0\leq\tau\leq\delta/\varepsilon$. Let $\alpha(\rho)$ be an infinitely differentiable function, which equals 0 for $\rho\geq 2\delta/3$, equals 1 for $\rho\leq\delta/3$, and $0\leq\alpha(\rho)\leq 1$ for $\delta/3<\rho<2\delta/3$. So, we can consider the functions $\alpha(\varepsilon\tau)c_j(\tau,p)$ as our new boundary layer functions, which are defined in the whole $\overline{D}$ and still satisfy the estimates above. This smooth continuation procedure will be used whenever we need it, without any special mention.

So, we have constructed an asymptotic expansion for $u_\varepsilon$. It is easily seen (see also [48], p. 86) that the partial sum

$$U_n(x,y,\varepsilon) = \sum_{j=0}^{n} \varepsilon^j \Big( u_j(x,y) + c_j(\tau,p) \Big)$$

is an asymptotic approximation in $D$ of $u_\varepsilon$ with respect to the sup norm, with an accuracy of the order of $\varepsilon^{n+1}$. Indeed, for a given $n$, $w_\epsilon = u_\varepsilon - U_n(\cdot,\cdot,\varepsilon)$ satisfies an equation of the form

$$-\varepsilon^2 \Delta w_\varepsilon(x,y) + w_\varepsilon(x,y) = h_\varepsilon(x,y),$$

with a homogeneous Dirichlet boundary condition, where $h_\varepsilon = \mathcal{O}(\varepsilon^{n+1})$. Now, the assertion follows from the fact that $\Delta w_\varepsilon \leq 0$ ($\geq 0$) at any maximum (respectively, minimum) point of $w_\varepsilon$.

On the other hand, since

$$\| c_0 \|_{L^p(D)} = \mathcal{O}(\varepsilon^{1/p}) \ \forall \ 1 \leq p < \infty,$$

we infer that $u_0$ is an asymptotic approximation in $D$ of $u_\varepsilon$ with respect to the norm $\| \cdot \|_{L^p(D)}$, with an accuracy of the order of $\varepsilon^{1/p}$, $\forall \ 1 \leq p < \infty$. In fact, $c_0$ is not important if we use this weaker norm.

We may ask ourselves what would happen if the data of a given $P_\varepsilon$ were not very regular. For example, let us consider the same Dirichlet $P_\varepsilon$ problem above, in a domain $D$ with a smooth boundary $\partial D$, but in which $f = f(x,y,\varepsilon)$ is no longer a series expansion with respect to $\varepsilon$. To be more specific, we consider the case in which $f$ admits a finite expansion of the form

$$f(x,y,\varepsilon) = \sum_{j=0}^{n} \varepsilon^j f_j(x,y) + \varepsilon^{n+1} g_\varepsilon(x,y),$$

for some given $n \in \mathbb{N}$, where $f_j$, $g_\varepsilon(\cdot,\cdot)$ are smooth functions defined on $\overline{D}$, and $\|g_\varepsilon(\cdot,\cdot)\|_{C(\overline{D})} \leq M$, for some constant $M$. In this case, we seek the solution of $P_\varepsilon$ in the form

$$u_\varepsilon(x,y) = \sum_{j=0}^{n} \varepsilon^j \Big( u_j(x,y) + c_j(\tau,p) \Big) + r_\varepsilon(x,y),$$

where $u_j$ and $c_j$ are defined as before, and $r_\varepsilon$ is given by

$$r_\varepsilon(x,y) = u_\varepsilon(x,y) - \sum_{j=0}^{n} \varepsilon^j \Big( u_j(x,y) + c_j(\tau,p) \Big),$$

and is called *remainder of the order* $n$. Using exactly the same argument as before, one can prove that

$$\|r_\varepsilon\|_{C(\overline{D})} = \mathcal{O}(\varepsilon^{n+1}).$$

Of course, this estimate is based on our assumptions on $f$. In general, an $n$th order remainder $r_\varepsilon$ will be a real remainder only if the norm of $\varepsilon^{-n} r_\varepsilon$ tends to zero, as $\varepsilon \to 0$. Note also that some of the corrections may not appear in the asymptotic expansion of $u_\varepsilon$. For example, if $f(x,y,0) = 0$ on the boundary, then $c_0(\tau,p)$ is the null function. We say that there is no boundary layer of the order zero. If, in addition, $c_1(\tau,p)$ is not identically zero (i.e., $f_1$ is not identically zero on $\partial D$), then we say that $P_\varepsilon$ is singularly perturbed of the order one (and there exists a first order boundary layer near $\partial D$).

The whole treatment above is essentially based on the assumption that the boundary $\partial D$ of the domain $D$ is smooth enough, so that the normal to $\partial D$ exists at any point of $\partial D$, allowing us to introduce the local variables $(\tau,p)$, etc. For a problem $P_\varepsilon$ in a domain $D$ whose boundary is no longer smooth, things get more complicated. For example, let us consider the same Dirichlet boundary value problem in a rectangle $D$. Again, this problem is in general singularly perturbed with respect to the sup norm. One can introduce new local coordinates and boundary layer functions for each of the four sides of $D$. But these boundary layer functions create new discrepancies at the four vertices of $D$. So one needs to introduce in the vicinities of these corner points new corrections (which are called corner boundary functions) which compensate for these new discrepancies. In our applications we will meet such domains, but the boundary layer will appear only at a single smooth part of the boundary (for example, at one of the four sides in the case of a rectangle $D$). In this case, instead of using corner boundary functions, one can assume additional conditions on the data to remove possible discrepancies at the corresponding corner points. Let us illustrate this on the problem $P_\varepsilon$ considered in Example 6 above. This problem admits a boundary layer near the side $\{(x,0);\ 0 \le x \le 1\}$ of the rectangle $D_T$ with respect to the sup norm. The existence of this boundary layer is suggested by the analogous ODE problem considered in Example 3 above. Indeed, $\varepsilon$ is present only in the first equation of system $(S)_\varepsilon$. For a given $x$ this has the same form as the equation discussed in Example 3, with $f_0(t) := f_1(x,t) - v_x(x,t)$. So we expect to have a singular behavior of the solution near the value $t = 0$ for all $x$. We will see that this is indeed the case. We will restrict ourselves to seeking an asymptotic expansion of the order zero for the solution of $P_\varepsilon$, i.e.,

$$\begin{aligned} U_\varepsilon(x,t) &= \big(u_\varepsilon(x,t), v_\varepsilon(x,t)\big) \\ &= \big(X(x,t), Y(x,t)\big) + \big(c_0(x,\tau), d_0(x,\tau)\big) + \big(R_{1\varepsilon}(x,t), R_{2\varepsilon}(x,t)\big), \end{aligned}$$

where $\big(X(x,t), Y(x,t)\big)$ is the regular term, $\tau = t/\varepsilon$ is the rapid variable for this problem, $\big(c_0(x,\tau), d_0(x,\tau)\big)$ is the correction (of order zero), and $R_\varepsilon := \big(R_{1\varepsilon}, R_{2\varepsilon}\big)$ is the remainder (of order zero). The form of the rapid variable $\tau$ is also suggested by the analogous problem in Example 3 above. In fact, according to this analogy, we expect a singular behavior near $t = 0$ only for $u_\varepsilon$, i.e., $d_0 = 0$. This will be indeed the case, but we have started with the above expansion, since our present problem is more complex and $u_\varepsilon$, $v_\varepsilon$ are connected to each other. We substitute the above expansion into $P_\varepsilon$ and identify the coefficients of the like powers of $\varepsilon$. Of course, we distinguish between the coefficients depending on $(x,t)$ and those depending on $(x,\tau)$. We should keep in mind that the remainder components are small as compared to the other terms. This idea comes from the case when we have a series expansion of the solution of $P_\varepsilon$. So, after substituting the above expansion into $(S)_\varepsilon$, we see that the only coefficient of $\varepsilon^{-1}$ in the second equation is $d_{0\tau}(x,\tau) = (\partial d_0/\partial\tau)(x,\tau)$. So, this should be zero, thus $d_0$ is a function depending on $x$ only. Taking also into account the fact that a boundary layer function should converge to zero as $\tau \to \infty$, we infer that $d_0$ is identically zero, as expected. From the first equation of $(S)_\varepsilon$, we derive the following boundary layer equation by identifying the coefficients of $\varepsilon^0$:

$$c_{0\tau}(x,\tau) + rc_0(x,\tau) = 0.$$

If we integrate this equation and use the usual condition $c_0(x,\tau) \to 0$ as $\tau \to \infty$, we get

$$c_0(x,\tau) = \alpha(x)e^{-r\tau},$$

where $\alpha(x)$ will be determined later. Applying the identification procedure to regular terms, one can see that $(X,Y)$ satisfies the reduced problem $P_0$ we have already indicated before. We have in mind that the remainder components should tend to 0 as $\varepsilon \to 0$. From the initial condition for $u_\varepsilon$, which reads

$$u_0(x) = X(x,0) + \alpha(x) + R_{1\varepsilon}(x,0),$$

we get

$$u_0(x) = X(x,0) + \alpha(x),$$

which shows how $c_0(x,\tau)$ compensate for the discrepancy in the corresponding initial condition. In fact, at this moment $c_0(x,\tau)$ is completely determined, since the last equation gives

$$\alpha(x) = u_0(x) - \frac{1}{r}\Big(f_1(x,0) - v_0'(x)\Big).$$

Let us now discuss the boundary conditions. From the equation

$$r_0\Big(X(0,t) + c_0(0,\tau) + R_{1\varepsilon}(0,t)\Big) + Y(0,t) + R_{2\varepsilon}(0,t) = 0$$

we derive

$$r_0X(0,t) + Y(0,t) = 0,$$

plus the condition $c_0(0,\tau) = 0$, i.e., $\alpha(0) = 0$. Now, let us discuss the nonlinear boundary condition, which reads

$$-\Big(X(1,t) + c_0(1,\tau) + R_{1\varepsilon}(1,t)\Big) + f_0\big(Y(1,t) + R_{2\varepsilon}(1,t)\big) = 0.$$

We obtain $c_0(1,\tau) = 0$, i.e., $\alpha(1) = 0$. For the regular part we get

$$-X(1,t) + f_0(Y(1,t)) = 0.$$

The same result is obtained if we consider a series (or Taylor) expansion for the solution of $P_\varepsilon$, expand $f_0\big(v_\varepsilon(1,t)\big)$ around $\varepsilon = 0$, and then equate the coefficients of $\varepsilon^0$. Of course, we need more regularity for $f_0$ to apply this method. Therefore it is indeed possible to show formally that $(u,v) = (X,Y)$ satisfies $P_0$. We have already determined the corrections of the order zero, and it is an easy matter to find the problem satisfied by the remainder components $R_{1\varepsilon}$, $R_{2\varepsilon}$. So, at this moment, we have a formal asymptotic expansion of the order zero. The next step would be to show that the expansion is well defined, in particular to show that, under some specific assumptions on the data, both $P_\varepsilon$ and $P_0$ have unique solutions in some function spaces. Finally, to show that the expansion is a real asymptotic expansion, it should be proved that the remainder tends to zero with respect to a given norm. In our applications (including the above hyperbolic $P_\varepsilon$) we are going to do even more, to establish error estimates for the remainder components (in most of the cases with respect to the sup norm).

Let us note that in the above formal derivation procedure we obtained two conditions for the correction $c_0(x,\tau)$, namely $\alpha(0) = 0$ and $\alpha(1) = 0$. These two conditions assure that $c_0$ does not create discrepancies at $(x,t) = (0,0)$ and $(x,t) = (1,0)$, which are corner boundary points of $D_T$. These conditions can be expressed in terms of our data:

$$v_0'(0) + ru_0(0) = f_1(0,0), \quad v_0'(1) + ru_0(1) = f_1(1,0).$$

In fact, we will see that these conditions are also necessary compatibility conditions for the existence of smooth solutions for problems $P_\varepsilon$ and $P_0$.

By the same technique, terms of the higher order approximations can be constructed as well.

For background material concerning the topics discussed in this chapter we refer the reader to [17], [18], [26], [27], [29], [32], [36], [37], [46], [47], [48], [50].

# Chapter 2

# Evolution Equations in Hilbert Spaces

In this chapter we are going to remind the reader of some basic concepts and results in the theory of evolution equations associated with monotone operators which will be used in the next chapters. The proofs of the theorems will be omitted, but appropriate references will be indicated.

## Function spaces and distributions

Let $X$ be a real Banach space with norm denoted by $\| \cdot \|_X$. If $\Omega \subset \mathbb{R}^n$ is a nonempty Lebesgue measurable set, we denote by $L^p(\Omega; X)$, $1 \leq p < \infty$, the space of all equivalence classes (with respect to the equality a.e. in $\Omega$) of strongly measurable functions $f : \Omega \to X$ such that $x \to \| f(x) \|_X^p$ is Lebesgue integrable over $\Omega$. In general, every class of $L^p(\Omega; X)$ is identified with one of its representatives. It is well known that $L^p(\Omega; X)$ is a real Banach space with the norm

$$\| u \|_{L^p(\Omega;X)} = \left( \int_\Omega \| u(x) \|_X^p \, dx \right)^{1/p}.$$

We will denote by $L^\infty(\Omega; X)$ the space of all equivalence classes of strongly measurable functions $f : \Omega \to X$ such that $x \to \| f(x) \|_X$ are essentially bounded in $\Omega$. Again, every class of $L^\infty(\Omega; X)$ is identified with one of its representatives. $L^\infty(\Omega; X)$ is also a real Banach space with the norm

$$\| u \|_{L^\infty(\Omega;X)} = \text{ess sup}_{x \in \Omega} \| u(x) \|_X .$$

In the case $X = \mathbb{R}$ we will simply write $L^p(\Omega)$ instead of $L^p(\Omega; \mathbb{R})$. On the other hand, if $\Omega$ is an interval of real numbers, say $\Omega = (a, b)$, then we will write $L^p(a, b; X)$ instead of $L^p((a, b); X)$.

In the following we assume that $\Omega$ is a nonempty open subset of $\mathbb{R}^n$. We will denote by $L^p_{\rm loc}(\Omega; X)$, $1 \le p \le \infty$, the space of all (equivalence classes of strongly) measurable functions $u : \Omega \to X$ such that the restriction of $u$ to every compact set $K \subset \Omega$ is in $L^p(K; X)$. We will use the notation $L^p_{\rm loc}(\Omega)$ instead of $L^p_{\rm loc}(\Omega; \mathbb{R})$.

Let $C(\Omega)$ be the set of all real-valued continuous functions defined on $\Omega$. As usual, we set

$$\begin{aligned} C^\infty(\Omega) &= \{\varphi \in C(\Omega); \varphi \text{ has continuous partial derivatives of any order}\}\,, \\ C_0^\infty(\Omega) &= \{\varphi \in C^\infty(\Omega);\ \operatorname{supp}\varphi \text{ is a bounded set included in } \Omega\}\,, \end{aligned}$$

where $\operatorname{supp}\varphi$ means the closure of the set of all points $x \in \Omega$ for which $\varphi(x) \neq 0$. When $C_0^\infty(\Omega)$ is endowed with the usual inductive limit topology, then it is denoted by $\mathcal{D}(\Omega)$.
A linear continuous functional $u : \mathcal{D}(\Omega) \to \mathbb{R}$ is said to be a *distribution* on $\Omega$. The linear space of all distributions on $\Omega$ is denoted by $\mathcal{D}'(\Omega)$.
Recall that if $u \in L^1_{\rm loc}(\Omega)$, then the functional defined by

$$\mathcal{D}(\Omega) \ni \varphi \longrightarrow \int_\Omega u(x)\varphi(x)dx$$

is a distribution on $\Omega$, called *regular distribution*. Such a distribution will always be identified with the corresponding function $u$ and so it will be denoted by $u$. Of course, there are distributions which are not regular, in particular the so-called Dirac distribution associated with some point $x_0 \in \Omega$, denoted by $\delta_{x_0}$ and defined by

$$\delta_{x_0}(\varphi) = \varphi(x_0) \ \ \forall \varphi \in \mathcal{D}(\Omega).$$

The partial derivative with respect to $x_j$ of a distribution $u \in \mathcal{D}'(\Omega)$ is defined by

$$\frac{\partial u}{\partial x_j}(\varphi) = -u\Big(\frac{\partial \varphi}{\partial x_j}\Big) \ \ \forall \varphi \in \mathcal{D}(\Omega),$$

hence

$$D^\alpha(\varphi) = (-1)^{|\alpha|} u(D^\alpha \varphi) \ \ \forall \varphi \in \mathcal{D}(\Omega),$$

where $\alpha = (\alpha_1, \alpha_2, \dots, \alpha_n) \in \mathbb{N}_0^n$ is a so-called multiindex and $\mid \alpha \mid = \alpha_1 + \alpha_2 + \cdots + \alpha_n$. Here $\mathbb{N}_0 = \{0\} \cup \mathbb{N}$, $\mathbb{N} = \{1, 2, \dots\}$, and

$$D^\alpha = \frac{\partial^{|\alpha|}}{\partial x_1^{\alpha_1} \cdots \partial x_n^{\alpha_n}}\,.$$

Let $1 \le p \le \infty$ and let $k \in \mathbb{N}$ be fixed. Then, the set

$$W^{k,p}(\Omega) = \{u : \Omega \to \mathbb{R};\ D^\alpha u \in L^p(\Omega) \ \ \forall \alpha \in \mathbb{N}_0^n,\ \mid \alpha \mid \le k\},$$

where $D^\alpha u$ are derivatives of $u$ in the sense of distributions, is said to be a Sobolev space of order $k$. Here $D^{(0,\dots,0)}u = u$.

For each $1 \le p < \infty$ and $k \in \mathbb{N}$, $W^{k,p}(\Omega)$ is a real Banach space with the norm

$$\|u\|_{W^{k,p}(\Omega)} = \left( \sum_{|\alpha| \le k} \|D^\alpha u\|^p_{L^p(\Omega)} \right)^{1/p}.$$

$W^{k,\infty}(\Omega)$ is also a real Banach space with the norm

$$\|u\|_{W^{k,\infty}(\Omega)} = \max_{|\alpha| \le k} \|D^\alpha u\|_{L^\infty(\Omega)}.$$

In the case $p = 2$ we have the notation $H^k(\Omega) := W^{k,2}(\Omega)$. This is a Hilbert space with respect to the scalar product

$$\langle u, v \rangle_k := \sum_{|\alpha| \le k} \int_\Omega D^\alpha u(x) D^\alpha v(x) dx .$$

We set for $k \in \mathbb{N}$, $1 \le p \le \infty$

$$W^{k,p}_{\text{loc}}(\Omega) = \{u : \Omega \to \mathbb{R};\ D^\alpha u \in L^p_{\text{loc}}(\Omega),\ \forall \alpha \in \mathbb{N}^n_0,\ \ | \alpha | \le k\}.$$

We will write $H^k_{\text{loc}}(\Omega)$ instead of $W^{k,2}_{\text{loc}}(\Omega)$. Let us also recall that the closure of $\mathcal{D}(\Omega)$ in $W^{k,p}(\Omega)$ is denoted by $W^{k,p}_0(\Omega)$, $k \in \mathbb{N}$, $1 \le p \le \infty$. In general this is a proper subspace of $W^{k,p}(\Omega)$. Note that $W^{k,2}_0(\Omega)$ is also denoted by $H^k_0(\Omega)$.
If $\Omega$ is an open interval of real numbers, say $\Omega = (a, b)$, $-\infty \le a < b \le +\infty$, then we will write $L^p(a, b)$, $L^p_{\text{loc}}(a, b)$, $W^{k,p}(a, b)$, $W^{k,p}_{\text{loc}}(a, b)$, $W^{k,p}_0(a, b)$, $H^k(a, b)$, $H^k_{\text{loc}}(a, b)$, $H^k_0(a, b)$ instead of $L^p\big((a, b)\big)$, $L^p_{\text{loc}}\big((a, b)\big)$, $W^{k,p}\big((a, b)\big)$, $W^{k,p}_{\text{loc}}\big((a, b)\big)$, $W^{k,p}_0\big((a, b)\big)$, $H^k\big((a, b)\big)$, $H^k_{\text{loc}}\big((a, b)\big)$, $H^k_0\big((a, b)\big)$, respectively.
If $a$, $b$ are finite numbers, then every element of the space $W^{k,p}(a, b)$, $k \in \mathbb{N}$, $1 \le p \le \infty$, can be identified with a function $f : [a, b] \to \mathbb{R}$ which is absolutely continuous, its derivatives $f^{(j)}$, $1 \le j \le k - 1$, exist and are absolutely continuous in $[a, b]$, and $f^{(k)}$ belongs to $L^p(a, b)$ (more precisely, the equivalence class of $f^{(k)}$ belongs to $L^p(a, b)$). Moreover, every element of $W^{k,p}_0(a, b)$ can be identified with such a function $f$ which satisfies in addition the following conditions:

$$f^{(j)}(a) = f^{(j)}(b) = 0, \quad 0 \le j \le k - 1.$$

Note also that $W^{k,1}(a, b)$ is continuously embedded into $C^{k-1}[a, b]$.

Now, let us recall basic information on vectorial distributions. So let $\Omega$ be an open interval $(a, b)$, $-\infty \le a < b \le \infty$. Denote by $\mathcal{D}'(a, b; X)$ the space of all linear continuous operators from $\mathcal{D}(a, b) := \mathcal{D}\big((a, b)\big)$ to $X$. The elements of $\mathcal{D}'(a, b; X)$ are called *vectorial distributions* on $(a, b)$ with values in $X$. Again, any $u \in L^1_{\text{loc}}(a, b; X) := L^1_{\text{loc}}\big((a, b); X\big)$ defines a distribution (which is identified with $u$) by

$$\mathcal{D}(a, b) \ni \varphi \longrightarrow \int_a^b \varphi(t) u(t) dt.$$

The distributional derivative of $u \in \mathcal{D}'(a,b;X)$ is the distribution $u'$ defined by

$$u'(\varphi) := -u(\varphi') \ \ \forall \varphi \in \mathcal{D}(a,b),$$

and hence

$$u^{(j)}(\varphi) := (-1)^j u(\varphi^{(j)}) \ \forall\ \varphi \in \mathcal{D}(a,b) \ \ \forall j = 1,2,\dots.$$

Now, for $k \in \mathbb{N}$ and $1 \le p \le \infty$, we set

$$W^{k,p}(a,b;X) = \{u \in L^p(a,b;X);\ u^{(j)} \in L^p(a,b;X),\ 1 \le j \le k\},$$

where $u^{(j)}$ is the $j$th distributional derivative of $u$. $W^{k,p}(a,b,X)$ is a Banach space with the norm

$$\|u\|_{W^{k,p}(a,b;X)} = \left( \sum_{j=0}^{k} \|u^{(j)}\|^p_{L^p(a,b;X)} \right)^{1/p}, 1 \le p < \infty,$$

$$\|u\|_{W^{k,\infty}(a,b;X)} = \max_{0 \le j \le k} \|u^{(j)}\|_{L^\infty(a,b;X)}.$$

For $p = 2$ we may use the notation $H^k(a,b;X)$ instead of $W^{k,2}(a,b;X)$. If $X$ is a real Hilbert space with scalar product denoted by $(\cdot,\cdot)_X$ then, for each $k \in \mathbb{N}$, $H^k(a,b;X)$ is also a real Hilbert space with respect to the scalar product

$$\langle u, v\rangle_{H^k(a,b;X)} := \sum_{j=0}^{k} \int_a^b (u^{(j)}(t), v^{(j)}(t))_X dt.$$

As usual, for $k \in \mathbb{N}$ and $1 \le p \le \infty$ we set

$$W^{k,p}_{\rm loc}(a,b;X) = \{u \in \mathcal{D}'(a,b;X);\ \text{for every bounded subinterval } (\alpha,\beta) \subset (a,b)$$
$$\text{the restriction of } u \text{ to } (\alpha,\beta) \in W^{k,p}(\alpha,\beta;X)\}.$$

Every $u \in W^{k,p}(a,b;X)$ has a representative $u_1$ which is an absolutely continuous function on $[a,b]$, such that its classic derivatives $d^j u_1/dt^j$, $1 \le j \le k-1$, are absolutely continuous functions on $[a,b]$, and the class of $d^k u_1/dt^k \in L^p(a,b;X)$. Usually, $u$ is identified with $u_1$.

A characterization of $W^{1,p}(a,b;X)$ is given by the following result:

**Theorem 2.0.3.** *Let $X$ be a real reflexive Banach space and let $u \in L^p(a,b;X)$, $a,b \in \mathbb{R}$, $a < b$, $1 < p < \infty$. Then, the following two conditions are equivalent:*

(i) *$u \in W^{1,p}(a,b;X)$;*

(ii) *There exists a constant $C > 0$ such that*

$$\int_a^{b-h} \|u(t+h) - u(t)\|_X^p dt \le Ch^p \ \forall\ h \in (0, b-a].$$

*If $p = 1$ then* (i) *implies* (ii). *Moreover,* (ii) *is true for $p = 1$ if one representative of $u \in L^1(a,b;X)$ is of bounded variation on $[a,b]$, where $X$ is a general Banach space, not necessarily reflexive.*

For background material concerning the topics discussed above we refer the reader to [1], [11], [30], [40], [39] and [52].

## Monotone operators

Let $H$ be a real Hilbert space with its scalar product and associated Hilbertian norm denoted by $\langle \cdot, \cdot \rangle$ and $\| \cdot \|$, respectively.

By a multivalued operator $A : D(A) \subset H \to H$ we mean a mapping that assigns to each $x \in D(A)$ a nonempty set $Ax \subset H$. In fact, $A$ is a mapping from $D(A)$ into $2^H$, but we prefer this notation. The situation in which some (or even all) of the sets $Ax$ are singletons is not excluded.
The graph of $A$ is defined as the following subset of $H \times H$

$$G(A) := \{(x, y) \in H \times H;\ x \in D(A),\ y \in Ax\}.$$

Obviously, for every $M \subset H \times H$ there exists a unique multivalued operator $A$ such that $G(A) = M$. More precisely,

$$\begin{aligned} D(A) &= \{x \in H;\ \text{there exists a } y \in H \text{ such that } (x, y) \in M\}, \\ Ax &= \{y \in H;\ (x, y) \in M\}\ \ \forall x \in D(A). \end{aligned}$$

So, every multivalued operator $A$ can be identified with $G(A)$ and we will simply write $(x, y) \in A$ instead of: $x \in D(A)$ and $y \in Ax$.

The range $R(A)$ of a multivalued operator $A : D(A) \subset H \to H$ is defined as the union of all $Ax$, $x \in D(A)$. For every multivalued operator $A$, there exists $A^{-1}$ which is defined as

$$A^{-1} = \{(y, x);\ (x, y) \in A\}.$$

Obviously, $D(A^{-1}) = R(A)$ and $R(A^{-1}) = D(A)$.

**Definition 2.0.4.** A multivalued operator $A : D(A) \subset H \to H$ is said to be monotone if

$$\langle x_1 - x_2, y_1 - y_2 \rangle \geq 0\ \ \forall\ (x_1, y_1), (x_2, y_2) \in A. \tag{2.1}$$

Sometimes, having in mind that $A$ can be identified with its graph, we will say that $A$ is a monotone subset of $H \times H$. If $Ax$ is a singleton then we will often identify $Ax$ with its unique element. So, if $A$ is single-valued (i.e., $Ax$ is a singleton for all $x \in D(A)$), then (2.1) can be written as

$$\langle x_1 - x_2, Ax_1 - Ax_2 \rangle \geq 0\ \ \forall\ x_1, x_2 \in D(A).$$

A very important concept is the following:

**Definition 2.0.5.** A monotone operator $A : D(A) \subset H \to H$ is called maximal monotone if $A$ has no proper monotone extension (in other words, $A$, viewed as a subset of $H \times H$, cannot be extended to any $A' \subset H \times H$, $A' \neq A$, such that the corresponding multivalued operator $A'$ is monotone).

We continue with the celebrated Minty's characterization for maximal monotonicity:

**Theorem 2.0.6.** (G. Minty) *Let $A : D(A) \subset H \to H$ be a monotone operator. It is maximal monotone if and only if $R(I + \lambda A) = H$ for some $\lambda > 0$ or, equivalently, for all $\lambda > 0$.*

Here $I$ denotes the identity operator on $H$. Recall also that for two multivalued operators $A$, $B$ and $r \in \mathbb{R}$, $A + B := \{(x, y + z);\ (x, y) \in A \text{ and } (x, z) \in B\}$, $rA := \{(x, ry);\ (x, y) \in A\}$.

**Theorem 2.0.7.** *Let $A : D(A) \subset H \to H$ be a maximal monotone operator. Assume in addition that $A$ is coercive with respect to some $x_0 \in H$, i.e.,*

$$\lim_{\substack{\|x\| \to \infty \\ (x,y) \in A}} \frac{\langle x - x_0, y \rangle}{\|x\|} = +\infty.$$

*Then $R(A) = H$, i.e., $A$ is surjective.*

In particular, if $A$ is strongly monotone, i.e., there exists a positive constant $a$ such that

$$\langle x_1 - x_2, y_1 - y_2 \rangle \geq a\|x_1 - x_2\|^2 \ \ \forall (x_1, y_1),\ (x_2, y_2) \in A,$$

then $A$ is coercive with respect to any $x_0 \in D(A)$. Therefore, if $A$ is maximal monotone and also strongly monotone, then $R(A) = H$.

**Theorem 2.0.8.** (R.T. Rockafellar) *If $A : D(A) \subset H \to H$ and $B : D(B) \subset H \to H$ are two maximal monotone operators such that $(Int\ D(A)) \cap D(B) \neq \emptyset$, then $A + B$ is maximal monotone, too.*

Now, recall that a single-valued operator $A : D(A) = H \to H$ is said to be *hemicontinuous* if for every $x, y, z \in H$

$$\lim_{t \to 0} \langle A(x + ty), z \rangle = \langle Ax, z \rangle.$$

**Theorem 2.0.9.** (G. Minty) *If $A : D(A) = H \to H$ is single-valued, monotone and hemicontinuous, then $A$ is maximal monotone.*

Now, for $A$ maximal monotone and $\lambda > 0$, we define the well-known operators:

$$J_\lambda = (I + \lambda A)^{-1}, \quad A_\lambda = (1/\lambda)(I - J_\lambda),$$

which are called the *resolvent* and the *Yosida approximation* of $A$, respectively. By Theorem 2.0.6, $D(J_\lambda) = D(A_\lambda) = H$. It is easily seen that $J_\lambda$, $A_\lambda$ are single-valued for all $\lambda > 0$.

Additional properties of $J_\lambda$ and $A_\lambda$ are recalled in the next result.

**Theorem 2.0.10.** *If $A : D(A) \subset H \to H$ is a maximal monotone operator, then for every $\lambda > 0$ we have:*

(i) *$J_\lambda$ is nonexpansive (i.e., Lipschitz continuous with the Lipschitz constant $= 1$);*
(ii) *$A_\lambda$ is monotone and Lipschitz continuous, with the Lipschitz constant $= 1/\lambda$;*
(iii) *$A_\lambda x \in AJ_\lambda x \ \ \forall\, x \in H$;*
(iv) *$\|A_\lambda x\| \le \|A^\circ x\| \ \forall\, x \in D(A)$;*
(v) *$A_\lambda x \to A^\circ x$ as $\lambda \to 0 \ \forall\, x \in D(A)$.*

*We have denoted by $A^\circ$ the so-called minimal section of $A$, which is defined by*

$$A^\circ x = Proj_{Ax} 0 \ \forall\, x \in D(A),$$

*i.e., $A^\circ x$ is the element of minimum norm of $Ax$ (which is a convex closed set).*

Let us continue with a brief presentation of the class of subdifferentials. First, let us recall that a function $\psi : H \to (-\infty, \infty]$ is said to be *proper* if $D(\psi) \neq \emptyset$, where $D(\psi) := \{x \in H;\ \psi(x) < \infty\}$. $D(\psi)$ is called the effective domain of $\psi$. Function $\psi$ is said to be *convex* if

$$\psi(\alpha x + (1-\alpha)y) \le \alpha\psi(x) + (1-\alpha)\psi(y) \ \ \forall \alpha \in (0,1),\ x, y \in H,$$

where usual conventions are used concerning operations which involve $\infty$.

A function $\psi : H \to (-\infty, \infty]$ is said to be *lower semicontinuous* at $x_0 \in H$ if $\psi(x_0) \le \liminf_{x \to x_0} \psi(x)$.
Let $\psi : H \to (-\infty, \infty]$ be a proper convex function. Its *subdifferential* at $x$ is defined by

$$\partial\psi(x) = \{y \in H;\ \psi(x) - \psi(v) \le \langle y, x - v\rangle,\ \forall\, v \in H\}.$$

The operator $\partial\psi \subset H \times H$ is called the subdifferential of $\psi$. Clearly, its domain $D(\partial\psi)$ is included in $D(\psi)$.

**Theorem 2.0.11.** *If $\psi : H \to (-\infty, \infty]$ is a proper convex lower semicontinuos (on $H$) function, then $\partial\psi$ is a maximal monotone operator and, furthermore,*

$$\overline{D(\partial\psi)} = \overline{D(\psi)}, \qquad Int\ D(\partial\psi) = Int\ D(\psi).$$

If $\psi : H \to (-\infty, \infty]$ is proper and convex, then operator $A = \partial\psi$ is *cyclically monotone*, i.e., for every $n \in \mathbb{N}$ we have

$$\langle x_0 - x_1, x_0^*\rangle + \langle x_1 - x_2, x_1^*\rangle + \cdots + \langle x_{n-1} - x_n, x_{n-1}^*\rangle + \langle x_n - x_0, x_n^*\rangle \ge 0,$$

for all $(x_i, x_i^*) \in A,\ i = 0, 1, \ldots, n$.

An operator $A : D(A) \subset H \to H$ is called *maximal cyclically monotone* if $A$ cannot be extended properly to any cyclically monotone operator. Obviously, if $\psi : H \to (-\infty, \infty]$ is a proper convex lower semicontinuous function, then $A = \partial\psi$ is maximal cyclically monotone. The converse implication is also true:

**Theorem 2.0.12.** *If $A : D(A) \subset H \to H$ is a maximal cyclically monotone operator, then there exists a proper convex lower semicontinuous function $\psi : H \to (-\infty, \infty]$, uniquely determined up to an additive constant, such that $A = \partial\psi$.*

The reader may find the detailed proofs of Theorems 2.0.6–2.0.12 in [7], [10] and [34].

In the following we give some examples of maximal monotone operators which will also be used later.

*Example* 1. We consider the following assumptions:

$(I_1)$ Let the function $g : [0,1] \times \mathbb{R} \to \mathbb{R}$ be such that $x \to g(x,\xi) \in L^2(0,1)$ for all $\xi \in \mathbb{R}$ and $\xi \to g(x,\xi)$ is continuous and non-decreasing for a.a. $x \in (0,1)$.

$(I_2)$ Let the mapping $\beta : D(\beta) \subset \mathbb{R}^2 \to \mathbb{R}^2$ be maximal monotone, where the space $\mathbb{R}^2$ is equipped with the usual scalar product and Euclidean norm.

Let $H = L^2(0,1)$ with the usual scalar product defined by

$$\langle u, v \rangle = \int_0^1 u(x)v(x)dx.$$

We define the operator $A : D(A) \subset H \to H$ by

$$\begin{aligned} D(A) &= \{u \in H^2(0,1);\, \big(u'(0), -u'(1)\big) \in \beta\big(u(0), u(1)\big)\}, \\ Au &= -u'' + g\big(\cdot, u(\cdot)\big) \quad \forall\, u \in D(A). \end{aligned}$$

**Proposition 2.0.13.** *Assume $(I_1)$, $(I_2)$. Then operator $A$ is maximal monotone. Moreover if $\beta$ is the subdifferential of a proper convex lower semicontinuous function $j : \mathbb{R}^2 \to (-\infty, +\infty]$, then $A$ is the subdifferential of the function $\varphi : H \to (-\infty, +\infty]$ defined by*

$$\varphi(u) = \begin{cases} \frac{1}{2}\int_0^1 u'(x)^2 dx + j\big(u(0), u(1)\big) + \int_0^1 dx \int_0^{u(x)} g(x,s)ds, \\ \textit{if } u \in H^1(0,1) \textit{ and } j\big(u(0), u(1)\big) < \infty, \\ +\infty, \quad \textit{otherwise.} \end{cases}$$

The proof of this result is given in [24], Chapter 2, and [34], Chapter 3, in more general contexts. However, the reader could easily reproduce the proof by using the following simple ideas:

*Step* 1. It is easily seen that $A$ is a monotone operator;

*Step* 2. For the case $g = 0$ one can prove that $A$ is maximal monotone by using Theorems 2.0.6–2.0.9;

*Step* 3. By virtue of Theorem 2.0.8 the operator $u \to -u'' + g_\lambda\big(\cdot, u(\cdot)\big)$, $\lambda > 0$, with domain $D(A)$, is maximal monotone in $H$, where $g_\lambda(x, \cdot)$ is the Yosida approximation of $g(x, \cdot)$. According to Theorem 2.0.6, this means that for each $\lambda > 0$ and $h \in H$ there exists a function $u_\lambda \in D(A)$ which satisfies the equation

$$u_\lambda - u_\lambda'' + g_\lambda\big(\cdot, u_\lambda\big) = h\,;$$

*Step* 4. Passing to the limit in the last equation, as $\lambda \longrightarrow 0$, we obtain that $R(I + A) = H$, i.e., $A$ is maximal monotone.

*Step* 5. If $\beta = \partial j$, where $j$ has the properties listed in the statement of our proposition, it is easily seen that $\varphi$ is proper, convex and lower semicontinuous, and $A \subset \partial\varphi$, which implies $A = \partial\varphi$.

*Example* 2. The some assumptions governing the function $g$ will be required, i.e., $(I_1)$. We further assume that
$(I_2')$ There are given two multivalued mappings $\beta_i : D(\beta_i) \subset \mathbb{R} \to \mathbb{R}$, $i = 1, 2$, which are both maximal monotone.

We consider the product space $H_1 = L^2(0,1) \times \mathbb{R}$, which is a real Hilbert space with respect to the inner product

$$\langle (p_1, a), (p_2, b) \rangle_{H_1} = \int_0^1 p_1(x) p_2(x) dx + ab.$$

Define the operator $A_1 : D(A_1) \subset H_1 \to H_1$ by

$$\begin{aligned} D(A_1) &= \{(p, a) \in H^2(0,1) \times \mathbb{R};\ a = p(1) \in D(\beta_2), \\ &\quad p(0) \in D(\beta_1),\ p'(0) = \beta_1(p(0))\}, \\ A_1(p, a) &= (-p'' + g(\cdot, p),\ \beta_2(a) + p'(1)). \end{aligned}$$

We have:

**Proposition 2.0.14.** ([24], p. 93) *Assume* $(I_1), (I_2')$. *Then the operator* $A_1$ *defined above is maximal cyclically monotone. More precisely,* $A_1$ *is the subdifferential of* $\varphi_1 : H \to (-\infty, +\infty]$ *defined by*

$$\varphi_1(p, a) = \begin{cases} \frac{1}{2} \int_0^1 p'(x)^2 dx + \int_0^1 k(x, p(x)) dx + j_1(p(0)) + j_2(a), \\ \text{if } p \in H^1(0,1), p(0) \in D(j_1),\ a = p(1) \in D(j_2), \\ +\infty, \quad \text{otherwise}, \end{cases}$$

*where* $j_1, j_2 : \mathbb{R} \to (-\infty, \infty]$ *are proper, convex and lower semicontinuous functions such that* $\beta_1 = \partial j_1$, $\beta_2 = \partial j_2$, *whilst* $k(x, \cdot)$ *is given by*

$$k(x, \zeta) = \int_0^\zeta g(x, s) ds.$$

Notice that the existence of $j_1$ and $j_2$ is not an assumption: any maximal monotone operator from $\mathbb{R}$ into $\mathbb{R}$ is the subdifferential of a proper, convex, lower semicontinuous function. The reader may either prove this result as an exercise, or consult, e.g., [10].

*Example* 3. We first formulate some hypotheses:

($j_1$) Let the functions $r, g : [0,1] \times \mathbb{R} \to \mathbb{R}$ be such that $r(\cdot,\xi)$ and $g(\cdot,\xi)$ belong to $L^2(0,1)$ for all $\xi \in \mathbb{R}$, and $\xi \to r(x,\xi)$, $\xi \to g(x,\xi)$ are continuous and nondecreasing for a.a. $x \in (0,1)$;

($j_2$) Let the functions $\alpha_1$, $\alpha_2$ belong to $L^\infty(0,1)$ such that both of them are bounded below by positive constants;

($j_3$) Let the operator $\beta : D(\beta) \subset \mathbb{R}^2 \to \mathbb{R}^2$ be maximal monotone.

We consider the Hilbert space $H_2 = L^2(0,1)^2$ with the scalar product defined by

$$\langle (p_1, q_1), (p_2, q_2) \rangle = \int_0^1 (\alpha_1 p_1 p_2 + \alpha_2 q_1 q_2) dx.$$

Define the operator $B : D(B) \subset H_2 \to H_2$ by

$$D(B) = \left\{ (p,q) \in H^1(0,1)^2;\, \big(-p(0), p(1)\big) \in \beta\Big(\big(q(0), q(1)\big)\Big) \right\},$$
$$B(p,q) = \left(\alpha_1^{-1}\left(q' + r(\cdot, p)\right),\ \alpha_2^{-1}\left(p' + g(\cdot, q)\right)\right),$$

where $\alpha_i^{-1}$ denotes the quotient function $1/\alpha_i$.

**Proposition 2.0.15.** ([24], Chapter 5) *If assumptions ($j_1$)–($j_3$) are satisfied, then operator $B$ is maximal monotone. In addition, $D(B)$ is dense in $H_2$.*

Although the proof can be found in [24], the reader is encouraged to reproduce it by using the steps indicated in Example 1.

*Example* 4. We consider the same assumptions ($j_1$), ($j_2$) formulated above and in addition:

($j_4$) Let the functions $r_0, f_0 : \mathbb{R} \to \mathbb{R}$ be continuous and nondecreasing, and let $c$ be a positive constant.

Consider the space $H_3 = L^2(0,1)^2 \times \mathbb{R}$ which is a Hilbert space with the scalar product

$$\langle (p_1, q_1, a_1), (p_2, q_2, a_2) \rangle = \int_0^1 (\alpha_1 p_1 p_2 + \alpha_2 q_1 q_2) dx + c a_1 a_2.$$

Define the operator $B_1 : D(B_1) \subset H_3 \to H_3$ by

$$D(B_1) = \left\{ (p,q,a) \in H^1(0,1)^2 \times \mathbb{R};\ a = q(1),\ r_0(q(0)) + p(0) = 0 \right\},$$
$$B_1(p,q,a) = \left(\alpha_1^{-1}\left(q' + r(\cdot,p)\right),\ \alpha_2^{-1}\left(p' + g(\cdot,q)\right),\ c^{-1}(-p(1) + f_0(a))\right).$$

**Proposition 2.0.16.** ([24], Chapter 6) *If ($j_1$), ($j_2$), ($j_4$) hold, then operator $B_1$ is maximal monotone.*

Note that operators $B$ and $B_1$ occur exactly in this form (involving in particular $\alpha_1^{-1}$, $\alpha_2^{-1}$) in some applications we are going to discuss later. Spaces $H_2$ and $H_3$ were chosen as adequate frameworks which guarantee the monotonicity of these operators.

# Evolution equations

Let $H$ be a real Hilbert space. Its scalar product and norm are again denoted by $\langle\cdot,\cdot\rangle$ and $\|\cdot\|$, respectively. Consider in $H$ the following Cauchy problem:

$$\begin{cases} u'(t) + Au(t) = f(t), & 0 < t < T, \\ u(0) = u_0, \end{cases} \tag{2.2}$$

where $A : D(A) \subset H \to H$ is a nonlinear single-valued operator and $f \in L^1(0,T;H)$. In fact, the most known existence results are valid for multivalued $A$'s, but we will consider only single-valued $A$'s. This is enough for our considerations.

**Definition 2.0.17.** A function $u \in C([0,T];H)$ is said to be a *strong solution* of equation $(2.2)_1$ if:

(i) $u$ is absolutely continuous on every compact subinterval of $(0,T)$,
(ii) $u(t) \in D(A)$ for a.a. $t \in (0,T)$,
(iii) $u$ satisfies $(2.2)_1$ for a.a. $t \in (0,T)$.

If in addition $u(0) = u_0$, then $u$ is called a strong solution of the Cauchy problem (2.2).

**Definition 2.0.18.** A function $u \in C([0,T];H)$ is called a *weak solution* of $(2.2)_1$ if there exist two sequences $\{u_n\} \subset W^{1,\infty}(0,T;H)$ and $\{f_n\} \subset L^1(0,T;H)$ such that:

$(k)$ $u_n'(t) + Au_n(t) = f_n(t)$ for a.a. $t \in (0,T)$, $n \in \mathbb{N}$,
$(kk)$ $u_n \to u$ in $C([0,T];H)$ as $n \to \infty$,
$(kkk)$ $f_n \to f$ in $L^1(0,T;H)$ as $n \to \infty$.

Again, if in addition $u(0) = u_0$, then $u$ is called a weak solution of the Cauchy problem (2.2).

We recall the following variant of Gronwall's lemma which will be used in the following (see [10], p. 157):

**Lemma 2.0.19.** *Let $a, b, c \in \mathbb{R}$, $a < b$, $g \in L^1(a,b)$ with $g \geq 0$ a.e. on $(a,b)$, and $h \in C[a,b]$ be such that*

$$\frac{1}{2}h^2(t) \leq \frac{1}{2}c^2 + \int_a^t g(s)h(s)ds \quad \forall t \in [a,b].$$

*Then*

$$| h(t) | \leq | c | + \int_a^t g(s)ds \quad \forall t \in [a,b].$$

**Theorem 2.0.20.** (see, e.g., [34], p. 48) *If $A : D(A) \subset H \to H$ is a maximal monotone operator, $u_0 \in D(A)$ and $f \in W^{1,1}(0,T;H)$, then the Cauchy problem (2.2) has a unique strong solution $u \in W^{1,\infty}(0,T;H)$. Moreover $u(t) \in D(A)$ for all $t \in [0,T]$, $u$ is differentiable on the right at every $t \in [0,T)$, and*

$$\frac{d^+u}{dt}(t) + Au(t) = f(t) \quad \forall\, t \in [0,T),$$

$$\left\| \frac{d^+u}{dt}(t) \right\| \leq \|f(0) - Au_0\| + \int_0^t \|f'(s)\| ds \quad \forall\, t \in [0,T). \tag{2.3}$$

*If $u$ and $\overline{u}$ are the strong solutions corresponding to $(u_0, f)$, $(\overline{u}_0, \overline{f}) \in D(A) \times W^{1,1}(0,T;H)$, then*

$$\|u(t) - \overline{u}(t)\| \leq \|u_0 - \overline{u}_0\| + \int_0^t \|f(s) - \overline{f}(s)\| ds, \quad 0 \leq t \leq T. \tag{2.4}$$

**Theorem 2.0.21.** (see, e.g., [34], p. 55) *If $A : D(A) \subset H \to H$ is a maximal monotone operator, $u_0 \in \overline{D(A)}$ and $f \in L^1(0,T;H)$, then the Cauchy problem (2.2) has a unique weak solution $u \in C([0,T];H)$. If $u$ and $\overline{u}$ are the weak solutions corresponding to $(u_0, f)$, $(\overline{u}_0, \overline{f}) \in \overline{D(A)} \times L^1(0,T;H)$, then $u$ and $\overline{u}$ still satisfy (2.4).*

*Remark* 2.0.22. The proof of the last result is straightforward. For the first part, it is enough to consider a sequence $(u_0^n, f_n) \in D(A) \times W^{1,1}(0,T;H)$ which approximates $(u_0, f)$ in $H \times L^1(0,T;H)$. By Theorem 2.0.20 there exists a unique strong solution, say $u_n$, for (2.2) with $u_0 := u_0^n$ and $f := f_n$. Writing (2.4) for $u_n$ and $u_m$, we see that $u_n$ is a Cauchy sequence in $C([0,T];H)$, thus it has a limit in this space, which is a weak solution of (2.2). A similar density argument leads us to estimate (2.4) for weak solutions. In particular this implies uniqueness for weak solutions.

*Remark* 2.0.23. Both the above theorems still hold in the case of Lipschitz perturbations, i.e., in the case in which $A$ is replaced by $A + B$, where $A$ is maximal monotone as before and $B : D(B) = H \to H$ is a Lipschitz operator (see [10], p. 105). The only modifications appear in estimates (2.3) and (2.4) :

$$\left\| \frac{d^+u}{dt}(t) \right\| \leq e^{\omega t} \Big( \|f(0) - Au_0\| + \int_0^t e^{-\omega s} \|f'(s)\| ds \Big), \quad 0 \leq t < T, \tag{2.5}$$

$$\|u(t) - \overline{u}(t)\| \leq e^{\omega t} \Big( \|u_0 - \overline{u}_0\| + \int_0^t e^{-\omega s} \|f(s) - \overline{f}(s)\| ds \Big), \quad 0 \leq t \leq T, \tag{2.6}$$

where $\omega$ is the Lipschitz constant of $B$.

**Theorem 2.0.24.** (H. Brézis, [10]) *If $A$ is the subdifferential of a proper convex lower semicontinuous function $\varphi : H \to (-\infty, +\infty]$, $u_0 \in \overline{D(A)}$ and $f \in L^2(0,T;H)$, then the Cauchy problem (2.2) has a unique strong solution $u$, such* *that* $t^{1/2}u' \in L^2(0,T;H)$, $t \to \varphi(u(t))$ *is integrable on* $[0,T]$ *and absolutely continuous on* $[\delta,T]$, $\forall \delta \in (0,T)$. *If, in addition,* $u_0 \in D(\varphi)$, *then* $u' \in L^2(0,T;H)$, $t \to \varphi(u(t))$ *is absolutely continuous on* $[0,T]$, *and*

$$\varphi(u(t)) \le \varphi(u_0) + \frac{1}{2}\int_0^T \|f(t)\|^2 dt \quad \forall t \in [0,T].$$

Before stating other existence results let us recall the following definition:

**Definition 2.0.25.** Let $C$ be a nonempty closed subset of $H$. A *continuous semigroup of contractions* on $C$ is a family of operators $\{S(t) : C \to C;\ t \ge 0\}$ satisfying the following conditions:

(j) $S(t+s)x = S(t)S(s)x \quad \forall x \in C, \quad t,s \ge 0$;

(jj) $S(0)x = x \quad \forall x \in C$;

(jjj) for every $x \in C$, the mapping $t \to S(t)x$ is continuous on $[0,\infty)$;

(jv) $\|S(t)x - S(t)y\| \le \|x-y\| \quad \forall t \ge 0, \quad \forall x,y \in C$.

The *infinitesimal generator* of a semigroup $\{S(t) : C \to C; t \ge 0\}$, say $G$, is given by:

$$G(x) = \lim_{\substack{h\to 0\\ h>0}} \frac{S(h)x - x}{h}, \quad x \in D(G),$$

where $D(G)$ consists of all $x \in C$ for which the above limit exists.

Let $A : D(A) \subset H \to H$ be a (single-valued) maximal monotone operator. From Theorem 2.0.20 we see that if $f \equiv 0$, then for every $u_0 \in D(A)$ there exists a unique strong solution $u(t)$, $t \ge 0$, of the Cauchy problem (2.2). We set $S(t)u_0 := u(t)$, $t \ge 0$. Then it is easily seen that $S(t)$ is a contraction (i.e., a non expansive operator) on $D(A)$ (see (2.4)) and so $S(t)$ can be extended as a contraction on $\overline{D(A)}$, for each $t \ge 0$. Moreover, it is obvious that the family $\{S(t) : \overline{D(A)} \to \overline{D(A)};\ t \ge 0\}$ is a continuous semigroup of contractions and its infinitesimal generator is $G = -A$. We will say that the semigroup is generated by $-A$. If $A$ is linear maximal monotone, then $\overline{D(A)} = H$, and the semigroup $\{S(t) : H \to H; t \ge 0\}$ generated by $-A$ is a $C_0$-semigroup of contractions. Recall that a family of linear continuous operators $\{S(t) : H \to H;\ t \ge 0\}$ is called a $C_0$-semigroup if conditions (j)–(jjj) above are satisfied with $C = H$. In fact, in this case, continuity at $t = 0$ of the function $t \to S(t)x$, $\forall x \in H$, combined with (j)–(jj) is enough to derive (jjj). For details on $C_0$-semigroups, we refer to [3], [22], [38].

**Definition 2.0.26.** Let $A : D(A) \subset H \to H$ be the infinitesimal generator of a $C_0$-semigroup $\{S(t) : H \to H;\ t \ge 0\}$, and let $f \in L^1(0,T;H)$, $u_0 \in H$. The function $u \in C([0,T];H)$ given by

$$u(t) = S(t)u_0 + \int_0^t S(t-s)f(s)ds \quad \forall t \in [0,T], \tag{2.7}$$

is called a *mild solution* of the Cauchy problem (2.2). Since $\overline{D(A)} = H$, the above formula (which is usually called *the variation of constants formula*) makes sense for any $u_0 \in H$.

Obviously, the mild solution has the uniqueness property.

**Theorem 2.0.27.** (see [21], [38], p. 109) *Let $A : D(A) \subset H \to H$ be the infinitesimal generator of a $C_0$-semigroup $\{S(t) : H \to H;\ t \geq 0\}$, and let $u_0 \in D(A)$, $f \in W^{1,1}(0,T;H)$. Then, the mild solution $u$ of (2.2) is also its unique strong solution. Moreover, $u \in C^1([0,T];H)$, and*

$$f(t) - Au(t) = u'(t) = S(t)\big(f(0) - Au_0\big) + \int_0^t S(t-s)f'(s)ds, \quad 0 \leq t \leq T. \tag{2.8}$$

We will apply Theorem 2.0.27 to the special case of a linear monotone (positive) operator $A$ for which $A + \omega I$ is maximal monotone for some $\omega > 0$ ($I$ is the identity of $H$).

Note that the mild solution of problem (2.2) given by the above definition is also a weak solution of the same problem. Indeed, let $(u_0^n, f_n) \in D(A) \times W^{1,1}(0,T;H)$ approximate $(u_0, f)$ in $H \times L^1(0,T;H)$. Denote by $u_n$ the strong solution of (2.2) with $u_0 := u_0^n$ and $f := f_n$. Then,

$$u_n(t) = S(t)u_0^n + \int_0^t S(t-s)f_n(s)\,ds, \ t \in [0,T],$$

which clearly implies our assertion.

The next two results deal with Cauchy problems associated with time-dependent Lipschitz perturbations of maximal monotone operators.

**Theorem 2.0.28.** *Let $A : D(A) \subset H \to H$ be a linear maximal monotone operator, $u_0 \in H$ and let $B : [0,T] \times H \to H$ satisfy the following two conditions:*

$$t \to B(t,z) \in L^1(0,T;H) \ \ \forall\, z \in H, \tag{2.9}$$

*and there exists some constant $\omega > 0$ such that*

$$\|B(t,x_1) - B(t,x_2)\| \leq \omega\|x_1 - x_2\| \ \ \forall\, t \in [0,T],\ x_1,\ x_2 \in H. \tag{2.10}$$

*Then there exists a unique function $u \in C([0,T];H)$ which satisfies the integral equation*

$$u(t) = S(t)u_0 - \int_0^t S(t-s)B(s,u(s))ds, \ t \in [0,T].$$

The solution of this integral equation is called a mild solution of the Cauchy problem

$$\begin{cases} u'(t) + Au(t) + B\big(t,u(t)\big) = 0, & 0 < t < T, \\ u(0) = u_0. \end{cases} \tag{2.11}$$

*Remark* 2.0.29. In the statement of the above theorem assumption (2.9) can be replaced by the following condition:
the function $t \to B(t,z)$ is strongly measurable for all $z \in H$, and $t \to B(t,0) \in L^1(0,T;H)$, which is equivalent with (2.9) due to (2.10).
The proof of this result is based on Banach's fixed point principle.

The next theorem is an extension of Theorem 2.0.24 to the case of time-dependent Lipschitz perturbations.

**Theorem 2.0.30.** (H. Brézis, [10], pp. 106–107) *If $A$ is the subdifferential of a proper convex lower semicontinuous function $\varphi : H \to (-\infty,+\infty]$, $u_0 \in \overline{D(A)} = \overline{D(\varphi)}$ and $B : [0,T] \times \overline{D(\varphi)} \to H$ satisfies the conditions:*

(i) *$\exists\, \omega \geq 0$ such that $\| B(t,x_1) - B(t,x_2) \| \leq \omega \| x_1 - x_2 \|$ $\forall t \in [0,T]$, $\forall x_1, x_2 \in \overline{D(A)}$;*

(ii) *$\forall x \in \overline{D(\varphi)}$, the function $t \to B(t,x)$ belongs to $L^2(0,T;H)$,*
*then, $\forall u_0 \in \overline{D(\varphi)}$, there exists a unique strong solution of the Cauchy problem*

$$\begin{cases} u'(t) + Au(t) + B(t,u(t)) = 0, & 0 < t < T, \\ u(0) = u_0, \end{cases}$$

*such that $t^{1/2}u'(t) \in L^2(0,T;H)$.*

Now, we are going to recall some existence results concerning fully nonlinear, non-autonomous (time-dependent) evolution equations.

**Theorem 2.0.31.** (T. Kato, [28]) *Let $A(t) : D \subset H \to H$ be a family of single-valued maximal monotone operators (with $D(A(t)) = D$ independent of $t$) satisfying the following condition*

$$\|A(t)x - A(s)x\| \leq L \mid t - s \mid (1 + \|x\| + \|A(s)x\|)$$

*for all $x \in D$, $s,t \in [0,T]$, where $L$ is a positive constant. Then, for every $u_0 \in D$, there exists a unique function $u \in W^{1,\infty}(0,T;H)$ such that $u(0) = u_0$, $u(t) \in D$ $\forall t \in [0,T]$, and*

$$u'(t) + A(t)u(t) = 0 \quad \textit{for a.a. } t \in (0,T).$$

In fact, Kato's result is valid in a general Banach space.

**Theorem 2.0.32.** (H. Attouch and A. Damlamian, [5]) *Let $A(t) = \partial\phi(t,\cdot)$, $0 \leq t \leq T$, where $\phi(t,\cdot) : H \to (-\infty,+\infty]$ are all proper, convex, and lower semicontinuous. Assume further that there exist some positive constants $C_1$, $C_2$ and a nondecreasing function $\gamma : [0,T] \to \mathbb{R}$ such that*

$$\phi(t,x) \leq \phi(s,x) + [\gamma(t) - \gamma(s)][\phi(s,x) + C_1\|x\|^2 + C_2], \tag{2.12}$$

*for all* $x \in H$, $0 \le s \le t \le T$. *Then, for every* $u_0 \in D(\phi(0,\cdot))$ *and* $f \in L^2(0,T;H)$, *there exists a unique function* $u \in W^{1,2}(0,T;H)$ *such that* $u(0) = u_0$ *and*

$$u'(t) + A(t)u(t) = f(t) \quad \textit{for a.a.} \quad t \in (0,T). \tag{2.13}$$

*Moreover, there exists a function* $h \in L^1(0,T)$ *such that*

$$\phi(t,u(t)) \le \phi(s,u(s)) + \int_s^t h(r)dr \quad \textit{for all} \ \ 0 \le s \le t \le T.$$

Assumption (2.12) implies that $D(\phi(t,\cdot))$ increases with $t$. In all our forthcoming applications of the above theorem $D(\phi(t,\cdot)) = D$, a constant set.

In the following we present some existence and uniqueness results for partial differential systems of the form

$$\begin{cases} \alpha_1(x)u_t + v_x + r(x,u) = f_1(x,t), \\ \alpha_2(x)v_t + u_x + g(x,v) = f_2(x,t), \ 0 < x < 1, \ 0 < t < T, \end{cases} \tag{2.14}$$

with which we associate the initial conditions

$$u(x,0) = u_0(x), \quad v(x,0) = v_0(x), \quad 0 < x < 1, \tag{2.15}$$

as well as one of the following types of boundary conditions:

$$(-u(0,t), u(1,t)) \in \beta(\ v(0,t), v(1,t)), \quad 0 < t < T \tag{2.16}$$

(algebraic boundary conditions),

$$\begin{cases} -u(0,t) = r_0\big(v(0,t)\big), \\ u(1,t) - cv_t(1,t) = f_0\big(v(1,t)\big) + e(t), \ 0 < t < T \end{cases} \tag{2.17}$$

(algebraic-differential boundary conditions).

The term "algebraic" indicates that the boundary conditions are algebraic relations involving the values of the unknowns on the boundary, i.e., in this specific case at $x = 0$ and $x = 1$.

The term "algebraic-differential" means that we have an algebraic boundary condition as well as a differential (or dynamic) boundary condition that involves the derivative of $v(1,t)$.

It should be mentioned that many practical problems arising from physics and engineering can be regarded as particular cases of problem (2.14)–(2.16) or (2.17). For details see the next chapter.

We are going to state an existence result for problem (2.14)–(2.16). The basic idea is to express this problem as a Cauchy problem in an appropriate functional framework. Indeed, by denoting $U(t) = (u(\cdot.t), v(\cdot,t))$, $U_0 = (u_0, v_0)$ and $f(t) =$

$(\alpha_1^{-1} f_1(\cdot,t),\, \alpha_2^{-1} f_2(\cdot,t))$, one can express this problem as the following Cauchy problem in $H_2$

$$\begin{cases} U'(t) + BU(t) = f(t), & 0 < t < T, \\ U(0) = U_0, \end{cases}$$

where $H_2$ and $B$ were defined in *Example* 3 above.

**Theorem 2.0.33.** (see [24], p. 115) *Assume* $(j_1)$–$(j_3)$ (*see Example* 3). *Let* $T > 0$ *be fixed. Then, for every* $f_1,\ f_2 \in L^1(0,T;L^2(0,1))$, *and* $u_0,\ v_0 \in L^2(0,1)$, *there exists a unique weak solution* $u, v \in C([0,T];L^2(0,1))$ *of problem* (2.14), (2.15), (2.16). *If in addition* $f_1,\ f_2 \in W^{1,1}(0,T;L^2(0,1))$, $u_0,\ v_0 \in H^1(0,1)$ *and* $(-u_0(0), u_0(1)) \in \beta((v_0(0), v_0(1))$, *then* $(u,v)$ *is a strong solution,* $(u,v) \in W^{1,\infty}(0,T;L^2(0,1))^2$ *and*

(i) $u, v \in L^\infty((0,1)\times(0,T))$, $u_x, v_x \in L^\infty(0,T;L^2(0,1))$;

(ii) $(u,v)$ *satisfies* (2.14) *for all* $t \in [0,T)$, *a.a.* $x \in (0,1)$, *where* $u_t, v_t$ *are replaced by* $\frac{\partial^+ u}{\partial t}$, $\frac{\partial^+ v}{\partial t}$;

(iii) $u, v$ *satisfy* (2.15) *and* (2.16).

All the conclusions of the above theorem follow from Theorems 2.0.20, 2.0.21 with the exception of the regularity property $u_x, v_x \in L^\infty(0,T;L^2(0,1))$, which can be proved in a standard manner.

On the other hand, problem (2.14), (2.15), (2.17) can be written as a Cauchy problem in the space $H_3$ :

$$\frac{d}{dt}\big(u(\cdot,t), v(\cdot,t), \xi(t)\big) + B_1(u(\cdot,t), v(\cdot,t), \xi(t)) = (f_1(\cdot,t), f_2(\cdot,t), e(t)),\ 0 < t < T,$$
$$\big(u(\cdot,0), v(\cdot,0), \xi(0)\big) = \big(u_0, v_0, v_0(1)\big),$$

where $H_3$ and $B_1$ were defined in *Example* 4.

**Theorem 2.0.34.** (see [24], p. 131) *Assume* $(j_1)$, $(j_2)$, $(j_4)$ (*see Examples* 3 *and* 4). *Let* $T > 0$ *be fixed. Then, for every* $f_1,\ f_2 \in L^1(0,T;L^2(0,1))$, *and* $u_0,\ v_0 \in L^2(0,1)$, $e \in L^1(0,T)$ *there exists a unique weak solution* $u,\ v \in C([0,T];L^2(0,1))$ *of problem* (2.14), (2.15), (2.17). *If, in addition,* $f_1,\ f_2 \in W^{1,1}(0,T;L^2(0,1))$, $e \in W^{1,1}(0,T)$, $u_0,\ v_0 \in H^1(0,1)$, *and* $-\,u_0(0) = r_0(v_0(0))$, *then* $(u,\ v) \in W^{1,\infty}(0,T;\ L^2(0,1))^2$ *and*

(i) $u, v \in L^\infty((0,1)\times(0,T))$, $u_x, v_x \in L^\infty(0,T;L^2(0,1))$, $v(1,\cdot) \in L^\infty(0,T)$;

(ii) $u,\ v$ *satisfy* (2.14) *for all* $t \in [0,T)$, *a.a.* $x \in (0,1)$, *where* $u_t,\ v_t$ *are replaced by* $\frac{\partial^+ u}{\partial t}$, $\frac{\partial^+ v}{\partial t}$;

(iii) $u,\ v$ *satisfy* (2.15) *and* (2.17) (*more precisely, the former equation of* (2.17) *is satisfied for all* $t \in [0,T]$, *while the latter is satisfied for all* $t \in [0,T)$, *where* $v_t(1,t)$ *is replaced by* $\frac{\partial^+ v}{\partial t}(1,t)$).

Again, the proof is based on Theorems 2.0.20, 2.0.21. For the strong solution we obtain that $\xi(t) = v(1,t)$, $0 \le t \le T$.

Note also that the last two results still hold if we consider uniform Lipschitz perturbations of the nonlinear functions $r(x,\cdot)$, $g(x,\cdot)$, and also a Lipschitz perturbation of $f_0$ in the case of a dynamic boundary condition.

The remainder of this chapter is meant for recalling some existence and regularity results for second order abstract differential equations.

Consider in $H$ the following boundary value problem

$$\begin{cases} u''(t) = Au(t) + f(t), & 0 < t < T, \\ u(0) = a, \quad u(T) = b. \end{cases} \tag{2.18}$$

As far as the problem (2.18) is concerned, we have the following result of V. Barbu (see [7], p. 310):

**Theorem 2.0.35.** *Let $A : D(A) \subset H \to H$ be a maximal monotone operator, $a, b \in D(A)$ and $f \in L^2(0,T;H)$. Then problem* (2.18) *has a unique strong solution $u \in W^{2,2}(0,T;H)$.*

The above result was extended by R.E. Bruck in [12] to the case $a, b \in \overline{D(A)}$ :

**Theorem 2.0.36.** *Let $A : D(A) \subset H \to H$ be a maximal monotone operator, $a, b \in \overline{D(A)}$ and $f \in L^2(0,T;H)$. Then problem* (2.18) *has a unique strong solution $u \in C([0,T];H) \bigcap W^{2,2}_{\text{loc}}(0,T;H)$ such that*

$$t^{1/2}(T-t)^{1/2}u' \in L^2(0,T;H), \quad t^{3/2}(T-t)^{3/2}u'' \in L^2(0,T;H).$$

Another extension has been established by A.R. Aftabizadeh and N.H. Pavel [2]. Specifically, let us consider in $H$ the problem

$$\begin{cases} p(t)u''(t) + r(t)u'(t) = Au(t) + f(t), & 0 < t < T, \\ u'(0) \in \beta_1(u(0) - a), \quad -u'(T) \in \beta_2(u(T) - b). \end{cases} \tag{2.19}$$

**Theorem 2.0.37.** *If $A$, $\beta_1$, $\beta_2$ are maximal monotone in $H$; $0$, $a$, $b \in D(A)$; $f \in L^2(0,T;H)$;*

$$\langle A_\lambda x - A_\lambda y, v\rangle \geq 0 \;\forall\, \lambda > 0 \;\forall\, v, x, y \in H, \; x - y \in D(\beta_1), \; v \in \beta_1(x-y),$$
$$\langle A_\lambda x - A_\lambda y, v\rangle \geq 0 \;\forall\, \lambda > 0 \;\forall\, v, x, y \in H, \; x - y \in D(\beta_2), \; v \in \beta_2(x-y),$$

*where $A_\lambda$ denotes the Yosida approximation of $A$; either $D(\beta_1)$ or $D(\beta_2)$ is bounded; $p, r \in W^{1,\infty}(0,T)$ and $p(t) \geq c > 0 \;\forall\, t \in [0,T]$, then problem* (2.19) *has at least one solution $u \in W^{2,2}(0,T;H)$.*

*If, in addition, at least one of the operators $A$, $\beta_1$, $\beta_2$ is injective, then the solution is unique.*

Note that the last three theorems hold true for multivalued $A$, and inclusion relation instead of $(2.18)_1$.

# Part II

# Singularly Perturbed Hyperbolic Problems

# Chapter 3

# Presentation of the Problems

In this chapter we present some singularly perturbed, hyperbolic, initial-boundary value problems which will be investigated in detail in the next chapters of Part II. More precisely, we are interested in some initial-boundary value problems associated with a partial differential system, known as the telegraph system. This system describes propagation phenomena in electrical circuits. In an attempt to get as close as possible to physical reality, we will associate with the telegraph system some linear or nonlinear boundary conditions which describe specific physical situations. We hope that the models we are going to discuss will be of interest to engineers and physicists. Although our theoretical investigations will be focused on specific models, this will not affect the generality of our methods. In fact, we are going to address some essential issues which could easily be adapted to further situations.

Note that the nonlinear character of the problems we are going to discuss creates many difficulties. Nonlinearities may occur either in the telegraph system or in the boundary conditions. Our aim here is to develop methods which work for such nonlinear problems.

Let $D_T := \big\{(x,t) \in \mathbb{R}^2;\ 0 < x < 1,\ 0 < t < T\big\}$, where $T > 0$ is a given time instant. Consider the telegraph system in $D_T$ :

$$\begin{cases} \varepsilon u_t + v_x + ru = f_1(x,t), \\ v_t + u_x + gv = f_2(x,t), \end{cases} \tag{$LS$}$$

with the initial conditions:

$$u(x,0) = u_0(x), \quad v(x,0) = v_0(x), \quad 0 \le x \le 1, \tag{$IC$}$$

and nonlinear algebraic-differential boundary conditions of the form:

$$\begin{cases} r_0 u(0,t) + v(0,t) = 0, \\ u(1,t) - kv_t(1,t) = f_0(v(1,t)) + e(t), \quad 0 \le t \le T. \end{cases} \tag{$BC.1$}$$

Problem $(LS)$, $(IC)$, $(BC.1)$ is a model for transmission (propagation) in electrical circuits. Specifically, let us consider a circuit of the form indicated in the figure below (see [15]).

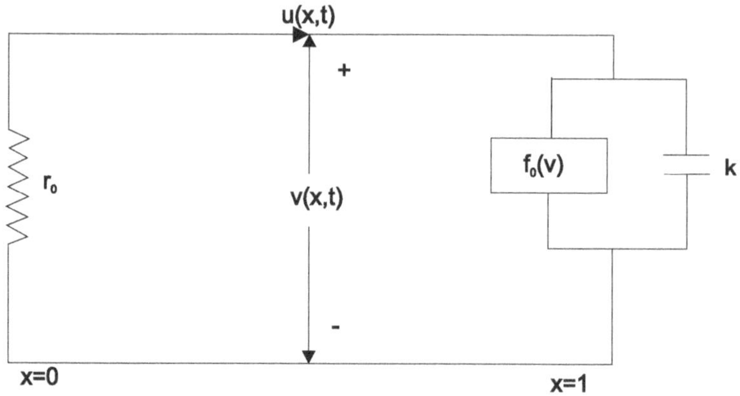


We denote by $u(x,t)$ the current at time instant $t$, at point $x$ of the line, $0 \leq x \leq 1$; $v(x,t)$ is the voltage at $t$ and $x$. It is well known that functions $u$ and $v$ satisfy the telegraph system:

$$\begin{cases} Lu_t + v_x + ru = E(x,t), \\ Cv_t + u_x + gv = 0, \end{cases}$$

where $r$, $g$, $L$, $C$ represent the resistance, conductance, inductance and capacitance per unit length, while $E$ is the voltage per unit length impressed along the line in series with it.

Obviously, the above system is of the form $(LS)$ in which we set $C=1$, since this assumption does not restrict the generality of the problem. Also, we re-denote $L$ by $\varepsilon$, since the inductance will be considered a small parameter. It is well known that the inductance of the line is small whenever the corresponding frequency is small.

At the end $x=0$ of the line, we have a resistance $r_0$ and we assume that Ohm's law applies, i.e., $(BC.1)_1$. At the other end of the line, i.e., at the point $x=1$, we have a capacitance $k$ and a nonlinear resistor whose characteristic is $f_0(v)$. Therefore, at $x=1$ condition $(BC.1)_2$ holds, with $e(t) \equiv 0$. This boundary condition is nonlinear due to the presence of the term $f_0(v(1,t))$. Also, this is a dynamic (or differential) condition due to the presence of $v_t(1,t)$.

Therefore, problem $(LS)$, $(IC)$, $(BC.1)$, which will be studied in the next chapters, has an obvious physical motivation.

We will also investigate a more general problem in which the latter equation of the telegraph system is nonlinear. More precisely, we consider in $D_T$ the following nonlinear hyperbolic system:

$$\begin{cases} \varepsilon u_t + v_x + ru = f_1(x,t), \\ v_t + u_x + g(x,v) = f_2(x,t), \end{cases} \tag{$NS$}$$

with initial conditions ($IC$) and boundary conditions ($BC$.1). As we will see, the treatment of this more general problem is more difficult and the results are weaker. The investigation of this nonlinear problem will show how to deal with the case where the partial differential system is also nonlinear.

Another problem which will be investigated consists of system ($NS$), initial conditions ($IC$), as well as two integro-differential boundary conditions of the form:

$$\begin{cases} u(0,t) + k_0 v_t(0,t) = -r_0(v(0,t)) - l_0 \int_0^t v(0,s)ds + e_0(t), \\ u(1,t) - k_1 v_t(1,t) = \ \ r_1(v(1,t)) + l_1 \int_0^t v(1,s)ds - e_1(t), \ 0 \le t \le T. \end{cases} \tag{$BC.2$}$$

Here, both $r_0$ and $r_1$ are nonlinear functions. Such boundary conditions also occur in the theory of electrical circuits. Indeed, let us consider the following electrical circuit (cf. [15]).

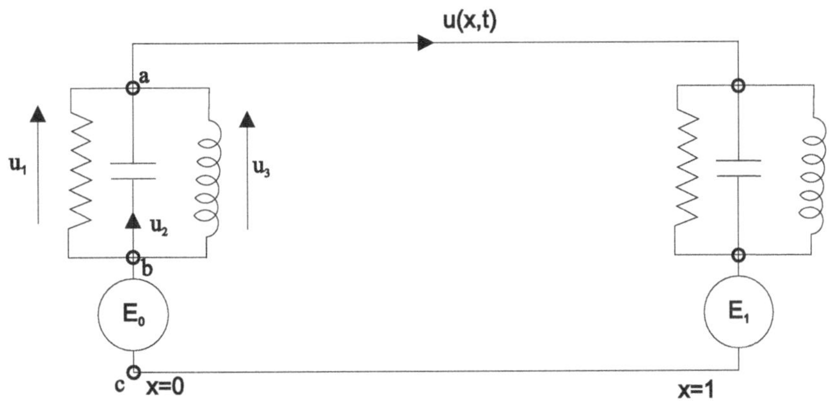


At the end $x$=0 we have the following obvious equations:

$$\begin{cases} v(0,t) = v_{ac}, \\ E_0(t) = v_{bc}, \\ u(0,t) = u_1 + u_2 + u_3, \end{cases}$$

where $v_{ac}$, $v_{bc}$ are the potential differences between the corresponding points, while $u_1$, $u_2$, $u_3$ are the currents passing through the three elements (the resistor, condenser, and inductor) located at $x$=0. Let us admit that the following equations connect the currents and $v_{ab}{=}v(0,t){-}E_0(t)$:

$$\begin{cases} -u_1 = r_0(v_{ab}), \\ -u_2 = k_0 \frac{d}{dt} v_{ab}, \\ -u_3 = l_0 \int_0^t v_{ab}(s)ds. \end{cases}$$

Note that the first equation is nonlinear, while the others are linear. We have denoted by $c_0$ the capacitance of the condenser and by $l_0$ the inverse of the inductance value. So we derive the following boundary condition:

$$\begin{aligned} u(0,t) &+ k_0[v_t(0,t) - E_0'(t)] \\ &= -r_0\big(v(0,t) - E_0(t)\big) - l_0 \int_0^t [v(0,s) - E_0(s)]ds. \end{aligned}$$

For the other end of the line, $x = 1$, we have a similar boundary condition:

$$\begin{aligned} u(1,t) &- k_1[v_t(1,t) - E_1'(t)] \\ &= r_1\big(v(1,t) - E_1(t)\big) + l_1 \int_0^t [v(1,s) - E_1(s)]ds. \end{aligned}$$

If $E_0$ and $E_1$ are null, then the last two equations become of the form $(BC.2)$ in which $e_0(t){\equiv}0$, $e_1(t){\equiv}0$. Otherwise, one may replace the unknown $v$ by $\tilde{v}$, where

$$\tilde{v}(x,t) = v(x,t) - xE_1(t) - (1-x)E_0(t).$$

Thus functions $u$, $\tilde{v}$ satisfy $(BC.2)$ in which $e_0$ and $e_1$ are zero.

Another case we will study is that in which the capacitance per unit length is negligible, while the inductance is sizeable. More precisely, system $(LS)$ is replaced by:

$$\begin{cases} u_t + v_x + ru = f_1(x,t), \\ \varepsilon v_t + u_x + gv = f_2(x,t), \end{cases} \tag{$(LS)'$}$$

with initial conditions $(IC)$ and boundary conditions of the form $(BC.1)$. Note that this new model also describes the situation in which the inductance is negligible and the two boundary conditions are reversed. In any case, we will see that this model requires separate analysis.

Also, we will consider $(LS)$ associated with boundary conditions at the two end points, which are both algebraic and nonlinear:

$$\begin{cases} u(0,t) + r_0(v(0,t)) = 0, \\ u(1,t) = f_0(v(1,t)), \quad 0 \le t \le T. \end{cases} \tag{$(BC.3)$}$$

The physical relevance of this problem is obvious. In fact, this will be the first model we are going to address. For each model we need specific methods of investigation.

Of course, there are further boundary conditions which may occur. However, the models discussed in this book basically cover all possible physical situations and the methods we use could easily be adapted to further problems, including more complex models. For example, in the case of integrated circuits, there are $n$ telegraph systems with $2n$ unknowns which are all connected by means of different boundary conditions. Such models could be investigated by similar methods and techniques. In some cases, there are negligible parameters (for instance, inductances) and so we are led to the idea of replacing the original perturbed models by reduced (unperturbed) models. In order to make sure that the reduced models still describe well the corresponding phenomena we have to develop an asymptotic analysis of the singular perturbation problems associated with such transmission processes in integrated circuits. This analysis can be done by using the same technique as for the case of a single telegraph system for which we will perform a complete investigation.

Mention should be made of the fact that the models presented above, or similar more complex models, also describe further physical problems, in particular problems which occur in hydraulics (see V. Barbu [8], V. Hara [23], V. Iftimie [25], I. Straskraba and V. Lovicar [41], V.L. Streeter and E.B. Wylie [42]).

# Chapter 4

# Hyperbolic Systems with Algebraic Boundary Conditions

In this chapter we investigate initial-boundary value problems of the form $(LS)$, $(IC)$, $(BC.3)$ presented in Chapter 3. The other problems will be discussed in the next chapter, since they require separate analysis.

In the first section we derive formally a zeroth order asymptotic expansion for the solution of $(LS)$, $(IC)$, $(BC.3)$ by using the method described in Chapter 1. This problem is singularly perturbed of the boundary layer type with respect to the sup norm. We determine the corresponding boundary layer function as well as the problems satisfied by the regular term and by the remainder components.

In the second section we prove some results on the existence, uniqueness and higher regularity of the solutions of both the perturbed and the reduced problem, under appropriate assumptions on the data. For the perturbed problem we derive the existence and uniqueness of the strong solution from the general theory for evolution equations. For higher regularity of the solution we use an approach based on d'Alembert like formulas and on Banach's contraction principle. For the reduced problem we use the general theory of evolution equations, including non-autonomous evolution equations. All these results guarantee the fact that our asymptotic expansion is well defined.
Moreover, they are used in the third and last section to derive estimates for the remainder components with respect to the uniform convergence norm. These estimates show that our expansion is a real asymptotic expansion.
In order to develop our treatment we need to assume higher smoothness of the data as well as adequate compatibility of the data with the boundary conditions.

## 4.1 A zeroth order asymptotic expansion

Let us consider in $D_T := \{(x,t) \in \mathbb{R}^2;\ 0 < x < 1,\ 0 < t < T\}$ the linear telegraph system:

$$\begin{cases} \varepsilon u_t + v_x + ru = f_1(x,t), \\ v_t + u_x + gv = f_2(x,t), \end{cases} \tag{$LS$}$$

with the initial conditions

$$u(x,0) = u_0(x), \quad v(x,0) = v_0(x), \quad 0 \le x \le 1, \tag{$IC$}$$

and nonlinear boundary conditions:

$$\begin{cases} u(0,t) + r_0(v(0,t)) = 0, \\ u(1,t) = f_0(v(1,t)), \quad 0 \le t \le T. \end{cases} \tag{$BC.3$}$$

We suppose that $f_1, f_2 : \overline{D}_T \to \mathbb{R}$, $u_0, v_0 : [0,1] \to \mathbb{R}$, $f_0, r_0 : \mathbb{R} \to \mathbb{R}$ are known functions while $r, g$ are given constants, $r > 0$, $g \ge 0$, $\varepsilon$ is a positive small parameter. Problem $(LS)$, $(IC)$, $(BC.3)$, which will also be called $P_\varepsilon$, is of the hyperbolic type with nonlinear algebraic boundary conditions. Denote the solution of $P_\varepsilon$ by $U_\varepsilon(x,t) = (u_\varepsilon(x,t),\ v_\varepsilon(x,t))$. Arguing as in Chapter 1, we see that $U_\varepsilon$ has a singular behavior with respect to the uniform convergence topology within a neighborhood of the segment $\{(x,0);\ 0 \le x \le 1\}$ which is the *boundary layer*. Indeed, if $U_\varepsilon$ would converge uniformly in $\overline{D}_T$, then necessarily

$$v_0'(x) + ru_0(x) - f_1(x,0) = 0, \quad 0 \le x \le 1,$$

and this condition is not satisfied in general. Using the classical singular perturbation theory (see Chapter 1) we can derive a zeroth order expansion for $U_\varepsilon$ in the form:

$$U_\varepsilon = U_0(x,t) + V_0(x,\tau) + R_\varepsilon(x,t), \tag{4.1}$$

where:

$U_0 = (X(x,t), Y(x,t))$ is the zeroth order term of the regular series;

$V_0 = (c_0(x,\tau), d_0(x,\tau)), \tau = t/\varepsilon$, is the boundary layer (vector) function;

$R_\varepsilon = (R_{1\varepsilon}(x,t), R_{2\varepsilon}(x,t))$ is the remainder of the order zero.

We substitute (4.1) into $P_\varepsilon$ and then identify the coefficients of $\varepsilon^k$, $k = -1, 0$, those depending on $t$ separately from those depending on $\tau$. In fact, we have already discussed the formal derivation of this asymptotic expansion in Chapter 1, but here we include additional details. First we substitute (4.1) into $(LS)$. We obtain that $X$ and $Y$ satisfy an algebraic equation and a parabolic equation, respectively:

$$X = (1/r)(f_1 - Y_x) \text{ in } D_T, \tag{4.2}$$

$$Y_t - (1/r)Y_{xx} + gY = f_2 - (1/r)f_{1x}, \text{ in } D_T. \tag{4.3}$$

For the boundary functions $c_0$, $d_0$ we get

$$d_0 \equiv 0,\ c_0(x,\tau) = \alpha(x)e^{-r\tau}, \tag{4.4}$$

where function $\alpha$ will be determined below from $(IC)$.
For the remainder components we formally obtain the following system:

$$\begin{cases} \varepsilon R_{1\varepsilon t} + R_{2\varepsilon x} + rR_{1\varepsilon} = -\varepsilon X_t, & \text{in } D_T, \\ R_{2\varepsilon t} + R_{1\varepsilon x} + gR_{2\varepsilon} = -c_{0x}, & \text{in } D_T. \end{cases} \tag{4.5}$$

Now, from $(IC)$ it follows $d_0(x,0) + Y(x,0) = v_0(x)$, so

$$Y(x,0) = v_0(x),\quad 0 \le x \le 1, \tag{4.6}$$

$$c_0(x,0) + X(x,0) = u_0(x) \Leftrightarrow$$

$$\alpha(x) = u_0(x) + (1/r)(v_0'(x) - f_1(x,0)),\quad 0 \le x \le 1, \tag{4.7}$$

$$R_{1\varepsilon}(x,0) = R_{2\varepsilon}(x,0) = 0,\quad 0 \le x \le 1. \tag{4.8}$$

Finally, substituting (4.1) into $(BC.3)$ we derive

$$X(0,t) + r_0\left(Y(0,t)\right) = 0,\quad X(1,t) - f_0\left(Y(1,t)\right) = 0,\quad 0 \le t \le T,$$

which can be written as (cf. (4.2))

$$\begin{cases} -(1/r)Y_x(0,t) + r_0\left(Y(0,t)\right) = -(1/r)f_1(0,t), \\ (1/r)Y_x(1,t) + f_0\left(Y(1,t)\right) = (1/r)f_1(1,t),\quad 0 \le t \le T. \end{cases} \tag{4.9}$$

Also we obtain

$$c_0(0,\tau) = c_0(1,\tau) = 0 \quad \Leftrightarrow \quad \alpha(0) = \alpha(1) = 0, \tag{4.10}$$

and the following boundary conditions for the remainder components

$$\begin{cases} R_{1\varepsilon}(0,t) + r_0\Big(R_{2\varepsilon}(0,t) + Y(0,t)\Big) - r_0\Big(Y(0,t)\Big) = 0, \\ R_{1\varepsilon}(1,t) - f_0\Big(R_{2\varepsilon}(1,t) + Y(1,t)\Big) + f_0\Big(Y(1,t)\Big) = 0,\quad 0 \le t \le T. \end{cases} \tag{4.11}$$

Summarizing, we see that the components of the regular part satisfy the reduced problem $P_0$, which comprises (4.2), (4.3), (4.6) and (4.9), while the remainder components satisfy the problem (4.5), (4.8), (4.11).

As for conditions (4.10), we have already pointed out at the end of Chapter 1 that they are needed to eliminate possible discrepancies which may occur due to $c_0$ at the corner points $(x,t) = (0,0)$ and $(x,t) = (1,0)$. As we will see in the next section, these conditions are also compatibility conditions that we need to derive results on the existence, uniqueness and smoothness of the solutions of the problems involved in our asymptotic analysis.

## 4.2 Existence, uniqueness and regularity of the solutions of problems $P_\varepsilon$ and $P_0$

Let us start with problem $P_\varepsilon$, which comprises $(LS)$, $(IC)$, $(BC.3)$. This is a particular case of problem (2.14), (2.15), (2.16) (see Chapter 2), so on account of Theorem 2.0.33 we have:

**Theorem 4.2.1.** *Assume that*

$$r, g \text{ are nonnegative constants;} \tag{4.12}$$

$$r_0, f_0 : \mathbb{R} \to \mathbb{R} \text{ are continuous nondecreasing functions;} \tag{4.13}$$

$$f_1, f_2 \in L^1\left(0,T; L^2(0,1)\right), \quad u_0, v_0 \in L^2(0,1).$$

*Then problem $P_\varepsilon$ has a unique weak solution $(u_\varepsilon, v_\varepsilon) \in C\big([0,T]; L^2(0,1)\big)^2$. If, in addition,*

$$f_1,\ f_2 \in W^{1,1}(0,T; L^2(0,1)), \tag{4.14}$$

*$u_0, v_0 \in H^1(0,1)$, and satisfy the zeroth order compatibility conditions:*

$$\begin{cases} u_0(0) + r_0\left(v_0(0)\right) = 0, \\ u_0(1) - f_0\left(v_0(1)\right) = 0, \end{cases} \tag{4.15}$$

*then the solution $(u_\varepsilon, v_\varepsilon)$ of problem $P_\varepsilon$ belongs to the space $W^{1,\infty}\big(0,T; L^2(0,1)\big)^2$ and $u_{\varepsilon x}, v_{\varepsilon x} \in L^\infty\big(0,T; L^2(0,1)\big)$.*

In the following we need higher regularity of the solution in order to validate the above asymptotic expansion, in particular to obtain estimates for the remainder components with respect to the uniform convergence norm. To this purpose, we will consider in a first stage the particular case $r = g = 0$, for which we can make use of some d'Alembert like formulae. We start with the following $C^1$ regularity result:

**Proposition 4.2.2.** *Assume that $r = g = 0$ and*

$$r_0, f_0 \in C^1(\mathbb{R}),\ \ r_0' \geq 0,\ \ f_0' \geq 0; \tag{4.16}$$

$$f_1, f_2 \in C^1([0,T]; C[0,1]); \tag{4.17}$$

*$u_0, v_0 \in C^1[0,1]$ satisfy (4.15), and the following first order compatibility conditions hold*

$$\begin{cases} f_1(0,0) - v_0'(0) + \varepsilon r_0'\left(v_0(0)\right)\left[f_2(0,0) - u_0'(0)\right] = 0, \\ f_1(1,0) - v_0'(1) - \varepsilon f_0'\left(v_0(1)\right)\left[f_2(1,0) - u_0'(1)\right] = 0. \end{cases} \tag{4.18}$$

*Then, the (strong) solution $(u_\varepsilon, v_\varepsilon)$ of problem $P_\varepsilon$ belongs to $C^1(\overline{D}_T)^2$.*

*Proof.* Without any loss of generality we suppose that $\varepsilon = 1$. For simplicity we will denote by $(u, v)$ the solution of $P_\varepsilon$ corresponding to $\varepsilon = 1$. It is easily seen that the general solution of system $(LS)$ with $r = g = 0$ is given by the following d'Alembert type formulae:

$$\begin{cases} u(x,t) = \frac{1}{2}\left[\mu(x-t) + \eta(x+t)\right] \\ +\frac{1}{2}\int_0^t [(f_1+f_2)(x-s,t-s) + (f_1-f_2)(x+s,t-s)]ds, \\ v(x,t) = \frac{1}{2}\left[\mu(x-t) - \eta(x+t)\right] \\ +\frac{1}{2}\int_0^t [(f_1+f_2)(x-s,t-s) - (f_1-f_2)(x+s,t-s)]ds, \end{cases} \tag{4.19}$$

where $\mu : [-T, 1] \to \mathbb{R}$, $\eta : [0, 1+T] \to \mathbb{R}$ are some arbitrary $C^1$ functions. We consider that in the above formulae $f_1$ and $f_2$ are extended to $\mathbb{R} \times [0, T]$ by

$$\begin{cases} f_i(x,t) := f_i(2-x,t), & \text{for } i = 1,2, \quad 1 < x \le 2, \\ f_i(x,t) := f_i(-x,t), & \text{for } i = 1,2, \quad -1 \le x < 0, \end{cases}$$

and so on. These extensions belong to $C^1([0,T]; C(\mathbb{R}))$. When necessary, every function defined on $\overline{D}_T$ will be extended in a similar manner. In the following we will show that the unknown functions $\mu$ and $\eta$ of system (4.19) can be determined from the conditions that $u$ and $v$ satisfy both $(IC)$ and $(BC.3)$.

First, we obtain from $(IC)$

$$\begin{cases} \mu(x) = u_0(x) + v_0(x), \\ \eta(x) = u_0(x) - v_0(x), \quad 0 \le x \le 1, \end{cases} \tag{4.20}$$

thus $\mu$ and $\eta$ are uniquely determined on the interval $[0, 1]$. By employing formulae (4.19) in $(BC.3)$ we obtain that

$$\begin{cases} (1/2)[\mu(-t) + \eta(t)] + \int_0^t f_1(s,t-s)ds \\ +r_0\left((1/2)[\mu(-t) - \eta(t)] + \int_0^t f_2(s,t-s)ds\right) = 0, \\ (1/2)[\mu(1-t) + \eta(1+t)] + \int_0^t f_1(1-s,t-s)ds \\ = f_0\left((1/2)[\mu(1-t) - \eta(1+t)] + \int_0^t f_2(1-s,t-s)ds\right), \end{cases} \tag{4.21}$$

for all $t \in [0, T]$. For simplicity we assume that $T \le 1$. It is easy to see that $\mu$ and $\eta$ can uniquely and completely be determined from (4.21). Indeed, if we denote by $z_1(t)$, $z_2(t)$ the above arguments of functions $r_0$, $f_0$, i.e.,

$$z_1(t) := (1/2)[\mu(-t) - \eta(t)] + \int_0^t f_2(s,t-s)ds,$$

$$z_2(t) := (1/2)[\mu(1-t) - \eta(1+t)] + \int_0^t f_2(1-s,t-s)ds,$$

then equations (4.21) can be written in the form:

$$\begin{cases} (I + r_0)(z_1(t)) = h_1(t), \\ (I + f_0)(z_2(t)) = h_2(t), \quad 0 \le t \le T, \end{cases} \tag{4.22}$$

where

$$\begin{cases} h_1(t) = -\eta(t) - \int_0^t (f_1 - f_2)(s, t-s)ds, \\ h_2(t) = \mu(1-t) + \int_0^t (f_1 + f_2)(1-s, t-s)ds, \quad 0 \le t \le T, \end{cases}$$

are known functions (see (4.20)). Since $r_0, f_0$ are maximal monotone, it follows that $z_1$ and $z_2$ are uniquely determined from (4.22) (see Theorem 2.0.9). Thus $\mu$ and $\eta$ are also uniquely determined. In addition, the compatibility conditions (4.15) and (4.18) imply that $\mu \in C^1[-T, 1]$ and $\eta \in C^1[0, 1+T]$. Therefore, on account of (4.19), we obtain that $(u, v) \in C^1([0, T]; C[0, 1])^2$. Since $(u, v)$ satisfies system $(LS)$ it follows that $(u, v) \in C^1(\overline{D}_T)^2$. □

*Remark* 4.2.3. Conditions (4.15), (4.18) are necessary conditions for the $C^1$ regularity of $(u_\varepsilon, v_\varepsilon)$. Indeed, by the continuity of $u$ and $v$ in $\overline{D}_T$, we can take $t = 0$ in $(BC.3)$ and obtain (4.15). Since $u$ and $v$ are continuously differentiable in $\overline{D}_T$, we can differentiate both the equations of $(BC.3)$ and then take $t = 0$, thus obtaining (4.18).

*Remark* 4.2.4. Note that, under the assumptions of Proposition 4.2.2 problem $P_\varepsilon$ is equivalent to system (4.19), (4.20), (4.21). But, this system may have a solution under weaker assumptions. In this situation, it is quite natural to call it a *generalized solution* of problem $P_\varepsilon$. For example, under the following weaker assumptions: (4.13), $f_1, f_2 \in C(\overline{D}_T)$, $u_0, v_0 \in C[0, 1]$ and satisfy the compatibility conditions (4.15), problem $P_\varepsilon$ has a unique generalized solution $(u_\varepsilon, v_\varepsilon) \in C(\overline{D}_T)^2$. See the proof of the above proposition.

*Remark* 4.2.5. Both Proposition 4.2.2 and the last statement of Remark 4.2.4 remain valid if $(BC.3)$ are replaced by boundary conditions of the form

$$\begin{cases} u(0, t) + \alpha_1(t)v(0, t) = 0, \\ u(1, t) - \alpha_2(t)v(1, t) = 0, \ 0 < t < T. \end{cases} \tag{$BC.3)'$$

Thus, if (4.17) hold, $\alpha_1, \alpha_2 \in C^1[0, T]$, $\alpha_1, \alpha_2 \ge 0$, $u_0, v_0 \in C^1[0, 1]$ and the following compatibility conditions hold

$$\begin{cases} u_0(0) + \alpha_1(0)v_0(0) = 0, \\ u_0(1) - \alpha_2(1)v_0(1) = 0, \end{cases} \tag{4.23}$$

$$\begin{cases} f_1(0, 0) - v_0'(0) + \varepsilon[\alpha_1'(0)v_0(0) + \alpha_1(0)(f_2(0, 0) - u_0'(0))] = 0, \\ f_1(1, 0) - v_0'(1) - \varepsilon[\alpha_2'(1)v_0(1) + \alpha_2(1)(f_2(1, 0) - u_0'(1))] = 0, \end{cases} \tag{4.24}$$

then $P_\varepsilon$ (with $(BC.3)'$ instead of $(BC.3)$) has a unique solution $(u_\varepsilon, v_\varepsilon) \in C^1(\overline{D}_T)^2$. If we assume the following weaker conditions: $\alpha_1, \alpha_2 \in C[0,T]$, $\alpha_1, \alpha_2 \geq 0$, $f_1, f_2 \in C(\overline{D}_T)$, $u_0, v_0 \in C[0,1]$, and (4.23) are satisfied, then $P_\varepsilon$ (with $(BC.3)'$ instead of $(BC.3)$) has a unique generalized solution $(u_\varepsilon, v_\varepsilon) \in C(\overline{D}_T)^2$.

We continue with another existence result which will be essential in our treatment.

**Proposition 4.2.6.** *Assume that* $r = g = 0$ *and* (4.13) *holds. If*

$$f_1, f_2 \in L^\infty(0,T; L^p(0,1)) \ (or\ f_1, f_2 \in C([0,T]; L^p(0,1))),$$

$u_0, v_0 \in L^p(0,1)$, *where* $p \in [1, \infty)$, *then problem* $P_\varepsilon$ *has a unique generalized solution* $(u_\varepsilon, v_\varepsilon) \in L^\infty(0,T; L^p(0,1))^2$ $(C([0,T]; L^p(0,1))^2$, *respectively*).

*Proof.* For simplicity and without any loss of generality, we assume again that $T \leq 1$. We suppose first that $f_1, f_2 \in L^\infty(0,T; L^p(0,1))$ $(C([0,T]; L^p(0,1)))$, and $u_0, v_0 \in L^p(0,1)$. Since both $(I + f_0)^{-1}$, $(I + r_0)^{-1}$ are Lipschitzian and

$$(t,s) \to f_i(s, t-s), \quad (t,s) \to f_i(1-s, t-s), \quad i = 1, 2,$$

are Lebesgue measurable, in view of (4.20) and (4.21) we obtain $\mu \in L^p(-T, 1)$ and $\eta \in L^p(0, 1+T)$. On the other hand, $\mu(x-t)$, $\eta(x+t) \in C([0,T]; L^p(0,1))$ and the integral terms in (4.19) belong to $L^\infty(0,T; L^p(0,1))$ $(C([0,T]; L^p(0,1))$, respectively). Thus $(u_\varepsilon, v_\varepsilon) \in L^\infty(0,T; L^p(0,1))^2$ $(C([0,T]; L^p(0,1))^2$, respectively). □

*Remark* 4.2.7. It is easily seen that Proposition 4.2.6 is also valid if $(BC.3)$ are replaced by boundary conditions of the form $(BC.3)'$ with appropriate assumptions on $\alpha_1$ and $\alpha_2$. Also, one can consider non-homogeneous boundary conditions and derive similar existence results.

Using the previous results, we are going to derive higher regularity for the solution of problem $P_\varepsilon$ in the case $r > 0$ and/or $g > 0$. The first step in this direction is the following proposition.

**Proposition 4.2.8.** *Assume that* (4.13) *holds, and* $r > 0$ *and/or* $g > 0$. *If* $f_1, f_2 \in C(\overline{D}_T)$, $u_0, v_0 \in C[0,1]$ *and satisfy* (4.15), *then problem* $P_\varepsilon$ *has a unique generalized solution* $(u_\varepsilon, v_\varepsilon) \in C(\overline{D}_T)^2$. *More precisely,* $(u_\varepsilon, v_\varepsilon)$ *satisfies* (4.19), (4.20), (4.21) *where* $f_1$, $f_2$ *are replaced by* $f_1 - ru_\varepsilon$, $f_2 - gv_\varepsilon$.
*If the following weaker assumptions are satisfied*

$$f_1, f_2 \in L^\infty(0,T; L^2(0,1)), \quad u_0, v_0 \in L^2(0,1),$$

*then problem* $P_\varepsilon$ *has a unique generalized solution* $(u_\varepsilon, v_\varepsilon) \in L^\infty(0,T; L^2(0,1))^2$.

*Proof.* In view of Theorem 4.2.1, there exists a unique weak solution $(u_\varepsilon, v_\varepsilon) \in C([0,T]; L^2(0,1))^2$ of problem $P_\varepsilon$. Using Banach's Fixed Point Theorem, one can show that the solution is more regular, more precisely it belongs to the space $C(\overline{D}_T)^2$, if $f_1, f_2 \in C(\overline{D}_T)$, $u_0, v_0 \in C[0,1]$ and satisfy (4.15). To this purpose, we consider the space $Z{=}C(\overline{D}_T)^2$ equipped with the norm $\|(u,v)\|_Z{:=}\max(\|u\|,\|v\|)$, where

$$\|u\| := \sup_{(x,t)\in\overline{D}_T} e^{-Lt} \mid u(x,t) \mid,$$

and $L$ is a large positive constant. Define the operator $\mathcal{S} : Z \to Z$, which assigns to each pair $\gamma = (\gamma_1, \gamma_2) \in C(\overline{D}_T)^2$ the unique solution of system (4.19), (4.20), (4.21), where $f_1$, $f_2$ are replaced by $f_1{-}r\gamma_1$, $f_2{-}g\gamma_2$. According to Remark 4.2.4 this solution belongs $C(\overline{D}_T)^2$, and hence operator $\mathcal{S}$ is well defined.
For a sufficiently large $L$, operator $\mathcal{S} : Z \to Z$ is a strict contraction, i.e.,

$$\|\mathcal{S}\gamma - \mathcal{S}\widehat{\gamma}\|_Z \leq k\|\gamma - \widehat{\gamma}\|_Z \ \forall\ \gamma = (\gamma_1, \gamma_2),\ \widehat{\gamma} = (\widehat{\gamma}_1, \widehat{\gamma}_2) \in Z$$

where $0 < k < 1$. Indeed, we have, for example,

$$\begin{aligned}\Big|\int_0^t &(f_1 - r\gamma_1 + f_2 - g\gamma_2)(x-s, t-s)ds \\ &- \int_0^t (f_1 - r\widehat{\gamma}_1 + f_2 - g\widehat{\gamma}_2)(x-s, t-s)ds\Big| \\ &\leq 2\ \max\{r, g\}\|\gamma - \widehat{\gamma}\|_Z \int_0^t e^{L(t-s)}ds \leq \frac{2}{L}\ \max\{r, g\}e^{Lt}\|\gamma - \widehat{\gamma}\|_Z,\end{aligned}$$

for all $t \in [0, T]$. Therefore

$$\begin{aligned}\Big\|\int_0^t &(f_1 - r\gamma_1 + f_2 - g\gamma_2)(x-s, t-s)ds \\ &- \int_0^t (f_1 - r\widehat{\gamma}_1 + f_2 - g\widehat{\gamma}_2)(x-s, t-s)ds\Big\|_Z \leq \frac{2}{L}\ \max\{r, g\}\|\gamma - \widehat{\gamma}\|_Z.\end{aligned}$$

Using (4.20) and (4.21), with $f_1 - r\gamma_1$, $f_2 - g\gamma_2$ instead of $f_1$, $f_2$, we obtain similar estimates for

$$\|\mu(x-t) - \widehat{\mu}(x-t)\|_Z,\quad \|\eta(x+t) - \widehat{\eta}(x+t)\|_Z.$$

In order to do that, we can make use of (4.22) and the fact that both $(I + f_0)^{-1}$, $(I + r_0)^{-1}$ are Lipschitz continuous. Thus, for a sufficiently large $L$, operator $\mathcal{S}$ is indeed a strict contraction. So it has a unique fixed point.

The last part of the proposition follows by similar arguments.
In this case, space $Z$ is replaced by $Z := L^\infty(0, T; L^2(0, 1))^2$ with the norm $\|(u,v)\|_Z{:=}\max(\|u\|,\|v\|)$, where

$$\|u\| := \text{ess sup}_{t\in(0,T)} e^{-Lt}\|u(\cdot, t)\|_{L^2(0,1)}. \qquad \square$$

*Remark* 4.2.9. Proposition 4.2.8 is also valid in the case of time-dependent boundary conditions of the type $(BC.3)'$. Moreover, the result can be extended to the case of non-homogeneous boundary conditions.

Finally, we can state and prove the following regularity result, which will be essential for our treatment:

**Theorem 4.2.10.** *Assume that* (4.12), (4.16), (4.17) *hold,* $u_0, v_0$ *belong to* $C^1[0,1]$ *and satisfy* (4.15), *and the following first order compatibility conditions are valid*

$$\begin{cases} f_1(0,0) - ru_0(0) - v_0'(0) + \varepsilon r_0'(v_0(0))\,[f_2(0,0) - gv_0(0) - u_0'(0)] = 0, \\ f_1(1,0) - ru_0(1) - v_0'(1) - \varepsilon f_0'(v_0(1))\,[f_2(1,0) - gv_0(1) - u_0'(1)] = 0. \end{cases} \tag{4.25}$$

*Then, solution* $(u_\varepsilon, v_\varepsilon)$ *of problem* $P_\varepsilon$ *belongs to* $C^1(\overline{D}_T)^2$.

*Proof.* We assume again, without loss of generality, that $\varepsilon = 1$ and, for simplicity, we re-denote $(u_\varepsilon, v_\varepsilon)$ by $(u, v)$. We know from Proposition 4.2.8 that $(u, v)$ belongs to $C(\overline{D}_T)^2$.

Now, we consider the following problem, which is obtained by formal differentiation with respect to $t$ of problem $P_\varepsilon$ :

$$\begin{cases} \tilde{u}_t + \tilde{v}_x = -r\tilde{u} + f_{1t}, \\ \tilde{v}_t + \tilde{u}_x = -g\tilde{v} + f_{2t} \ \text{ in } \ D_T, \end{cases} \tag{4.26}$$

$$\begin{cases} \tilde{u}(x,0) = f_1(x,0) - ru_0(x) - v_0'(x), \\ \tilde{v}(x,0) = f_2(x,0) - gv_0(x) - u_0'(x), \quad 0 < x < 1, \end{cases} \tag{4.27}$$

$$\begin{cases} \tilde{u}(0,t) + r_0'(v(0,t))\,\tilde{v}(0,t) = 0, \\ \tilde{u}(1,t) - f_0'(v(1,t))\,\tilde{v}(1,t) = 0, \quad 0 < t < T. \end{cases} \tag{4.28}$$

It follows from Remark 4.2.9 that this problem has a unique generalized solution $(\tilde{u}, \tilde{v}) \in C(\overline{D}_T)^2$.

Let us prove that $(\tilde{u}, \tilde{v}) = (u_t, v_t)$. First, we note that $(u, v)$ satisfies system (4.19), (4.20), (4.21) with $f_1 - ru$ and $f_2 - gv$ instead of $f_1$ and $f_2$. By differentiating this system with respect to $t$, we see that $(u_t, v_t)$, which belongs to $L^\infty(0,T; L^2(0,1))^2$ (see Theorem 4.2.1), satisfies (4.26), (4.27), (4.28) in the generalized sense. By the uniqueness property of the solution to (4.26), (4.27), (4.28) in the class of $L^\infty(0,T; L^2(0,1))^2$ (see Proposition 4.2.8 and Remark 4.2.9), we get $(u_t, v_t) = (\tilde{u}, \tilde{v})$, and hence $(u_t, v_t) \in C(\overline{D}_T)^2$. Finally, using system $(LS)$, it follows that $(u, v) \in C^1(\overline{D}_T)^2$. □

*Remark* 4.2.11. It should be noted that more smoothness of the data and additional compatibility conditions imply even more regularity of the solution of problem $P_\varepsilon$. For a discussion of this issue see [24], p. 126. In fact, any degree of regularity can be reached under appropriate assumptions.

The rest of this section is focused on the reduced problem $P_0$, which consists of the algebraic equation (4.2) and the boundary value problem (4.3), (4.6), (4.9). Let us consider the Hilbert space $H_0 := L^2(0,1)$ endowed with the usual scalar product and the associated norm denoted by $\|\cdot\|_0$.
We define the operator $A(t) : D(A(t)) \subset H_0 \to H_0$,

$$\begin{aligned} D(A(t)) = \{p \in H^2(0,1);\ & r^{-1}p'(0) + \sigma_1(t) = r_0(p(0)), \\ & -r^{-1}p'(1) + \sigma_2(t) = f_0(p(1))\},\quad A(t)p = -(1/r)p'' + gp, \\ \sigma_1(t) = -r^{-1}f_1(0,t),\ & \sigma_2(t) = r^{-1}f_1(1,t). \end{aligned}$$

Denoting $\widetilde{y}(t) = Y(\cdot,t)$, $\widetilde{h}(t) = h(\cdot,t) = (f_2 - r^{-1}f_{1x})(\cdot,t)$, $0<t<T$, it is obvious that problem (4.3), (4.6), (4.9) can be expressed as the following Cauchy problem in $H_0$:

$$\begin{cases} \widetilde{y}'(t) + A(t)\widetilde{y}(t) = \widetilde{h}(t),\ 0<t<T, \\ \widetilde{y}(0) = v_0. \end{cases} \tag{4.29}$$

Since we plan to prove high regularity properties of the solution of the reduced problem, we note that by formal differentiation with respect to $t$ of problem (4.3), (4.6), (4.9) we obtain the following problem:

$$\begin{cases} z_t - r^{-1}z_{xx} + gz = h_t \ \text{ in } D_T, \\ z(x,0) = z_0(x),\quad 0 \le x \le 1, \\ r^{-1}z_x(0,t) + \sigma_1'(t) = \alpha(t)z(0,t), \\ -r^{-1}z_x(1,t) + \sigma_2'(t) = \beta(t)z(1,t),\quad 0 \le t \le T, \end{cases} \tag{4.30}$$

where $z = Y_t$, $z_0 = h(0) - A(0)v_0$, $\alpha = r_0'(Y(0,\cdot))$, $\beta = f_0'(Y(1,\cdot))$. We see that this problem can be expressed as the following Cauchy problem in the Hilbert space $H_0$:

$$\begin{cases} \widetilde{z}'(t) + A_1(t)\widetilde{z}(t) = h_t(\cdot,t),\quad 0<t<T, \\ \widetilde{z}(0) = z_0, \end{cases} \tag{4.31}$$

with $\widetilde{z}(t) := z(\cdot,t)$ and $A_1(t) : D(A_1(t)) \subset H_0 \to H_0$,

$$\begin{aligned} D(A_1(t)) = \{p \in H^2(0,1);\ & r^{-1}p'(0) + \sigma_1'(t) = \alpha(t)p(0), \\ & -r^{-1}p'(1) + \sigma_2'(t) = \beta(t)p(1)\},\quad A_1(t)p = -r^{-1}p'' + gp. \end{aligned}$$

We continue with a result about the existence, uniqueness, and regularity of the solution to problem (4.29).

**Theorem 4.2.12.** *Assume that*

$$h \in W^{2,2}(0,T;L^2(0,1)),\ \sigma_1,\ \sigma_2 \in H^2(0,T); \tag{4.32}$$

$$r>0,\ g \ge 0,\ r_0, f_0 \in C^2(\mathbb{R}),\ r_0',\ f_0' \ge 0; \tag{4.33}$$

$$v_0 \in D(A(0)),\ z_0 := h(0) - A(0)v_0 \in D(A_1(0)). \tag{4.34}$$

*Then, problem* (4.29) *has a unique strong solution* $\widetilde{y}(t) = Y(\cdot, t)$, *and*

$$Y \in W^{2,2}(0,T; H^1(0,1)) \bigcap W^{1,2}(0,T; H^2(0,1)).$$

*Proof.* First we note that under assumptions (4.33), operator $A(t)$ defined above is maximal monotone for all $t \in [0,T]$ (see Example 1 of Chapter 2). In addition, $A(t)$ is the subdifferential of the function $\phi(t,\cdot){:}H_0 \to (-\infty, +\infty]$,

$$\phi(t,p) = \begin{cases} \frac{1}{2r}\int_0^1 p'(x)^2 dx + \frac{g}{2}\int_0^1 p^2(x)dx + j_1(p(0)) + j_2(p(1)) \\ -\sigma_1(t)p(0) - \sigma_2(t)p(1), \text{ if } p \in H^1(0,1), \\ +\infty, \text{ otherwise.} \end{cases}$$

where $j_1,\ j_2$ are primitives for $r_0,\ f_0$, respectively. For every $t \in [0,T]$, the effective domain $D(\phi(t,\cdot)) = H^1(0,1)$. Thus, $D(\phi(t,\cdot))$ does not depend on $t$. Let us show that condition (2.12) of Theorem 2.0.32 is satisfied. Indeed, for every $p \in H^1(0,1)$ and $0 \le s \le t \le T$, we have

$$\phi(t,p) - \phi(s,p) \le (\mid p(0) \mid + \mid p(1) \mid) \int_s^t (\mid \sigma_1'(\tau) \mid + \mid \sigma_2'(\tau) \mid) d\tau. \tag{4.35}$$

Since $j_1,\ j_2$ are bounded from below by some affine functions, we have

$$\phi(s,p) \ge \frac{1}{2r} \parallel p' \parallel_0^2 - C_1 \mid p(0) \mid - C_2 \mid p(1) \mid - C_3,\ \forall s \in [0,T],\ p \in H^1(0,1),$$

where $C_1,\ C_2,\ C_3$ are some positive constants. Thus, it is easily seen that

$$\mid p(0) \mid + \mid p(1) \mid \le \phi(s,p) + M_1 \parallel p \parallel_0^2 + M_2,$$

where $M_1,\ M_2$ are positive constants, and therefore, (4.35) implies (2.12) with $\gamma(t) = \int_0^t (\mid \sigma_1'(\tau) \mid + \mid \sigma_2'(\tau) \mid) d\tau$. We have used the following obvious inequality

$$p^2(x) \le (1+\eta) \parallel p \parallel_0^2 + \eta^{-1} \parallel p' \parallel_0^2,\ \forall x \in [0,1],\ \eta > 0,\ p \in H^1(0,1). \tag{4.36}$$

So, according to Theorem 2.0.32, problem (4.29) has a unique strong solution $\widetilde{y}(t) := Y(\cdot,t)$,

$$Y \in W^{1,2}(0,T; H_0) \bigcap L^2(0,T; H^2(0,1)).$$

Now, we are going to prove that $Y \in W^{1,2}(0,T; H^1(0,1))$. We start from the obvious inequalities

$$\begin{aligned} &\frac{1}{2}\frac{d}{dt}\|\widetilde{y}(t+\delta) - \widetilde{y}(t)\|_0^2 + \frac{1}{r}\|Y_x(\cdot,t+\delta) - Y_x(\cdot,t)\|_0^2 \\ &\le \mid \sigma_1(t+\delta) - \sigma_1(t) \mid \cdot \mid Y(0,t+\delta) - Y(0,t) \mid \\ &\quad + \mid \sigma_2(t+\delta) - \sigma_2(t) \mid \cdot \mid Y(1,t+\delta) - Y(1,t) \mid \\ &\quad + \parallel \widetilde{h}(t+\delta) - \widetilde{h}(t)\|_0 \cdot \|\widetilde{y}(t+\delta) - \widetilde{y}(t)\|_0, \end{aligned}$$

for a.a. $0 \leq t \leq t+\delta \leq T$, and

$$\frac{1}{2}\frac{d}{d\delta}\|\widetilde{y}(\delta)-v_0\|_0^2+\frac{1}{r}\|Y_x(\cdot,\delta)-v_0'\|_0^2 \leq \mid \sigma_1(\delta)-\sigma_1(0)\mid \cdot \mid Y(0,\delta)-v_0(0)\mid$$
$$+\mid \sigma_2(\delta)-\sigma_2(0)\mid \cdot \mid Y(1,\delta)-v_0(1)\mid + \| \widetilde{h}(\delta)-A(0)v_0\|_0 \cdot \|\widetilde{y}(\delta)-v_0\|_0,$$

for a.a. $\delta \in (0,T)$. Since $H^1(0,1)$ is continuously embedded into $C[0,1]$, we infer from the above inequalities the following estimates

$$\begin{aligned}
&\frac{1}{2}\frac{d}{dt}\|\widetilde{y}(t+\delta)-\widetilde{y}(t)\|_0^2+\frac{1}{r}\|Y_x(\cdot,t+\delta)-Y_x(\cdot,t)\|_0^2 \\
&\quad\leq C_4\Big(\mid \sigma_1(t+\delta)-\sigma_1(t)\mid^2+\mid \sigma_2(t+\delta)-\sigma_2(t)\mid^2\Big) \\
&\qquad+\frac{1}{2r}\|Y_x(\cdot,t+\delta)-Y_x(\cdot,t)\|_0^2+\Big(\| \widetilde{h}(t+\delta)-\widetilde{h}(t)\|_0 \\
&\qquad+C_5(\mid \sigma_1(t+\delta)-\sigma_1(t)\mid+\mid \sigma_2(t+\delta)-\sigma_2(t)\mid)\Big)\|\widetilde{y}(t+\delta)-\widetilde{y}(t)\|_0,
\end{aligned}$$

for a.a. $0 \leq t \leq t+\delta \leq T$, and

$$\begin{aligned}
&\frac{1}{2}\frac{d}{d\delta}\|\widetilde{y}(\delta)-v_0\|_0^2+\frac{1}{r}\|Y_x(\cdot,\delta)-v_0'\|_0^2 \leq \frac{1}{2r}\|Y_x(\cdot,\delta)-v_0'\|_0^2 \\
&\qquad+C_6\Big(\mid \sigma_1(\delta)-\sigma_1(0)\mid^2+\mid \sigma_2(\delta)-\sigma_2(0)\mid^2\Big)+\Big(\| \widetilde{h}(\delta)-A(0)v_0\|_0 \\
&\qquad+C_7(\mid \sigma_1(\delta)-\sigma_1(0)\mid+\mid \sigma_2(\delta)-\sigma_2(0)\mid)\Big)\|\widetilde{y}(\delta)-v_0\|_0,
\end{aligned}$$

for a.a. $\delta \in (0,T)$, where $C_4$, $C_5$, $C_6$, $C_7$ are some positive constants. Integration over $[0,T-\delta]$ and $[0,\delta]$ of the above inequalities leads us to

$$\begin{aligned}
\frac{1}{2r}\int_0^{T-\delta} & \|Y_x(\cdot,t+\delta)-Y_x(\cdot,t)\|_0^2\,dt \leq \frac{1}{2}\parallel \widetilde{y}(\delta)-v_0\parallel_0^2 \\
&+\int_0^{T-\delta}\Big[C_4\Big(\mid \sigma_1(t+\delta)-\sigma_1(t)\mid^2+\mid \sigma_2(t+\delta)-\sigma_2(t)\mid^2\Big) \\
&+\Big(C_5(\mid \sigma_1(t+\delta)-\sigma_1(t)\mid+\mid \sigma_2(t+\delta)-\sigma_2(t)\mid) \\
&+\| \widetilde{h}(t+\delta)-\widetilde{h}(t)\|_0\Big)\|\widetilde{y}(t+\delta)-\widetilde{y}(t)\|_0\Big]dt,
\end{aligned} \tag{4.37}$$

for all $\delta \in (0,T]$, and

$$\begin{aligned}
\frac{1}{2}\parallel \widetilde{y}(\delta)-v_0\parallel_0^2 \leq & \int_0^{\delta}\Big[C_6\Big(\mid \sigma_1(s)-\sigma_1(0)\mid^2+\mid \sigma_2(s)-\sigma_2(0)\mid^2\Big) \\
&+\Big(C_7(\mid \sigma_1(s)-\sigma_1(0)\mid+\mid \sigma_2(s)-\sigma_2(0)\mid) \\
&+\| \widetilde{h}(s)-A(0)v_0\|_0\Big)\|\widetilde{y}(s)-v_0\|_0\Big]ds,
\end{aligned} \tag{4.38}$$

for all $\delta \in (0,T]$. Applying Lemma 2.0.19 to inequality (4.38) and using $\sigma_1,\ \sigma_2 \in H^2(0,T),\ \widetilde{h},\ \widetilde{y} \in W^{1,2}(0,T;H_0)$, we infer that

$$\|\widetilde{y}(\delta) - v_0\|_0 \le C_8\delta, \text{ for all } \delta \in (0,T],$$

and then, by (4.37), we get

$$\int_0^{T-\delta} \|Y_x(\cdot,t+\delta) - Y_x(\cdot,t)\|_0^2 dt \le C_9\delta^2,$$

from which it follows that $Y \in W^{1,2}(0,T;H^1(0,1))$ (see Theorem 2.0.3).

Now, let us prove that $Y \in W^{2,2}(0,T;H^1(0,1))$. Set $V := H^1(0,1)$ and denote its dual by $V^*$. We will prove that $Y_t$ satisfies problem (4.30) derived by formal differentiation with respect to $t$ of problem (4.29). First, let us check that $Y_t \in W^{1,2}(0,T;V^*)$. Note that

$$\begin{aligned}\langle \varphi, Y_t(\cdot,t+\delta) - Y_t(\cdot,t)\rangle &= \langle \varphi, r^{-1}(Y_{xx}(\cdot,t+\delta) - Y_{xx}(\cdot,t))\rangle \\ &- g\langle \varphi, Y(\cdot,t+\delta) - Y(\cdot,t)\rangle + \langle \varphi, h(\cdot,t+\delta) - h(\cdot,t)\rangle\end{aligned} \tag{4.39}$$

for a.a. $t \in (0,T-\delta)$ and all $\varphi \in V$, where $\langle\cdot,\cdot\rangle$ denotes the pairing between $V$ and $V^*$. Integration by parts in (4.39) yields

$$\begin{aligned}&\langle \varphi, Y_t(\cdot,t+\delta) - Y_t(\cdot,t)\rangle + \langle \varphi', r^{-1}(Y_x(\cdot,t+\delta) - Y_x(\cdot,t))\rangle \\ &\quad + [(r_0(Y(0,t+\delta)) - r_0(Y(0,t)) - (\sigma_1(t+\delta) - \sigma_1(t))]\varphi(0) \\ &\quad + [(f_0(Y(1,t+\delta)) - f_0(Y(1,t))) - (\sigma_2(t+\delta) - \sigma_2(t))]\varphi(1) \\ &\quad + g\langle \varphi, Y(\cdot,t+\delta) - Y(\cdot,t)\rangle = \langle \varphi, h(\cdot,t+\delta) - h(\cdot,t)\rangle\end{aligned}$$

for a.a. $t \in (0,T-\delta)$, $\forall \varphi \in V$. This equation together with (4.32) and (4.33) implies

$$\| Y_t(\cdot,t+\delta) - Y_t(\cdot,t) \|_{V^*} \le C_{10}(\| \widetilde{y}(t+\delta) - \widetilde{y}(t) \|_V + \| h(\cdot,t+\delta) - h(\cdot,t) \|_0 + \delta),$$

where $C_{10}$ is a positive constant. Thus,

$$\int_0^{T-\delta} \| Y_t(\cdot,t+\delta) - Y_t(\cdot,t) \|_{V^*}^2 \, dt \le C_{11}\delta^2, \ \forall \delta \in (0,T],$$

and so it follows by Theorem 2.0.3 that $Y_t \in W^{1,2}(0,T;V^*)$. Therefore, one can differentiate with respect to $t$ the equation in $Y$:

$$\begin{aligned}&\langle \varphi, Y_t(\cdot,t)\rangle + \langle \varphi', r^{-1}Y_x(\cdot,t)\rangle + g\langle \varphi, Y(\cdot,t)\rangle + (r_0(Y(0,t)) - \sigma_1(t))\varphi(0) \\ &\qquad + (f_0(Y(1,t)) - \sigma_2(t))\varphi(1) = \langle \varphi, h(\cdot,t)\rangle, \ \forall \varphi \in V,\end{aligned}$$

thus obtaining

$$\begin{aligned}&\langle \varphi, z_t(\cdot,t)\rangle + \langle \varphi', r^{-1}z_x(\cdot,t)\rangle + g\langle \varphi, z(\cdot,t)\rangle + (\alpha(t)z(0,t) - \sigma_1'(t))\varphi(0) \\ &\qquad + (\beta(t)z(1,t) - \sigma_2'(t))\varphi(1) = \langle \varphi, h_t(\cdot,t)\rangle, \ \forall \varphi \in V,\end{aligned} \tag{4.40}$$

where $z = Y_t$, $\alpha(t) = r_0'(Y(0,t))$, $\beta(t) = f_0'(Y(1,t))$. Obviously, $\alpha, \beta \in H^1(0,T)$, $\alpha \geq 0$, $\beta \geq 0$, and

$$z(\cdot, 0) = z_0. \tag{4.41}$$

One can see that $z$ is the unique solution of problem (4.40), (4.41). Indeed, if we choose $z_0 = 0$, $\sigma_1' = \sigma_2' = 0$, $\varphi = z(\cdot,t)$ and $h_t \equiv 0$ in (4.40), (4.41), we see that

$$\frac{d}{dt} \| z(\cdot,t) \|_0^2 \leq 0 \ \text{ for } a.a.\ t \in (0,T) \Rightarrow z \equiv 0.$$

Thus $z := Y_t$ is a variational solution of problem (4.30), which can be also written in the form of the Cauchy problem (4.31).

In fact, problem (4.31) has a unique strong solution. To show this, we notice that operator $A_1(t)$ is a maximal monotone operator for all $t \in [0,T]$. Moreover, $A_1(t)$ is the subdifferential of the function $\phi_1(t,\cdot){:}H_0 \to (-\infty, +\infty]$,

$$\phi_1(t,p) = \begin{cases} \frac{1}{2r}\int_0^1 p'(x)^2 dx + \frac{g}{2}\int_0^1 p(x)^2 dx + \frac{\alpha(t)}{2} p(0)^2 + \frac{\beta(t)}{2} p(1)^2 \\ -\sigma_1'(t)p(0) - \sigma_2'(t)p(1), \quad \text{if } p \in H^1(0,1), \\ +\infty, \quad \text{otherwise.} \end{cases}$$

For every $t \in [0,T]$, $D(\phi_1(t,\cdot)) = H^1(0,1)$. Let us show that condition (2.12) of Theorem 2.0.32 is satisfied by $\phi_1$. Indeed, for every $p \in H^1(0,1)$ and $0 \leq s \leq t \leq T$, we have

$$\begin{aligned} \phi_1(t,p) - \phi_1(s,p) \leq & \frac{1}{2}(p(0)^2 + p(1)^2) \int_s^t (| \alpha' | + | \beta' |)(\tau) d\tau \\ & + (| p(0) | + p(1) |) \int_s^t (| \sigma_1' | + | \sigma_2' |)(\tau) d\tau. \end{aligned} \tag{4.42}$$

Since

$$\phi_1(s,p) \geq \frac{1}{2r} \| p' \|_0^2 - C_{12} | p(0) | - C_{13} | p(1) |,$$

where $C_{12}$, $C_{13}$ are some positive constants, one can easily derive (2.12) from (4.42). So, according to Theorem 2.0.32, problem (4.31) has a unique strong solution $\tilde{z}(t) := z(\cdot,t)$, $z(\cdot,t) \in D(\phi_1(t,\cdot)) = H^1(0,1) \, \forall t \in [0,T]$,

$$z \in W^{1,2}(0,T;H_0) \bigcap L^2(0,T;H^2(0,1)),$$

and there exists a function $\vartheta \in L^1(0,T)$ such that

$$\phi_1(t,\tilde{z}(t)) \leq \phi_1(0,\tilde{z}(0)) + \int_0^t \vartheta(s) ds, \ \forall\, t \in [0,T],$$

which implies

$$\begin{aligned} \frac{1}{2r}\int_0^1 z_x(x,t)^2 dx & \leq C_{14} + \sigma_1'(t) z(0,t) + \sigma_2'(t) z(1,t) \\ & \leq C_{15} + C_{16} \| z(\cdot,t) \|_0^2 + \frac{1}{4r} \| z_x(\cdot,t) \|_0^2, \ \forall\, t \in [0,T], \end{aligned}$$

from which we get $\| z_x(\cdot,t) \|_0 \leq C_{17}$. Therefore, $z \in L^\infty(0,T;H^1(0,1))$. We have used $\sigma_1,\ \sigma_2 \in H^2(0,T),\ \ z(\cdot,t) \in H^1(0,1)\,\forall\, t \in [0,T]$, as well as inequality (4.36). As $z = Y_t$, we have already proved that

$$Y \in W^{2,2}(0,T;L^2(0,1)) \bigcap W^{1,2}(0,T;H^2(0,1)) \bigcap W^{1,\infty}(0,T;H^1(0,1)).$$

It remains to show that $Y_{tt} = z_t \in L^2(0,T;H^1(0,1))$. To this purpose one can apply a reasoning similar to that used in the first part of this proof. We need only some slight modifications which we are going to point out. Starting from equations $(4.31)_1$ and

$$(z - z_0)_t - r^{-1}(z_{xx} - z_0'') + g(z - z_0) = h_t - A_1(0)z_0,$$

one gets the following estimates

$$\begin{aligned}
&\frac{1}{2}\frac{d}{dt}\|\widetilde{z}(t+\delta) - \widetilde{z}(t)\|_0^2 + \frac{1}{r}\|z_x(\cdot,t+\delta) - z_x(\cdot,t)\|_0^2\\
&\leq \Big(\mid \sigma_1'(t+\delta) - \sigma_1'(t) \mid + \mid z(0,t) \mid \cdot \mid \alpha(t+\delta) - \alpha(t) \mid\Big) \mid z(0,t+\delta) - z(0,t) \mid\\
&\quad + \Big(\mid \sigma_2'(t+\delta) - \sigma_2'(t) \mid + \mid z(1,t) \mid \cdot \mid \beta(t+\delta) - \beta(t) \mid\Big) \mid z(1,t+\delta) - z(1,t) \mid\\
&\quad + \| \widetilde{h}'(t+\delta) - \widetilde{h}'(t)\|_0 \cdot \|\widetilde{z}(t+\delta) - \widetilde{z}(t)\|_0,
\end{aligned}$$

for a.a. $0 \leq t \leq t+\delta \leq T$, and

$$\begin{aligned}
&\frac{1}{2}\frac{d}{d\delta}\|\widetilde{z}(\delta) - z_0\|_0^2 + \frac{1}{r}\|z_x(\cdot,\delta) - z_0'\|_0^2\\
&\qquad \leq \Big(\mid \sigma_1'(\delta) - \sigma_1'(0) \mid + \mid z_0(0) \mid \cdot \mid \alpha(\delta) - \alpha(0) \mid\Big)\\
&\qquad \times \mid z(0,\delta) - z_0(0) \mid + \Big(\mid \sigma_2'(\delta) - \sigma_2'(0) \mid + \mid z_0(1) \mid \cdot \mid \beta(\delta) - \beta(0) \mid\Big)\\
&\qquad \times \mid z(1,\delta) - z_0(1) \mid + \| \widetilde{h}'(\delta) - A_1(0)z_0\|_0 \cdot \|\widetilde{z}(\delta) - z_0(0)\|_0,
\end{aligned}$$

for a.a. $\delta \in (0,T)$. We have used the conditions $\alpha \geq 0,\ \beta \geq 0$.
As $z(0,\cdot),\ z(1,\cdot) \in L^\infty(0,T)$, one can use the fact that $H^1(0,1)$ is continuously embedded into $C[0,1]$ to derive the following estimates

$$\begin{aligned}
&\frac{1}{2}\frac{d}{dt}\|\widetilde{z}(t+\delta) - \widetilde{z}(t)\|_0^2\\
&\quad + \frac{1}{r}\|z_x(\cdot,t+\delta) - z_x(\cdot,t)\|_0^2 \leq \frac{1}{2r}\|z_x(\cdot,t+\delta) - z_x(\cdot,t)\|_0^2\\
&\quad + C_{18}\Big((\sigma_1'(t+\delta) - \sigma_1'(t))^2 + (\sigma_2'(t+\delta) - \sigma_2'(t))^2\\
&\quad + (\alpha(t+\delta) - \alpha(t))^2 + (\beta(t+\delta) - \beta(t))^2\Big)\\
&\quad + \Big(C_{19}(\mid \sigma_1'(t+\delta) - \sigma_1'(t) \mid + \mid \sigma_2'(t+\delta) - \sigma_2'(t) \mid\\
&\quad + \mid \alpha(t+\delta) - \alpha(t) \mid + \mid \beta(t+\delta) - \beta(t) \mid)\\
&\quad + \| \widetilde{h}_t(t+\delta) - \widetilde{h}_t(t)\|_0\Big)\|\widetilde{z}(t+\delta) - \widetilde{z}(t)\|_0,
\end{aligned} \tag{4.43}$$

for a.a. $0 \le t \le t+\delta \le T$, and

$$\begin{aligned}\frac{1}{2}\frac{d}{d\delta}\|\widetilde{z}(\delta)-z_0\|_0^2 &+ \frac{1}{r}\|z_x(\cdot,\delta)-z_0'\|_0^2 \le \frac{1}{2r}\|z_x(\cdot,\delta)-z_0'\|_0^2 \\ &+ C_{20}\Big((\sigma_1'(\delta)-\sigma_1'(0))^2+(\sigma_2'(\delta)-\sigma_2'(0))^2 \\ &+ (\alpha(\delta)-\alpha(0))^2+(\beta(\delta)-\beta(0))^2\Big) \\ &+ \Big(C_{21}(\mid \sigma_1'(\delta)-\sigma_1'(0)\mid + \mid \sigma_2'(\delta)-\sigma_2'(0)\mid \\ &+ \mid \alpha(\delta)-\alpha(0)\mid + \mid \beta(\delta)-\beta(0)\mid) \\ &+ \|\,\widetilde{h}'(\delta)-A_1(z_0)\|_0\Big)\|\widetilde{z}(\delta)-z_0\|_0,\end{aligned} \tag{4.44}$$

for a.a. $\delta \in (0,T)$, where $C_{18},\dots,\ C_{21}$ are some positive constants. Note that every $\zeta \in H^1(0,T)$ satisfies the following estimate

$$\mid \zeta(s_1)-\zeta(s_2)\mid = \left|\int_{s_1}^{s_2}\zeta'(\tau)d\tau\right| \le \|\,\zeta'\,\|_{L^2(0,T)}\cdot \mid s_1-s_2\mid^{1/2} \quad \forall\, s_1,s_2 \in [0,T].$$

Thus, since $\alpha,\ \beta,\ \sigma_1',\ \sigma_2' \in H^1(0,T)$ and $\widetilde{h}',\ \widetilde{z} \in W^{1,2}(0,T;H_0)$, we can derive from (4.44)

$$\|\,\widetilde{z}(\delta)-z_0\,\|_0^2 \le C_{22}\delta^2 + C_{23}\int_0^\delta \|\,\widetilde{z}(s)-z_0(0)\,\|_0\, ds.$$

Then, Lemma 2.0.19 yields $\|\,\widetilde{z}(\delta)-z_0\,\|_0 \le C_{24}\delta$. This together with (4.43) implies the desired conclusion, i.e., $z \in W^{1,2}(0,T;H^1(0,1))$. □

*Remark* 4.2.13. In fact we can also prove that $Y \in W^{2,\infty}(0,T;L^2(0,1))$ (for details, see [24, p. 82]). In [24] a thorough discussion is also included concerning different regularity issues, such as: regularity under more relaxed assumptions, more regularity with respect to $x$, and higher regularity in the case of a more general nonlinear parabolic equation. Here, we have included only facts which are absolutely necessary for our treatment.

Summarizing what we have done so far, we can state the following concluding result:

**Corollary 4.2.14.** *Assume that*

$$\begin{gathered} r>0,\ g\ge 0,\ r_0, f_0 \in C^2(\mathbb{R}),\ r_0' \ge 0,\ f_0' \ge 0; \\ f_1,\ f_2 \textit{ are sufficiently regular;} \\ u_0 \in H^2(0,1),\ v_0 \in H^4(0,1); \end{gathered}$$

*and the following compatibility conditions are fulfilled*

$$\begin{cases} u_0(0)+r_0(v_0(0))=0, \\ u_0(1)-f_0(v_0(1))=0, \end{cases}$$

$$\begin{cases} f_1(0,0) - ru_0(0) - v_0'(0) = 0, \\ r_0'(v_0(0))\,(f_2(0,0) - gv_0(0) - u_0'(0)) = 0, \\ f_1(1,0) - ru_0(1) - v_0'(1) = 0, \\ f_0'(v_0(1))\,(f_2(1,0) - gv_0(1) - u_0'(1)) = 0, \end{cases}$$

$$\begin{cases} r^{-1}v_0^{(3)}(0) - gv_0'(0) + f_{2x}(0,0) - r^{-1}f_{1xx}(0,0) - rr_0'(v_0(0))\times \\ \times\left(r^{-1}v_0''(0) - gv_0(0) + f_2(0,0) - r^{-1}f_{1x}(0,0)\right) = f_{1t}(0,0), \\ r^{-1}v_0^{(3)}(1) - gv_0'(1) + f_{2x}(1,0) - r^{-1}f_{1xx}(1,0) + rf_0'(v_0(1))\times \\ \times\left(r^{-1}v_0''(1) - gv_0(1) + f_2(1,0) - r^{-1}f_{1x}(1,0)\right) = f_{1t}(1,0). \end{cases}$$

*Then, all the conclusions of Theorems* 4.2.10 *and* 4.2.12 *hold.*

*Remark* 4.2.15. It should be noted that in the statement of the above corollary, the compatibility conditions are independent of $\varepsilon$, and (4.25) are satisfied for all $\varepsilon > 0$. Note also that (4.15), (4.25) as well as the compatibility conditions required for the reduced problem $P_0$ form a compatible system of conditions. These conditions include in particular our previous conditions (4.10), which were required to eliminate possible discrepancies caused by correction $c_0$ at the corner points $(x,t) = (0,0)$ and $(x,t) = (1,0)$. So, we have created an adequate framework for what we are going to do in the next section.

## 4.3 Estimates for the remainder components

Under the assumptions of Corollary 4.2.14, our expansion (4.1) is well defined, in the sense that all its terms exist. We are now going to show it is a real asymptotic expansion, that is the remainder tends to zero with respect to the sup norm. We will show even more, that the remainder components satisfy certain estimates, as seen in the following result:

**Theorem 4.3.1.** *Assume that all the assumptions of Corollary* 4.2.14 *are fulfilled. Then, for every* $\varepsilon > 0$*, the solution of problem* $P_\varepsilon$ *admits an asymptotic expansion of the form* (4.1) *and the following estimates are valid:*

$$\|R_{1\varepsilon}\|_{C(\overline{D}_T)} = \mathcal{O}(\varepsilon^{1/8}), \quad \|R_{2\varepsilon}\|_{C(\overline{D}_T)} = \mathcal{O}(\varepsilon^{3/8}).$$

*Proof.* Throughout this proof we denote by $K_1, K_2, \ldots$ some positive constants which depend on the data, but are independent of $\varepsilon$. By Corollary 4.2.14, problem (4.5), (4.8),(4.11) has a unique smooth solution $(R_{1\varepsilon}, R_{2\varepsilon})$.

In order to establish the desired estimates, we consider the Hilbert space $H := L^2(0,1)^2$, endowed with the scalar product

$$\langle p, q\rangle := \varepsilon \int_0^1 p_1(x)q_1(x)dx + \int_0^1 p_2(x)q_2(x)dx, \; p = (p_1,p_2), \; q = (q_1,q_2) \in H.$$

Denote by $\|\cdot\|$ the corresponding Hilbertian norm. Note that both the scalar product and the norm depend on $\varepsilon$. Also, we consider the operator $\mathcal{B}_\varepsilon(t) : D\big(\mathcal{B}_\varepsilon(t)\big) \subset H \to H$,

$$\begin{aligned} D\big(\mathcal{B}_\varepsilon(t)\big) = \Big\{&(p,q) \in \big(H^1(0,1)\big)^2,\ p(0) + r_0\big(q(0) + Y(0,t)\big) = r_0\big(Y(0,t)\big), \\ &p(1) - f_0\big(q(1) + Y(1,t)\big) = -f_0\big(Y(1,t)\big)\Big\}, \\ \mathcal{B}_\varepsilon(t)(p,q) &= \big(\varepsilon^{-1}q' + r\varepsilon^{-1}p,\ p' + gq\big) \ \text{ for all } t \in [0,T]. \end{aligned}$$

Obviously, problem (4.5), (4.8), (4.11) can be written as the following Cauchy problem in $H$:

$$\begin{cases} R_\varepsilon'(t) + \mathcal{B}_\varepsilon(t)R_\varepsilon(t) = F_\varepsilon(t), & 0 < t < T, \\ R_\varepsilon(0) = 0, \end{cases} \tag{4.45}$$

where $R_\varepsilon(t) := (R_{1\varepsilon}(\cdot,t), R_{2\varepsilon}(\cdot,t))$, $F_\varepsilon(t) := \big( - X_t(\cdot,t), -c_{0x}(\cdot,\tau)\big)$, $0<t<T$.

Taking the scalar product in $H$ of $(4.45)_1$ and $R_\varepsilon(t)$ and integrating the resulting equation over $[0,t]$, we get

$$\frac{1}{2}\|R_\varepsilon(t)\|^2 + \int_0^t \langle \mathcal{B}_\varepsilon(s)R_\varepsilon(s), R_\varepsilon(s)\rangle ds = \int_0^t \langle F_\varepsilon(s), R_\varepsilon(s)\rangle ds \tag{4.46}$$

for all $t \in [0,T]$. We have denoted by $\|\cdot\|_0$ the usual norm of $L^2(0,1)$. An easy computation involving assumptions $r_0', f_0' \geq 0$ shows that

$$\langle \mathcal{B}_\varepsilon(s)R_\varepsilon(s), R_\varepsilon(s)\rangle \geq r\big\|R_{1\varepsilon}(\cdot,s)\big\|_0^2 + g\big\|R_{2\varepsilon}(\cdot,s)\big\|_0^2 \ \text{ for all } s \in [0,T]. \tag{4.47}$$

Therefore,

$$\frac{1}{2}\|R_\varepsilon(t)\|^2 \leq \int_0^t \|F_\varepsilon(s)\| \cdot \|R_\varepsilon(s)\| ds \quad \text{for all } t \in [0,T].$$

By Lemma 2.0.19, it follows that

$$\|R_\varepsilon(t)\| \leq \int_0^t \|F_\varepsilon(s)\| ds \ \text{ for all } t \in [0,T]. \tag{4.48}$$

On the other hand,

$$\|F_\varepsilon(s)\|^2 = \varepsilon \int_0^1 X_s(x,s)^2 dx + \int_0^1 c_{0x}\big(x, s/\varepsilon\big)^2 dx \leq K_1\varepsilon$$

for all $s \in [0,T]$. Thus,

$$\int_0^T \|F_\varepsilon(s)\| ds = \mathcal{O}(\varepsilon^{1/2}). \tag{4.49}$$

Now, it is easily seen that (4.46), (4.47), (4.48) and (4.49) imply

$$\left\|R_\varepsilon(t)\right\|^2 = \varepsilon\left\|R_{1\varepsilon}(\cdot,t)\right\|_0^2 + \left\|R_{2\varepsilon}(\cdot,t)\right\|_0^2 \le K_2\varepsilon \quad \text{for all } t \in [0,T], \tag{4.50}$$

$$\int_0^T \left(r\left\|R_{1\varepsilon}(\cdot,s)\right\|_0^2 + g\left\|R_{2\varepsilon}(\cdot,s)\right\|_0^2\right)ds = \mathcal{O}(\varepsilon). \tag{4.51}$$

In order to continue the proof, we need the following auxiliary result:

**Lemma 4.3.2.** *If the assumptions of Theorem* 4.3.1 *hold, then*

$$\|v_\varepsilon\|_{C(\overline{D}_T)} = \mathcal{O}(1).$$

*Proof.* We define the operator $\mathcal{B}_{1\varepsilon} : D(\mathcal{B}_{1\varepsilon}) \subset H \to H$,

$$\begin{aligned} D(\mathcal{B}_{1\varepsilon}) &= \left\{(p,q) \in H^1(0,1)^2,\ p(0) + r_0\big(q(0)\big) = 0,\ p(1) - f_0\big(q(1)\big) = 0\right\}, \\ \mathcal{B}_{1\varepsilon}(p,q) &= \left(\varepsilon^{-1}(q' + rp),\ p' + gq\right). \end{aligned}$$

Obviously, problem $P_\varepsilon$ can be written as the following Cauchy problem in $H$:

$$\begin{cases} U_\varepsilon'(t) + \mathcal{B}_{1\varepsilon}U_\varepsilon(t) = G_\varepsilon(t), \quad 0 < t < T, \\ U_\varepsilon(0) = (u_0, v_0), \end{cases} \tag{4.52}$$

where $U_\varepsilon(t) := \big(u_\varepsilon(\cdot,t), v_\varepsilon(\cdot,t)\big)$, $G_\varepsilon(t) := \big(\varepsilon^{-1}f_1(\cdot,t), f_2(\cdot,t)\big)$, $0 < t < T$. The assumptions of Theorem 2.0.20 are satisfied, therefore we have:

$$\|U_\varepsilon'(t)\| \le \|G_\varepsilon(0) - \mathcal{B}_{1\varepsilon}(u_0,v_0)\| + \int_0^t \|G_\varepsilon'(s)\|ds \le K_3\varepsilon^{-1/2}$$

for all $t \in [0,T]$, which implies

$$\varepsilon\|u_{\varepsilon t}(\cdot,t)\|_0^2 + \|v_{\varepsilon t}(\cdot,t)\|_0^2 \le K_3^2\varepsilon^{-1} \quad \forall\, t \in [0,T]. \tag{4.53}$$

On the other hand, by (4.50) we obtain the estimates:

$$\|v_\varepsilon(\cdot,t)\|_0 \le \|Y(\cdot,t)\|_0 + \|R_{2\varepsilon}(\cdot,t)\|_0 \le K_4, \tag{4.54}$$

$$\|u_\varepsilon(\cdot,t)\|_0 \le \|X(\cdot,t)\|_0 + \|c_0(\cdot,\tau)\|_0 + \|R_{1\varepsilon}(\cdot,t)\|_0 \le K_5, \tag{4.55}$$

for all $t \in [0,T]$. By making use of (4.53) and (4.55), we derive from the first equation of $(LS)$

$$\|v_{\varepsilon x}(\cdot,t)\|_0 \le K_6, \quad 0 \le t \le T.$$

Together with (4.54), this last inequality shows that $v_\varepsilon(\cdot,t)$ is uniformly bounded with respect to $\varepsilon > 0$ and $t \in [0,T]$ in $H^1(0,1)$. Since $H^1(0,1)$ is continuously embedded into $C[0,1]$, our conclusion follows. □

Let us continue the proof of the theorem. We are going to prove some estimates for $R_{1\varepsilon t}$, $R_{2\varepsilon t}$. Denote $R'_\varepsilon(t) = \Big(R_{1\varepsilon t}(\cdot,t), R_{2\varepsilon t}(\cdot,t)\Big)$ and write system (4.5) in $t$, $t+\delta \in [0,T]$, $\delta > 0$, then subtract one system from the other, and finally take the scalar product in $H$ of the resulting system with $R_\varepsilon(t+\delta) - R_\varepsilon(t)$. Thus, we get

$$\begin{aligned} \frac{1}{2}\frac{d}{dt} &\parallel R_\varepsilon(t+\delta) - R_\varepsilon(t) \parallel^2 \\ &+ r \parallel R_{1\varepsilon}(\cdot,t+\delta) - R_{1\varepsilon}(\cdot,t) \parallel_0^2 + E_{1\varepsilon}(t,\delta) + E_{2\varepsilon}(t,\delta) \\ &\leq \parallel F_\varepsilon(t+\delta) - F_\varepsilon(t) \parallel \cdot \parallel R_\varepsilon(t+\delta) - R_\varepsilon(t) \parallel \end{aligned} \tag{4.56}$$

for all $0 \leq t < t+\delta \leq T$, where

$$\begin{aligned} E_{1\varepsilon}(t,\delta) = &\Big[\big(f_0(v_\varepsilon(1,t+\delta)) - f_0(Y(1,t+\delta))\big) - \big(f_0(v_\varepsilon(1,t)) - f_0(Y(1,t))\big)\Big] \\ &\times \Big[R_{2\varepsilon}(1,t+\delta) - R_{2\varepsilon}(1,t)\Big], \\ E_{2\varepsilon}(t,\delta) = &\Big[\big(r_0(v_\varepsilon(0,t+\delta)) - r_0(Y(0,t+\delta))\big) - \big(r_0(v_\varepsilon(0,t)) - r_0(Y(0,t))\big)\Big] \\ &\times \Big[R_{2\varepsilon}(0,t+\delta) - R_{2\varepsilon}(0,t)\Big]. \end{aligned}$$

Integration of (4.56) over $[0,t]$ yields

$$\begin{aligned} \frac{1}{2} &\parallel R_\varepsilon(t+\delta) - R_\varepsilon(t) \parallel^2 + r\int_0^t \parallel R_{1\varepsilon}(\cdot,s+\delta) - R_{1\varepsilon}(\cdot,s) \parallel_0^2 ds \\ &+ \int_0^t \big(E_{1\varepsilon}(s,\delta) + E_{2\varepsilon}(s,\delta)\big) ds \leq \frac{1}{2} \parallel R_\varepsilon(\delta) \parallel^2 \\ &+ \int_0^t \parallel F_\varepsilon(s+\delta) - F_\varepsilon(s) \parallel \cdot \parallel R_\varepsilon(s+\delta) - R_\varepsilon(s) \parallel ds \end{aligned} \tag{4.57}$$

for all $0 \leq t < t+\delta \leq T$.
Now, we divide (4.57) by $\delta^2$. Then, if we take into account that $R_\varepsilon \in C^1([0,T];H)$, $r_0,\ f_0 \in C^2(\mathbb{R})$, $F_\varepsilon \in W^{1,2}(0,T;H)$, and let $\delta \to 0$, we infer that

$$\begin{aligned} \frac{1}{2} \parallel R'_\varepsilon(t) \parallel^2 + r\int_0^t \parallel R_{1\varepsilon t}(\cdot,s) \parallel_0^2 ds \leq K_7 + \frac{1}{2} \parallel R'_\varepsilon(0) \parallel^2 \\ + \int_0^t \parallel F'_\varepsilon(s) \parallel \cdot \parallel R'_\varepsilon(s) \parallel ds \ \forall\, t \in [0,T]. \end{aligned} \tag{4.58}$$

We have used the following inequality

$$\begin{aligned} \lim_{\delta\to 0}\int_0^t \frac{E_{1\varepsilon}(s,\delta)}{\delta^2} ds &= \int_0^t \frac{d}{ds}\Big[\big(f_0(v_\varepsilon(1,s)) - f_0(Y(1,s))\big)\Big] \cdot \frac{d}{ds}\big(R_{2\varepsilon}(1,s)\big) ds \\ &= \int_0^t \big(f_0'(v_\varepsilon(1,s))v_{\varepsilon s}(1,s) - f_0'(Y(1,s))Y_s(1,s)\big)\big(v_{\varepsilon s}(1,s) - Y_s(1,s)\big) ds \end{aligned}$$

$$
\begin{aligned}
&\geq -\int_0^t \big(f_0'(v_\varepsilon(1,s))v_{\varepsilon s}(1,s)Y_s(1,s) + f_0'(Y(1,s))v_{\varepsilon s}(1,s)Y_s(1,s)\big)ds \\
&= -f_0(v_\varepsilon(1,s))Y_s(1,s)\Big|_0^t - f_0'(Y(1,s))Y_s(1,s)v_\varepsilon(1,s)\Big|_0^t \\
&\quad + \int_0^t f_0(v_\varepsilon(1,s))Y_{ss}(1,s)ds + \int_0^t \frac{d^2}{ds^2}\Big(f_0(Y(1,s))\Big)v_\varepsilon(1,s)ds \geq -K_8
\end{aligned}
$$

for all $t \in [0,T]$, as well as the similar one

$$
\lim_{\delta\to 0}\int_0^t \frac{E_{2\varepsilon}(s,\delta)}{\delta^2}ds \geq -K_9 \quad \text{for all } t \in [0,T].
$$

For the last two inequalities we have used Lemma 4.3.2. Now, if we combine inequality (4.58) and the obvious estimates

$$
\|R_\varepsilon'(0)\| = \|F_\varepsilon(0)\| \leq K_{10},
$$

$$
\int_0^t \| F_\varepsilon'(s) \| \, ds \leq K_{11} \quad \text{for all } t \in [0,T],
$$

we get by Lemma 2.0.19

$$
\|R_\varepsilon'(t)\|^2 = \varepsilon\|R_{1\varepsilon t}(\cdot,t)\|_0^2 + \|R_{2\varepsilon t}(\cdot,t)\|_0^2 \leq K_{12}, \tag{4.59}
$$

for all $t \in [0,T]$. Therefore,

$$
\varepsilon\|R_{1\varepsilon t}(\cdot,t)\|_0^2 \leq K_{12}, \ \|R_{2\varepsilon t}(\cdot,t)\|_0 \leq K_{12}, \tag{4.60}
$$

for all $t \in [0,T]$. Now, combining (4.58) with (4.59) we find

$$
\|R_{1\varepsilon t}\|_{L^2(D_T)} \leq K_{13}. \tag{4.61}
$$

From (4.51) and (4.61), we derive

$$
\begin{aligned}
\|R_{1\varepsilon}(\cdot,t)\|_0^2 &= 2\int_0^t \langle R_{1\varepsilon s}(\cdot,s), R_{1\varepsilon}(\cdot,s)\rangle_0 ds \\
&\leq 2\int_0^t \|R_{1\varepsilon s}(\cdot,s)\|_0 \cdot \|R_{1\varepsilon}(\cdot,s)\|_0 ds \\
&\leq 2\|R_{1\varepsilon t}\|_{L^2(D_T)} \cdot \|R_{1\varepsilon}\|_{L^2(D_T)} = \mathcal{O}(\varepsilon^{1/2}) \ \ \forall\, t \in [0,T].
\end{aligned} \tag{4.62}
$$

Using (4.5), (4.50), (4.60) and (4.62), we obtain

$$
\|R_{1\varepsilon x}(\cdot,t)\|_0 \leq K_{14}, \ \|R_{2\varepsilon x}(\cdot,t)\|_0 \leq K_{15}\varepsilon^{1/4} \ \text{ for all } t \in [0,T]. \tag{4.63}
$$

By the mean value theorem, for every $t\in[0,T]$ and $\varepsilon > 0$ there exists a point $x_{t\varepsilon}\in[0,1]$ such that $\|R_{1\varepsilon}(\cdot,t)\|_0^2 = R_{1\varepsilon}(x_{t\varepsilon},t)^2$. Since

$$
\begin{aligned}
R_{1\varepsilon}(x,t)^2 &= R_{1\varepsilon}(x_{t\varepsilon},t)^2 + 2\int_{x_{t\varepsilon}}^x R_{1\varepsilon\xi}(\xi,t)R_{1\varepsilon}(\xi,t)d\xi \\
&\leq R_{1\varepsilon}(x_{t\varepsilon},t)^2 + 2\|R_{1\varepsilon}(\cdot,t)\|_0 \cdot \|R_{1\varepsilon x}(\cdot,t)\|_0,
\end{aligned}
$$

we obtain by (4.62) and (4.63) that

$$R_{1\varepsilon}(x,t)^2 \leq K_{16}\varepsilon^{1/4} \text{ for all } (x,t) \in \overline{D}_T.$$

Similarly, we can show that

$$R_{2\varepsilon}(x,t)^2 \leq K_{17}\varepsilon^{3/4} \text{ for all } (x,t) \in \overline{D}_T.$$

The proof is complete. □

*Remark* 4.3.3. We suspect that the above estimates could be proved under weaker assumptions on the data. Also, estimates in weaker norms are expected under even more relaxed assumptions on the data, including less compatibility. Furthermore, one may investigate only simple convergence to zero of the remainder components with respect to different norms. From a practical point of view, it is important to relax our requirements. This seems to be possible, at the expense of getting weaker approximation results. Note that the set of regular data $(u_0, v_0, f_1, f_2)$, as required in Corollary 4.2.14, is dense in the space $V_1 := L^2(0,1) \times H^1(0,1) \times H^1(D_T) \times L^2(D_T)$. It is easily seen that for $(u_0, v_0, f_1, f_2) \in V_1$, both problems $P_\varepsilon$ and $P_0$ have unique weak solutions (i.e., limits of strong solutions in $C([0,T]; L^2(0,1))^2$, and $L^2(D_T) \times \{C([0,T]; L^2(0,1)) \cap L^2(0,T; H^1(0,1))\}$, respectively), provided that $f_0$ and $r_0$ are smooth nondecreasing functions. This remark could be a starting point in proving further approximation results.

# Chapter 5

# Hyperbolic Systems with Dynamic Boundary Conditions

In this chapter we investigate the first four problems presented in Chapter 3. All these problems include dynamic boundary conditions (which involve the derivatives of $v(0,t)$, $v(1,t)$; $(BC.2)$ also include integrals of these functions). Note that all the four problems are singularly perturbed of the boundary layer type with respect to the sup norm.

The chapter consists of four sections, each of them addressing one of the four problems.

As a first step in our treatment, we construct a formal asymptotic expansion for each of the four problems, by employing the method presented in Chapter 1. For three of the four problems we construct zeroth order asymptotic expansions. In the case of problem $(LS)$, $(IC)$, $(BC.1)$, we construct a first order expansion of the solution in order to offer an example of a higher order asymptotic expansion. It should be pointed out that first or even higher order asymptotic expansions can be constructed for all the problems considered in this chapter but additional assumptions on the data should be required and much more laborious computations are needed.

Once a formal asymptotic expansion is determined, we will continue with its validation. More precisely, as a second step in our analysis, we will formulate and prove results concerning the existence, uniqueness, and higher regularity of the terms which occur in each of the previously determined asymptotic expansions. As in the previous chapter, we need higher regularity to show that our asymptotic expansions are well defined and to derive estimates for the remainder components. Our investigations here are mainly based on classic methods in the theory of evolution equations in Hilbert spaces associated with monotone operators as well as on linear semigroup theory. It should be pointed out that each of the four problems requires a different framework and separate analysis. All the operators associated with the corresponding reduced (unperturbed) problems are subdifferentials, except for the reduced problem in Subsection 5.4.2. When the PDE system under investigation is nonlinear (see Subsections 5.2.2 and 5.3.2), the treatment becomes much more complex.

The third and final step in our asymptotic analysis is to derive estimates for the remainders of the asymptotic expansions, in order to validate completely these expansions (see Subsections 5.1.3–5.4.3). In particular, note that our estimates from Subsection 5.1.3 are of the order of $\varepsilon^{\rho}, \rho > 1$, which enables us to say that the corresponding asymptotic expansion is a real first order expansion with respect to the sup norm.

Note also that Subsection 5.2.3 includes two different results: the former gives estimates in a weaker norm, in which the boundary layer is not visible (and thus the problem is regularly perturbed with respect to this norm), while the latter result provides estimates in the sup norm (and the boundary layer is visible in this norm). Obviously, the latter result is possible at the expense of stronger assumptions on the data (i.e., higher smoothness and more compatibility conditions). Note also that estimates in weaker norms are possible, under weaker assumptions, for all the problems under investigation. The reader is encouraged to derive such estimates under minimal assumptions.

## 5.1 A first order asymptotic expansion for the solution of problem $(LS)$, $(IC)$, $(BC.1)$

In this section we will investigate the following initial-boundary value problem, called $P_\varepsilon$ as usual:

$$\begin{cases} \varepsilon u_t + v_x + ru = f_1, \\ v_t + u_x + gv = f_2 \text{ in } D_T, \end{cases} \tag{$LS$}$$

with initial conditions

$$u(x,0) = u_0(x), \quad v(x,0) = v_0(x), \quad 0 \le x \le 1, \tag{$IC$}$$

and boundary conditions of the form

$$\begin{cases} r_0 u(0,t) + v(0,t) = 0, \\ u(1,t) - kv_t(1,t) = f_0\big(v(1,t)\big) + e_0(t), \quad 0 \le t \le T. \end{cases} \tag{$BC.1$}$$

The present data $f_1, f_2, u_0, v_0, r, g$ have the same meanings as in the previous chapter; $r_0 \ge 0$, $k > 0$ are given constants and $e_0 : [0,T] \to \mathbb{R}$ is a known function. Note that $(BC.1)$ are boundary conditions of algebraic-differential type, the latter condition being a (nonlinear) dynamic one (since it involves $v_t(1,t)$).
A reasoning similar to that used at the end of Chapter 1 reveals the existence of a boundary layer phenomenon in a neighborhood of the segment $\{(x,0); 0 \le x \le 1\}$.

### 5.1.1 Formal expansion

Here we construct a first order asymptotic expansion for the solution $U_\varepsilon = (u_\varepsilon(x,t), v_\varepsilon(x,t))$ of problem $P_\varepsilon$ introduced above.
We are going to apply the usual technique described in Chapter 1. More precisely, we will try to find a first order asymptotic expansion in the form

$$U_\varepsilon = U_0(x,t) + \varepsilon U_1(x,t) + V_0(x,\tau) + \varepsilon V_1(x,\tau) + R_\varepsilon(x,t), \quad (x,t) \in D_T, \tag{5.1}$$

where:

$\tau = t/\varepsilon$ is the fast variable;
$U_0 = (X_0(x,t), Y_0(x,t))$,
$U_1 = (X_1(x,t), Y_1(x,t))$ are the regular terms;
$V_0 = (c_0(x,\tau), d_0(x,\tau))$,
$V_1 = (c_1(x,\tau), d_1(x,\tau))$ are the boundary layer functions;
$R_\varepsilon = (R_{1\varepsilon}(x,t), R_{2\varepsilon}(x,t))$ is the remainder of the order one.

Now, we require $U_\varepsilon$ given by (5.1) to satisfy $P_\varepsilon$ formally and identify the coefficients of like powers of $\varepsilon^k$, $k = -1, 0, 1$.
For the components of the zeroth order regular term we obtain as in the previous chapter

$$\begin{cases} X_0 = (1/r)(f_1 - Y_{0x}), \\ Y_{0t} - (1/r)Y_{0xx} + gY_0 = f_2 - (1/r)f_{1x} \text{ in } D_T, \end{cases} \tag{5.2}$$

$$Y_0(x,0) = v_0(x), \ 0 \le x \le 1. \tag{5.3}$$

We have re-denoted the equations for the convenience of the reader. For the zeroth order boundary layer functions we obtain the formulae

$$c_0(x,\tau) = \alpha(x)e^{-r\tau}, \ d_0(x,\tau) \equiv 0, \tag{5.4}$$

where function $\alpha$ has the known expression which is derived from $(IC)$,

$$\alpha(x) = u_0(x) + (1/r)\big(v_0'(x) - f_1(x,0)\big).$$

For the components of the first order regular term we can easily derive the system

$$\begin{cases} X_{0t} + Y_{1x} + rX_1 = 0, \\ Y_{1t} + X_{1x} + gY_1 = 0 \text{ in } D_T, \end{cases} \tag{5.5}$$

which can be written in the following equivalent form

$$X_1 = -(1/r)(X_{0t} + Y_{1x}), \tag{5.6}$$

$$Y_{1t} - (1/r)Y_{1xx} + gY_1 = (1/r)X_{0xt} \text{ in } D_T. \tag{5.7}$$

The components of the first order remainder satisfy formally

$$\begin{cases} \varepsilon R_{1\varepsilon t} + R_{2\varepsilon x} + rR_{1\varepsilon} = -\varepsilon^2 X_{1t}, \\ R_{2\varepsilon t} + R_{1\varepsilon x} + gR_{2\varepsilon} = -\varepsilon(c_{1x} + gd_1) \text{ in } D_T. \end{cases} \tag{5.8}$$

The first order boundary layer functions satisfy the system

$$\begin{cases} c_{1\tau} + d_{1x} + rc_1 = 0, \\ d_{1\tau} + c_{0x} = 0, \end{cases}$$

and so it is easily seen that

$$\begin{cases} c_1(x,\tau) = \beta(x)e^{-r\tau} - (\alpha''(x)/r)\tau e^{-r\tau}, \\ d_1(x,\tau) = (1/r)\alpha'(x)e^{-r\tau}, \end{cases}$$

where function $\beta$ can be determined from $(IC)$. Indeed, from $(IC)$ we get

$$d_1(x,0) + Y_1(x,0) = 0 \Leftrightarrow$$

$$Y_1(x,0) = -(1/r)\alpha'(x), \ 0 \le x \le 1, \tag{5.9}$$

$$X_1(x,0) + c_1(x,0) = 0 \Leftrightarrow$$

$$\beta(x) = (1/r)X_{0t}(x,0) - (1/r^2)\alpha''(x), \quad 0 \le x \le 1. \tag{5.10}$$

For the remainder components we obtain the initial conditions:

$$R_{1\varepsilon}(x,0) = R_{2\varepsilon}(x,0) = 0, \quad 0 \le x \le 1. \tag{5.11}$$

Now, let us use $(BC.1)$. We first derive the equation

$$r_0 X_0(0,t) + Y_0(0,t) = 0, \quad 0 \le t \le T,$$

which yields (see (5.2)):

$$Y_0(0,t) - (r_0/r)Y_{0x}(0,t) = -(r_0/r)f_1(0,t), \quad 0 \le t \le T. \tag{5.12}$$

Next, we derive a condition similar to $(4.10)_1$ of the previous chapter, namely $c_0(0,\tau) = 0$, i.e.,

$$\alpha(0) = 0 \Leftrightarrow ru_0(0) = f_1(0,0) - v_0'(0). \tag{5.13}$$

We can also derive from $(BC.1)$ :

$$r_0 X_1(0,t) + Y_1(0,t) = 0, \quad 0 \le t \le T,$$

and, consequently (see (5.6)),

$$Y_1(0,t) - (r_0/r)Y_{1x}(0,t) = (r_0/r)X_{0t}(0,t), \quad 0 \le t \le T. \tag{5.14}$$

In addition, we obtain the equation

$$r_0 c_1(0,\tau) + d_1(0,\tau) = 0,$$

i.e.,

$$\begin{cases} \alpha''(0) = 0, \\ \alpha'(0) + r r_0 \beta(0) = 0. \end{cases} \tag{5.15}$$

From the latter boundary condition we get

$$X_0(1,t) - kY_{0t}(1,t) = f_0(Y_0(1,t)) + e_0(t), \quad 0 \le t \le T,$$
$$X_1(1,t) - kY_{1t}(1,t) = f_0'(Y_0(1,t))Y_1(1,t), \quad 0 \le t \le T.$$

Clearly, these equations can be rewritten (see $(5.2)_1$, (5.6)) as:

$$kY_{0t}(1,t) + (1/r)Y_{0x}(1,t) + f_0(Y_0(1,t)) = (1/r)f_1(1,t) - e_0(t), \tag{5.16}$$

$$kY_{1t}(1,t) + (1/r)Y_{1x}(1,t) + f_0'(Y_0(1,t))Y_1(1,t) = -(1/r)X_{0t}(1,t), \tag{5.17}$$

for all $t \in [0,T]$. Another equation given by the identification procedure is

$$c_0(1,\tau) - kd_{1\tau}(1,\tau) = 0,$$

which is satisfied if and only if

$$\alpha(1) + k\alpha'(1) = 0. \tag{5.18}$$

Note that equations (5.13), (5.15) and (5.18) guarantee that our boundary layer functions $c_0, c_1$ and $d_1$ do not introduce any discrepancies at the corner points $(0,0)$ and $(1,0)$ of $D_T$. We will see in the next subsection that these equations are also among the compatibility conditions which we will require to get enough smoothness for the terms of expansion (5.1).

Let us also point out that, unlike $d_0$, $d_1$ is not zero. This means that the boundary layer phenomenon comes also into effect for the latter component of the solution, but starting with the second term (first order term) of the asymptotic expansion of order $N$, $N \ge 1$.

Finally, it is easily seen that $R_\varepsilon$ should satisfy the following boundary conditions:

$$\begin{cases} r_0 R_{1\varepsilon}(0,t) + R_{2\varepsilon}(0,t) = 0, \\ R_{1\varepsilon}(1,t) - kR_{2\varepsilon t}(1,t) = f_0(v_\varepsilon(1,t)) - f_0(Y_0(1,t)) \\ -\varepsilon f_0'(Y_0(1,t))Y_1(1,t) - \varepsilon c_1(1,\tau), \ 0 < t < T. \end{cases} \tag{5.19}$$

It is worth noting that the first term of expansion (5.1), $U_0 = (X_0(x,t), Y_0(x,t))$, coincides with the solution of the reduced problem $P_0$ which comprises the algebraic equation $(5.2)_1$ and the boundary value problem $(5.2)_2$, (5.3), (5.12), (5.16).

In conclusion, if there are smooth solutions for $P_\varepsilon$, $\varepsilon > 0$, $P_0$, and for problem (5.6), (5.7), (5.9), (5.14), (5.17), denoted by $P_1$, then $U_\varepsilon$ can be represented by (5.1), where $R_\varepsilon = (R_{1\varepsilon}, R_{2\varepsilon})$ satisfies (5.8), (5.11), (5.19).

Note that a zeroth order asymptotic expansion for the solution of $P_\varepsilon$ had previously been constructed in [35] under weaker assumptions on the data.

### 5.1.2 Existence, uniqueness and regularity of the solutions of problems $P_\varepsilon$, $P_0$ and $P_1$

We start this subsection with problem $P_\varepsilon$. Note that this problem is a particular case of problem (2.14), (2.15), (2.17) presented in Chapter 2. Therefore, we can apply Theorem 2.0.34. However, we plan to obtain higher regularity results for the solution $U_\varepsilon$ of $P_\varepsilon$, so we have to pay more attention to this problem.

Let us consider the Hilbert space $H_1 = \left(L^2(0,1)\right)^2 \times \mathbb{R}$, endowed with the scalar product

$$\langle h_1, h_2 \rangle = \varepsilon \int_0^1 p_1(x)p_2(x)dx + \int_0^1 q_1(x)q_2(x)dx + ka_1a_2,$$
$$\text{for all } h_i = (p_i, q_i, a_i) \in H_1, \quad i = 1,2,$$

and the associated Hilbertian norm, denoted by $\|\cdot\|$.

Define the operator $B_\varepsilon : D(B_\varepsilon) \subset H_1 \to H_1$ by:

$$D(B_\varepsilon) = \left\{(p,q,a);\ p,q \in H^1(0,1),\ r_0p(0) + q(0) = 0,\ a = q(1)\right\},$$
$$B_\varepsilon(p,q,a) = \left(\varepsilon^{-1}\left(q' + rp\right),\ p' + gq,\ k^{-1}\left(-p(1) + f_0\left(a\right)\right)\right).$$

Obviously, $B_\varepsilon$ is a special case of operator $B_1$ considered in *Example* 4, Chapter 2. Therefore, by Proposition 2.0.16, $B_\varepsilon$ is a maximal monotone operator. We associate with problem $P_\varepsilon$ the following Cauchy problem in $H_1$:

$$\begin{cases} w_\varepsilon'(t) + B_\varepsilon w_\varepsilon(t) = F_\varepsilon(t), & 0 < t < T, \\ w_\varepsilon(0) = w_0, \end{cases} \tag{5.20}$$

where

$$w_0 = (u_0, v_0, \xi_0), \quad w_\varepsilon(t) = \left(u_\varepsilon(\cdot,t), v_\varepsilon(\cdot,t), \xi_\varepsilon(t)\right),$$
$$F_\varepsilon(t) = \left(\varepsilon^{-1}f_1(\cdot,t), f_2(\cdot,t), -k^{-1}e_0(t)\right),\ 0 < t < T.$$

Let us reformulate Theorem 2.0.34 for the present particular case:

**Theorem 5.1.1.** *Assume that*

$$g, r \text{ are nonnegative constants, and } r_0, k > 0; \tag{5.21}$$

$$f_0 : \mathbb{R} \to \mathbb{R} \quad \text{is a continuous nondecreasing function;} \tag{5.22}$$

$$u_0, v_0 \in H^1(0,1),\quad r_0u_0(0) + v_0(0) = 0,\ \xi_0 = v_0(1),\quad (f_1, f_2, e) \in W^{1,1}(0,T;H_1).$$

*Then, problem* (5.20) *has a unique strong solution* $w_\varepsilon \in W^{1,\infty}(0,T;H_1)$, *with* $\xi_\varepsilon(t) = v_\varepsilon(1,t)$ *for all* $t \in [0,T]$. *In addition,* $u_\varepsilon, v_\varepsilon \in L^\infty\left(0,T;H^1(0,1)\right)$.

*Remark* 5.1.2. Obviously, the assumptions on $u_0, v_0$ in Theorem 5.1.1 say nothing else but that

$$w_0 = (u_0, v_0, \xi_0) \in D(B_\varepsilon).$$

The equation $r_0 u_0(0) + v_0(0) = 0$ is called a *zeroth order compatibility condition.*

In order to obtain higher regularity results we can use the classical method: first we differentiate formally our problem $P_\varepsilon$ with respect to $t$, then we establish some regularity of the solution of the resulting problem, and then we return to the original problem to derive higher regularity for its solution. This method provides immediate results in the case of linear problems. Nonlinear problems require much more work. Now, we prove the following result:

**Theorem 5.1.3.** *Assume* (5.21) *and*

$$f_0 \in C^2(\mathbb{R}),\ f_0' \geq 0, \quad f_1, f_2 \in W^{2,\infty}\left(0,T; L^2(0,1)\right),\ e_0 \in W^{2,\infty}(0,T); \tag{5.23}$$

$$w_0 \in D(B_\varepsilon), \quad w_{01} := F_\varepsilon(0) - B_\varepsilon w_0 \in D(B_\varepsilon). \tag{5.24}$$

*Then the strong solution $w_\varepsilon$ of the Cauchy problem* (5.20) *satisfies*

$$w_\varepsilon \in C^2([0,T]; H_1),\ u_\varepsilon, v_\varepsilon \in C^1([0,T]; H^1(0,1)).$$

*Proof.* By Theorem 5.1.1, we know that problem (5.20) has a unique strong solution $w_\varepsilon \in W^{1,\infty}(0,T;H_1)$, with $\xi_\varepsilon(t) = v_\varepsilon(1,t)$ for all $t\in[0,T]$. Next, we define the operators $C_\varepsilon(t) : D(C_\varepsilon(t)) = H_1 \to H_1$,

$$C_\varepsilon(t)(p,q,a) = d_\varepsilon(t)(0,0,a) - F_\varepsilon'(t),$$

where $d_\varepsilon(t) := k^{-1} f_0'\left(v_\varepsilon(1,t)\right),\ d_\varepsilon(t) \geq 0$ for all $t \in [0,T]$. Denote by $B_{0\varepsilon}$ the operator obtained by taking $f_0 \equiv 0$ in the definition of $B_\varepsilon$.

Obviously, operator $E_\varepsilon(t) = B_{0\varepsilon} + C_\varepsilon(t)$, with $D\left(E_\varepsilon(t)\right) = D(B_\varepsilon)$, is maximal monotone for all $t \in [0,T]$.

Moreover, since $d_\varepsilon' \in L^\infty(0,T)$ and $F_\varepsilon'' \in L^\infty(0,T;H_1)$, there exists a positive constant $L_\varepsilon$ such that

$$\|E_\varepsilon(t)x - E_\varepsilon(s)x\| \leq L_\varepsilon \mid t-s \mid (1 + \|x\|),$$

for all $x \in D(B_\varepsilon)$ and all $s,t \in [0,T]$. Therefore, according to Theorem 2.0.31, the following Cauchy problem

$$\begin{cases} z_\varepsilon'(t) + E_\varepsilon(t) z_\varepsilon(t) = 0, \quad 0 < t < T, \\ z_\varepsilon(0) = w_{01}, \end{cases} \tag{5.25}$$

has a unique strong solution

$$z_\varepsilon(t) = (\tilde{u}_\varepsilon(\cdot,t), \tilde{v}_\varepsilon(\cdot,t), \tilde{\xi}_\varepsilon(t)) \in W^{1,\infty}(0,T;H_1),$$

such that $z_\varepsilon(t) \in D(B_\varepsilon)$ for all $t \in [0,T]$, which implies $\tilde{\xi}_\varepsilon(t) = \tilde{v}_\varepsilon(1,t)$ for all $t \in [0,T]$. On the other hand, operator $B_{0\varepsilon}$ is linear and maximal monotone. If we denote by $\{S_\varepsilon(t),\ t \geq 0\}$ the continuous semigroup of contractions generated by $-B_{0\varepsilon}$, then the solutions of problems (5.20) and (5.25) satisfy

$$w_\varepsilon(t) = S_\varepsilon(t)w_0 + \int_0^t S_\varepsilon(t-s)F_{1\varepsilon}(s)ds, \quad 0 \leq t \leq T, \tag{5.26}$$

$$z_\varepsilon(t) = S_\varepsilon(t)w_{01} + \int_0^t S_\varepsilon(t-s)F_{2\varepsilon}(s)ds, \quad 0 \leq t \leq T, \tag{5.27}$$

where:

$$F_{1\varepsilon}(t) = F_\varepsilon(t) - \left(0,0,k^{-1}f_0\left(v_\varepsilon(1,t)\right)\right),$$
$$F_{2\varepsilon}(t) = F_\varepsilon'(t) - \left(0,0,d_\varepsilon(t)\tilde{v}_\varepsilon(1,t)\right),\ 0 \leq t \leq T.$$

Since $w_0 \in D(B_{0\varepsilon}) = D(B_\varepsilon)$ and $F_{1\varepsilon} \in W^{1,\infty}(0,T;H_1)$, we can differentiate (5.26) (see Theorem 2.0.27) to obtain

$$w_\varepsilon'(t) = S_\varepsilon(t)w_{01} + \int_0^t S_\varepsilon(t-s)F_{1\varepsilon}'(s)ds,\ 0 \leq t \leq T. \tag{5.28}$$

From (5.27) and (5.28) it follows that

$$\begin{aligned} \|w_\varepsilon'(t) - z_\varepsilon(t)\| &\leq \int_0^t \|F_{1\varepsilon}'(s) - F_{2\varepsilon}(s)\|ds \\ &\leq K_\varepsilon \int_0^t \|w_\varepsilon'(s) - z_\varepsilon(s)\|ds, \quad 0 \leq t \leq T, \end{aligned} \tag{5.29}$$

where $K_\varepsilon = \sup\{\mid d_\varepsilon(t) \mid;\ t \in [0,T]\}$. Gronwall's lemma applied to (5.29) yields $w_\varepsilon'(t) = z_\varepsilon(t)$ for all $t \in [0,T]$.
Therefore, $w_\varepsilon \in W^{2,\infty}(0,T;H_1)$, i.e.,

$$u_\varepsilon, v_\varepsilon \in W^{2,\infty}(0,T;L^2(0,1)),\ v_\varepsilon(1,\cdot) \in W^{2,\infty}(0,T).$$

In fact, $w_\varepsilon$ is even more regular. Indeed, $z_\varepsilon = w_\varepsilon'$ satisfies the problem

$$\begin{cases} z_\varepsilon'(t) + B_{0\varepsilon}z_\varepsilon(t) = F_{3\varepsilon}(t), & 0 < t < T, \\ z(0) = w_{01}, \end{cases}$$

where $F_{3\varepsilon}(t) = F_\varepsilon'(t) - (0,0,k^{-1}f_0'(v_\varepsilon(1,t))v_{\varepsilon t}(1,t))$ for all $t \in [0,T]$.
Since $w_{01} \in D(B_{0\varepsilon}) = D(B_\varepsilon)$ and $F_{3\varepsilon} \in W^{1,\infty}(0,T;H_1)$, it follows by Theorem 2.0.27 that $z_\varepsilon \in C^1([0,T];H_1)$, $B_{0\varepsilon}z_\varepsilon \in C([0,T];H_1)$, i.e.,

$$w_\varepsilon \in C^2([0,T];H_1),\ u_\varepsilon, v_\varepsilon \in C^1([0,T];H^1(0,1)). \qquad \square$$

*Remark* 5.1.4. Hypotheses (5.24) hold if the following sufficient conditions (expressed in terms of the data) are fulfilled: $u_0, v_0 \in H^2(0,1)$,

$$\begin{cases} r_0 u_0(0) + v_0(0) = 0, \\ f_1(0,0) = v_0'(0) + r u_0(0), \\ f_2(0,0) = u_0'(0) + g v_0(0), \\ k[f_2(1,0) - u_0'(1) - g v_0(1)] = -e_0(0) + u_0(1) - f_0\left(v_0(1)\right). \end{cases} \tag{5.30}$$

Let us point out that all these conditions are independent of $\varepsilon$, and so they are good for any value of $\varepsilon > 0$. This is important for our asymptotic analysis.

In the following we are going to investigate the reduced problem $P_0$, which comprises the algebraic equation $(5.2)_1$ and the boundary value problem $(5.2)_2$, (5.3), (5.12), (5.16). In this case, we choose as a basic setup the space $H_2 = L^2(0,1) \times \mathbb{R}$, which is a Hilbert space with the scalar product

$$\langle (p,a),(q,b) \rangle_{H_2} := \int_0^1 p(x)q(x)dx + kab,$$

and we denote by $\| \cdot \|_{H_2}$ the associated norm. Let us define an operator $A(t) : D(A(t)) \subset H_2 \to H_2$, by

$$D(A(t)) = \left\{(p,a) \in H_2;\ p \in H^2(0,1),\ a = p(1),\ r^{-1}p'(0) + \alpha_1(t) = r_0^{-1}p(0)\right\},$$
$$A(t)(p,a) = \left(-r^{-1}p'' + gp,\ k^{-1}[r^{-1}p'(1) + f_0(a)]\right),\ \alpha_1(t) = -r^{-1}f_1(0,t).$$

It is easily seen that $A(t) = \partial\phi(t,\cdot),\ t \in [0,T]$, where

$$\phi(t,\cdot) : H_2 \to (-\infty, +\infty],$$

$$\phi(t,(p,a)) = \begin{cases} \frac{1}{2r}\int_0^1 p'(x)^2 dx + \frac{g}{2}\int_0^1 p(x)^2 dx + \frac{1}{2r_0}p(0)^2 \\ -\alpha_1(t)p(0) + l_0(a), \quad \text{if } p \in H^1(0,1) \text{ and } a = p(1), \\ +\infty, \qquad \text{otherwise}, \end{cases}$$

in which $l_0 : \mathbb{R} \to \mathbb{R}$ is a primitive of $f_0$ (see also [24], p. 93). Obviously, for every $t \in [0,T]$, $D(\phi(t,\cdot)) = \{(p,a);\ p \in H^1(0,1),\ a = p(1)\} =: V$, which is independent of $t$. In addition, $V$ is a Hilbert space with the scalar product:

$$\langle (p_1,a_1),(p_2,a_2) \rangle_V = \int_0^1 (p_1'p_2' + p_1p_2)dx + ka_1a_2,\ \forall\ (p_i,a_i) \in V,\ i = 1,2.$$

We associate with $P_0$ the following Cauchy problem in $H_2$ :

$$\begin{cases} Z'(t) + A(t)Z(t) = h(t),\ 0 < t < T, \\ Z(0) = Z_0, \end{cases} \tag{5.31}$$

where

$$Z(t) = (Y_0(\cdot,t), \zeta(t)), \ Z_0 = (v_0, \zeta_0), \ h(t) = (h_1(\cdot,t), h_2(t)),$$
$$h_1(x,t) = (f_2 - r^{-1} f_{1x})(x,t), \ h_2(t) = k^{-1}(r^{-1} f_1(1,t) - e_0(t)).$$

In the following we are interested in higher order regularity of the solution. Thus we will formally differentiate several times problem (5.31). Consequently, we need to introduce several operators and notations. First, let $B(t) : D(B(t)) \subset H_2 \to H_2$ be defined by

$$D(B(t)) = \{(p,a) \in H_2; \ p \in H^2(0,1), \ a = p(1), \ r^{-1}p'(0) + \alpha_1'(t) = r_0^{-1}p(0)\},$$
$$B(t)(p,a) = \left(-r^{-1}p'' + gp, k^{-1}[r^{-1}p'(1) + \alpha_2(t)a]\right), \text{with } \alpha_2(t) = f_0'(Y_0(1,t)).$$

We will use the notations $B'(t)$ and $B''(t)$ for the operators which are obtained from the definition of $B(t)$ by replacing $\alpha_1'$ by $\alpha_1''$ and $\alpha_1^{(3)}$, respectively. We also define the function

$$\overline{h}(t) = h''(t) - k^{-1}(0, \alpha_2'(t) Y_{0t}(1,t)),$$

and denote

$$Z_{01} = h(0) - A(0)Z_0, \ Z_{02} = h'(0) - B(0)Z_{01},$$
$$Z_{03} = \overline{h}(0) - B'(0)Z_{02}.$$

**Theorem 5.1.5.** *Assume that*

$$r, \ r_0, \ k \text{ are some positive constants, and } g \geq 0; \tag{5.32}$$

$$f_0 \in C^3(\mathbb{R}), \ f_0' \geq 0, \ h \in W^{3,2}(0,T;H_2), \ \alpha_1 \in H^4(0,T); \tag{5.33}$$

$$Z_0 \in D(A(0)), \ Z_{01} \in D(B(0)), \ Z_{02} \in D(B'(0)), \ Z_{03} \in V. \tag{5.34}$$

*Then, problem* (5.31) *has a unique strong solution* $Z \in W^{4,2}(0,T;H_2)$, $\zeta(t) = Y_0(1,t)$ *for all* $t \in [0,T]$, *and* $Y_{0tttxx} \in L^2(D_T)$.

*Proof.* We will apply Theorem 2.0.32. As noted above, $D(\phi(t,\cdot) = V$ for all $t \in [0,T]$. We are going to show that condition (2.12) is fulfilled. Indeed, for every $p \in H^1(0,1)$, $0 \leq s \leq t \leq T$, we have

$$\begin{aligned} \phi(t,(p,p(1))) - \phi(s,(p,p(1))) &= -(\alpha_1(t) - \alpha_1(s))p(0) \\ &\leq \mid p(0) \mid \int_s^t \mid \alpha_1'(\tau) \mid d\tau. \end{aligned} \tag{5.35}$$

On the other hand, it follows from the definition of $\phi$ that

$$\phi(s,(p,p(1))) \geq K_1 \parallel p' \parallel_0^2 - K_2 \mid p(1) \mid - K_3 \mid p(0) \mid - K_4,$$

where $\| \cdot \|_0$ denotes the usual norm of $L^2(0,1)$), and thus

$$| p(0) | \leq \phi(s, (p, p(1))) + K_5 \| (p, p(1)) \|_{H_2}^2 + K_6,$$

where $K_i$, $i = 1, \dots, 6$, are different positive constants. This inequality together with (5.35) implies (2.12) with $\gamma(t) = \int_0^t | \alpha_1'(s) | \, ds$. Therefore, according to Theorem 2.0.32, problem (5.31) has a unique strong solution $Z \in W^{1,2}(0,T;H_2)$, such that $\zeta(t) = Y_0(1,t)$ for almost all $t \in (0,T)$, and $Y_0 \in L^\infty(0,T;H^1(0,1))$. By a reasoning similar to that used in the proof of Theorem 4.2.12, it follows that $Y_0 \in W^{1,2}(0,T;H^1(0,1))$.

To prove higher regularity of the solution, we will formally differentiate problem (5.31) with respect to $t$.
First, let us check whether function $t \to (Y_{0t}(\cdot,t), Y_{0t}(1,t)) \in W^{1,2}(0,T;V^*)$, where $V^*$ is the dual of $V$. We will reason as in the proof of Theorem 4.2.12. We start with the following obvious equation which is satisfied by the solution of problem (5.31):

$$\begin{aligned} &\langle (\varphi, \varphi(1)), Z'(t) \rangle + r^{-1} \langle \varphi', Y_{0x}(\cdot,t) \rangle_0 + g \langle \varphi, Y_0(\cdot,t) \rangle_0 \\ &\qquad - \alpha_1(t)\varphi(0) + r_0^{-1} Y_0(0,t)\varphi(0) + f_0(Y_0(1,t))\varphi(1) \\ &\qquad = \langle (\varphi, \varphi(1)), h(t) \rangle_{H_2} \ \ \forall \ (\varphi, \varphi(1)) \in V. \end{aligned} \tag{5.36}$$

We have denoted by $\langle \cdot, \cdot \rangle_0$ the usual scalar product of $L^2(0,1)$ and by $\langle \cdot, \cdot \rangle$ the pairing between $V$ and $V^*$. It follows from (5.36) that

$$\begin{aligned} &\langle (\varphi, \varphi(1)), Z'(t+\delta) - Z'(t) \rangle + r^{-1} \langle \varphi', Y_{0x}(\cdot,t+\delta) - Y_{0x}(\cdot,t) \rangle_0 \\ &\qquad - (\alpha_1(t+\delta) - \alpha_1(t))\varphi(0) + g\langle \varphi, Y_0(\cdot,t+\delta) - Y_0(\cdot,t) \rangle_0 \\ &\qquad + r_0^{-1}(Y_0(0,t+\delta) - Y_0(0,t))\varphi(0) \\ &\qquad + (f_0(Y_0(1,t+\delta)) - f_0(Y_0(1,t)))\varphi(1) \\ &\qquad = \langle (\varphi, \varphi(1)), h(t+\delta) - h(t) \rangle_{H_2}, \end{aligned}$$

for all $(\varphi, \varphi(1)) \in V$, a.a. $t$, $t + \delta \in (0,T)$, $\delta > 0$. Therefore,

$$\begin{aligned} \| Z'(t+\delta) - Z'(t) \|_{V^*}^2 \leq K_7 (&| \alpha_1(t+\delta) - \alpha_1(t) |^2 \\ &+ \| Z(t+\delta) - Z(t) \|_V^2 + \| h(t+\delta) - h(t) \|_{H_2}^2), \end{aligned}$$

where $K_7$ is a positive constant. We have taken into account that $H^1(0,1)$ is continuously embedded into $C[0,1]$ and that $f_0$ is Lipschitz continuous on bounded sets. The above inequality leads us to

$$\int_0^{T-\delta} \| (Y_{0t}(\cdot,t+\delta) - Y_{0t}(\cdot,t), Y_{0t}(1,t+\delta) - Y_{0t}(1,t)) \|_{V^*}^2 \, dt \leq K_8 \delta^2 \ \forall \ \delta \in (0,T],$$

where $K_8$ is another positive constant. Therefore, according to Theorem 2.0.3, $t \to (Y_{0t}(\cdot,t), Y_{0t}(1,t))$ belongs to $W^{1,2}(0,T;V^*)$. Thus, one can differentiate equation

(5.36) and see that $Z'(t) =: \overline{Z}(t) = (\overline{Y}_0(\cdot,t), \overline{Y}_0(1,t))$ satisfies the equation

$$\begin{aligned}&\langle(\varphi,\varphi(1)),\overline{Z}'(t)\rangle + r^{-1}\langle\varphi',\overline{Y}_{0x}(\cdot,t)\rangle_0 + g\langle\varphi,\overline{Y}_0(\cdot,t)\rangle_0\\ &\quad - \alpha_1'(t)\varphi(0) + r_0^{-1}\overline{Y}_0(0,t)\varphi(0) + \alpha_2(t)\overline{Y}_0(1,t)\varphi(1)\\ &\quad = \langle(\varphi,\varphi(1)),h'(t)\rangle_{H_2} \quad \forall\ (\varphi,\varphi(1)) \in V\end{aligned} \tag{5.37}$$

as well as the initial condition

$$\overline{Z}(0) = Z_{01}. \tag{5.38}$$

It is easy to check that problem (5.37), (5.38) has at most one solution. This problem is a weak form of the following Cauchy problem in $H_2$

$$\begin{cases} \overline{Z}'(t) + B(t)\overline{Z}(t) = h'(t), \ 0 < t < T,\\ \overline{Z}(0) = Z_{01}. \end{cases} \tag{5.39}$$

Obviously, $\alpha_2 \in H^1(0,T)$, $\alpha_2 \geq 0$. On the other hand, operator $B(t)$ is the subdifferential of the function $\phi_1(t,\cdot) : H_2 \to (-\infty,+\infty]$,

$$\phi_1(t,(p,a)) = \begin{cases} \frac{1}{2r}\int_0^1 p'(x)^2dx + \frac{g}{2}\int_0^1 p(x)^2dx + \frac{1}{2}\alpha_2(t)a^2 + \frac{1}{2r_0}p(0)^2\\ -\alpha_1'(t)p(0), \text{ if } p \in H^1(0,1) \text{ and } a = p(1),\\ +\infty, \text{ otherwise,} \end{cases}$$

with $D(\phi_1(t,\cdot)) = V$ for all $t \in [0,T]$. One can apply again Theorem 2.0.32 (since condition (2.12) is fulfilled). Therefore problem (5.39) has a unique strong solution $\overline{Z}$, $\overline{Z} \in W^{1,2}(0,T;H_2)$, $\overline{Y}_{0x} \in L^\infty(0,T;L^2(0,1))$.
Moreover, since $Z_{01} \in D(B(0))$, it follows by a standard reasoning (see Theorem 4.2.12) that $\overline{Y}_0 \in W^{1,2}(0,T;H^1(0,1))$. As $\overline{Z}$ satisfies problem (5.37), (5.38), one can see that $\overline{Z} = Z'$ and so

$$Y_0 \in W^{2,2}(0,T;H^1(0,1)), \quad Y_0(1,\cdot) \in H^2(0,T).$$

Now, using similar arguments, we can differentiate equation (5.37) and see that $\overline{Z}'(t) = Z''(t) =: \overline{\overline{Z}}(t) = (\overline{\overline{Y}}_0(\cdot,t), \overline{\overline{Y}}_0(1,t))$ satisfies a similar equation,

$$\begin{aligned}&\langle(\varphi,\varphi(1)),\overline{\overline{Z}}'(t)\rangle + r^{-1}\langle\varphi',\overline{\overline{Y}}_{0x}(\cdot,t)\rangle_0 + g\langle\varphi,\overline{\overline{Y}}_0(\cdot,t)\rangle_0\\ &\quad - \alpha_1''(t)\varphi(0) + r_0^{-1}\overline{\overline{Y}}_0(0,t)\varphi(0) + \alpha_2(t)\overline{\overline{Y}}_0(1,t)\varphi(1)\\ &\quad = \langle(\varphi,\varphi(1)),\overline{h}(t)\rangle_{H_2} \quad \forall\ (\varphi,\varphi(1)) \in V,\end{aligned} \tag{5.40}$$

as well as the initial condition

$$\overline{\overline{Z}}(0) = Z_{02}. \tag{5.41}$$

In fact, $\overline{\overline{Z}}(t) = Z''(t)$ is the unique strong solution of the Cauchy problem in $H_2$

$$\begin{cases} \overline{\overline{Z}}'(t) + B'(t)\overline{\overline{Z}}(t) = \overline{h}(t), \ 0 < t < T,\\ \overline{\overline{Z}}(0) = Z_{02}. \end{cases} \tag{5.42}$$

Since $Z_{02} \in D(B'(0))$, $\overline{h} \in W^{1,2}(0,T;H_2)$, $\alpha_1'' \in H^1(0,T)$, we get by known arguments

$$\overline{\overline{Z}} \in W^{1,2}(0,T;H_2),\ \overline{\overline{Y}}_0 \in W^{1,2}(0,T;H^1(0,1)),$$

and thus

$$Y_0 \in W^{3,2}(0,T;H^1(0,1)),\ Y_0(1,\cdot) \in H^3(0,T).$$

To conclude the proof, it is sufficient to notice that under our assumptions $\widetilde{Z}(t) = (Y_{0ttt}(\cdot,t), Y_{0ttt}(1,t))$ is the unique strong solution of the time-dependent Cauchy problem

$$\begin{cases} \widetilde{Z}'(t) + B''(t)\widetilde{Z}(t) = \widetilde{h}(t),\ 0 < t < T, \\ \widetilde{Z}(0) = Z_{03}, \end{cases} \tag{5.43}$$

where $\widetilde{h}(t) = \overline{h}'(t) - k^{-1}(0, \alpha_2'(t)Y_{0tt}(1,t))$. Since $\widetilde{h} \in L^2(0,T;H_2)$, $\alpha_1^{(3)} \in H^1(0,1)$, and $Z_{03} \in V$, one can apply again Theorem 2.0.32 which yields $\widetilde{Z} \in W^{1,2}(0,T;H_2)$. From (5.43) we derive $Y_{0tttxx} \in L^2(D_T)$, which completes the proof. □

*Remark* 5.1.6. Our assumptions $Z_0 \in D(A(0))$, $Z_{01} \in D(B(0))$ in the above Theorem hold if the following sufficient conditions are fulfilled: $v_0$, $f_1$ and $f_2$ are smooth enough, and

$$\begin{cases} rv_0(0) - r_0v_0'(0) = -r_0f_1(0,0),\ \zeta_0 = v_0(1), \\ (v_0''(0) - rgv_0(0) + rf_2(0,0) - f_{1x}(0,0)) - r_0[r^{-1}v_0^{(3)}(0) \\ -gv_0'(0) + f_{2x}(0,0) - r^{-1}f_{1xx}(0,0)] = -r_0f_{1t}(0,0), \\ v_0'(1) + rf_0(v_0(1)) - f_1(1,0) + re_0(0) \\ = -k[v_0''(1) - rgv_0(1) - f_{1x}(1,0) + rf_2(1,0)], \end{cases} \tag{5.44}$$

We encourage the reader to formulate additional conditions for data which guarantee that $Z_{02} \in D(B'(0))$, and $Z_{03} \in V$.

Now, we are going to investigate problem $P_1$. We associate with $P_1$ the Cauchy problem in $H_2$ :

$$\begin{cases} Z_1'(t) + C(t)Z_1(t) = l(t),\ \ 0 < t < T, \\ Z(0) = Z_{10}, \end{cases} \tag{5.45}$$

where $C(t) : D(C(t)) \subset H_2 \to H_2$,

$$\begin{aligned} D(C(t)) &= \{(p,a) \in H_2;\ p \in H^2(0,1),\ a = p(1), \\ &\qquad r^{-1}p'(0) + \alpha_3(t) = r_0^{-1}p(0)\}, \\ C(t)\,((p,a)) &= \left(-r^{-1}p'' + gp, k^{-1}[r^{-1}p'(1) + \alpha_2(t)a]\right), \\ Z_1(t) &= (Y_1(\cdot,t), \zeta_1(t)),\ Z_{10} = -r^{-1}(\alpha'(\cdot), \alpha'(1)), \\ l(t) &= (r^{-1}X_{0xt}(\cdot,t), -(kr)^{-1}X_{0t}(1,t)),\ \alpha_3(t) = r^{-1}X_{0t}(0,t). \end{aligned}$$

Obviously, if in addition to the assumptions of the previous theorem we require $f_1 \in W^{3,2}(0,T;H^1(0,1))$, we can see (if we also take into account the equation $X_0 = r^{-1}(f_1 - Y_{0x})$), that $l \in W^{2,2}(0,T;H_2)$ and $\alpha_3 \in H^2(0,T)$. If we suppose in addition that

$$Z_{10} \in D(C(0)),\ l(0) - C(0)Z_{10} \in D(C'(0)), \tag{5.46}$$

where $C'(t)$ is the operator obtained by replacing $\alpha_3$ with $\alpha_3'$ in the definition of $C(t)$, then a device similar to that used in the proof of the preceding theorem shows that problem $P_1$ has a unique solution $(X_1, Y_1) \in W^{2,2}(0,T;L^2(0,1)) \times W^{2,2}(0,T;H^1(0,1))$.

*Remark* 5.1.7. Assumption $(5.46)_1$ reads $\left(-r^{-1}\alpha'(\cdot), -r^{-1}\alpha'(1)\right) \in D(C(0))$. Clearly, this condition is satisfied iff

$$\alpha \in H^3(0,1),\quad -r_0\alpha''(0) + r_0 r X_{0t}(0,0) + r\alpha'(0) = 0. \tag{5.47}$$

If we take into account (5.10), we see that (5.47) is equivalent to $(5.15)_2$, which was required in the previous subsection.

On the other hand,

$$l(0) - C(0)Z_{10}$$
$$= \begin{pmatrix} r^{-1}X_{0xt}(\cdot,0) - r^{-2}\alpha^{(3)} + r^{-1}g\alpha' \\ -k^{-1}r^{-1}X_{0t}(1,0) + r^{-2}k^{-1}\alpha''(1) + r^{-1}k^{-1}\alpha'(1)f_0'(v_0(1)) \end{pmatrix}^T.$$

Thus, if $u_0$, $v_0$, $f_1$, $f_2$ and $e_0$ are smooth functions, then one can indicate explicit sufficient conditions in terms of the data such that (5.46) are satisfied.

So, we are able to formulate the following concluding result:

**Corollary 5.1.8.** *Assume that* (5.32) *holds. If* $u_0$, $v_0$, $f_1$, $f_2$ *and* $e_0$ *are smooth enough,*

$$f_0 \in C^3(\mathbb{R}),\ f_0' \geq 0, \tag{5.48}$$

*and the compatibility conditions* (5.30), (5.34), (5.46) *are satisfied, then the conclusions of Theorems* 5.1.3, 5.1.5 *hold and problem* $P_1$ *has a unique solution*

$$(X_1, Y_1) \in W^{2,2}(0,T;L^2(0,1)) \times W^{2,2}(0,T;H^1(0,1)). \tag{5.49}$$

*Remark* 5.1.9. Concerning the above compatibility conditions (5.34), (5.46), we have formulated just a partial system of sufficient conditions (see (5.44) and (5.47)), but it is not difficult to give a complete system of conditions in terms of the data. It is worth mentioning that such a system is compatible. For example, equations $(5.30)_{1,2}$, $(5.44)_1$,

$$\begin{cases} r_0 u_0(0) + v_0(0) = 0, \\ f_1(0,0) = v_0'(0) + r u_0(0), \\ r v_0(0) - r_0 v_0'(0) = -r_0 f_1(0,0), \end{cases}$$

are compatible, the third one being a combination of the first two equations.

Note also that all the conditions which have been introduced to avoid discrepancies at the corner points $(0,0)$ and $(1,0)$ (i.e., (5.13), (5.15) and (5.18)) are incorporated into our set of assumptions on the data. Indeed, (5.13) is exactly the second equation above; condition $(5.15)_2$ follows from (5.47), as remarked before; in fact, in view of (5.2) and (5.12), we can see that equation $(5.30)_3$ implies that $r_0 X_{0t}(0,0) + \alpha'(0) = 0$ and so (5.47) also yields $(5.15)_1$; (5.18) follows from $(5.30)_4$ and $(5.44)_3$.

### 5.1.3 Estimates for the remainder components

We are going to show that, under appropriate assumptions, the components of the first order remainder $R_\varepsilon$ are of the order of $\varepsilon^\alpha$, with $\alpha = 9/8$ and $\alpha = 11/8$, respectively, with respect to the uniform convergence norm. This is more than enough to guarantee that (5.1) is a real first order asymptotic expansion.

**Theorem 5.1.10.** *Assume that all the assumptions of Corollary* 5.1.8 *hold. Then, for every $\varepsilon > 0$, the solution of problem $P_\varepsilon$ admits an asymptotic expansion of the form* (5.1) *and the following estimates are valid:*

$$\|R_{1\varepsilon}\|_{C(\overline{D}_T)} = \mathcal{O}(\varepsilon^{9/8}), \qquad \|R_{2\varepsilon}\|_{C(\overline{D}_T)} = \mathcal{O}(\varepsilon^{11/8}),$$
$$\|R_{1\varepsilon t}\|_{C([0,T];L^2(0,1))} = \mathcal{O}(\varepsilon^{1/2}), \qquad \|R_{2\varepsilon t}\|_{C([0,T];L^2(0,1))} = \mathcal{O}(\varepsilon),$$
$$\|R_{1\varepsilon x}\|_{C([0,T];L^2(0,1))} = \mathcal{O}(\varepsilon), \qquad \|R_{2\varepsilon x}\|_{C([0,T];L^2(0,1))} = \mathcal{O}(\varepsilon^{5/4}).$$

*Proof.* Throughout this proof we denote by $M_k$ $(k = 1, 2, \ldots, 19)$ different positive constants which depend on the data, but are independent of $\varepsilon$. Using the conclusions of Corollary 5.1.8 as well as the definition of $(R_{1\varepsilon}, R_{2\varepsilon})$ (see (5.1)), we obtain that, for every $\varepsilon > 0$, problem (5.8), (5.11), (5.19) has a unique solution $R_\varepsilon \in W^{2,2}(0,T;L^2(0,1))^2$. In fact,

$$R_\varepsilon(t) = (R_{1\varepsilon}(\cdot,t), R_{2\varepsilon}(\cdot,t), R_{2\varepsilon}(1,t)), \quad 0 < t < T,$$

satisfies the Cauchy problem in $H_1$ (the Hilbert space $H_1$ as well as operator $B_{0\varepsilon}$ were defined in Subsection 5.1.2):

$$\begin{cases} R_\varepsilon'(t) + B_{0\varepsilon} R_\varepsilon(t) + (0,0,\sigma_\varepsilon(t)) = L_\varepsilon(t), \ 0 < t < T, \\ R_\varepsilon(0) = 0, \end{cases} \tag{5.50}$$

where $L_\varepsilon(t) = (L_{1\varepsilon}(\cdot,t), L_{2\varepsilon}(\cdot,t), L_{3\varepsilon}(t))$,

$$\begin{aligned} L_{1\varepsilon}(x,t) &= -\varepsilon X_{1t}(x,t), \\ L_{2\varepsilon}(x,t) &= -\varepsilon(c_{1x}(x,\tau) + g d_1(x,\tau)), \\ L_{3\varepsilon}(t) &= k^{-1}(\varepsilon c_1(1,\tau) + \varepsilon f_0'(Y_0(1,t))Y_1(1,t) + f_0(Y_0(1,t)) - f_0(\theta_\varepsilon(t))), \\ \sigma_\varepsilon(t) &= k^{-1}(f_0(R_{2\varepsilon}(1,t) + \theta_\varepsilon(t)) - f_0(\theta_\varepsilon(t)), \\ \theta_\varepsilon(t) &= Y_0(1,t) + \varepsilon Y_1(1,t) + \varepsilon d_1(1,\tau). \end{aligned}$$

Let us take the scalar product in $H_1$ of $(5.50)_1$ and $R_\varepsilon(t)$. Integrating the resulting equation over $[0,t]$ we get

$$\begin{aligned}\frac{1}{2}\|R_\varepsilon(t)\|^2 + \int_0^t \langle B_{0\varepsilon}R_\varepsilon(s), R_\varepsilon(s)\rangle ds + k\int_0^t \sigma_\varepsilon(s)R_{2\varepsilon}(1,s)ds \\ = \int_0^t \langle L_\varepsilon(s), R_\varepsilon(s)\rangle ds,\end{aligned} \tag{5.51}$$

for all $t \in [0,T]$. Therefore

$$\frac{1}{2}\|R_\varepsilon(t)\|^2 \le \int_0^t \|L_\varepsilon(s)\| \cdot \|R_\varepsilon(s)\| ds \quad \text{for all } t \in [0,T].$$

We have used the monotonicity of $B_{0\varepsilon}$ and $f_0' \ge 0$ which implies the obvious inequality

$$\int_0^t \sigma_\varepsilon(s)R_{2\varepsilon}(1,s)ds \ge 0 \quad \forall\, t \in [0,T].$$

If we apply Lemma 2.0.19 to the previous inequality we get

$$\|R_\varepsilon(t)\| \le \int_0^t \|L_\varepsilon(s)\| ds, \quad \text{for all } t \in [0,T]. \tag{5.52}$$

On the other hand,

$$\begin{aligned}\|L_\varepsilon(s)\|^2 \le& \varepsilon^3 \int_0^1 X_{1s}(x,s)^2 dx \\ &+ M_1\varepsilon^2\Big[\int_0^1 \Big(c_1\big(x,(s/\varepsilon)\big)^2 + g^2 d_1\big(x,(s/\varepsilon)\big)^2\Big)dx \\ &+ c_1\big(1,(s/\varepsilon)\big)^2 + d_1\big(1,(s/\varepsilon)\big)^2 + \varepsilon^2\Big],\end{aligned}$$

for all $s \in [0,T]$, because

$$\begin{aligned}&|\ f_0\big(\theta_\varepsilon(t)\big) - f_0\big(Y_0(1,t)\big) - \varepsilon f_0'\big(Y_0(1,t)\big)Y_1(1,t)\ | \\ &\qquad \le M_2\varepsilon(d_1(1,\tau) + \varepsilon).\end{aligned}$$

Therefore,

$$\begin{aligned}\|L_\varepsilon(s)\|^2 \le& M_3\varepsilon^3 + M_4\varepsilon^4 \\ &+ M_5\varepsilon^2 e^{-2rs/\varepsilon}\Big(1 + \frac{s^2}{\varepsilon^2}\Big) \quad \text{for all } s \in [0,T].\end{aligned}$$

An elementary computation yields

$$\int_0^T \Big(1 + \frac{s}{\varepsilon}\Big)e^{-\frac{rs}{\varepsilon}} ds = \mathcal{O}(\varepsilon),$$

and thus

$$\int_0^T \|L_\varepsilon(s)\|ds \le M_6\varepsilon^{3/2}. \tag{5.53}$$

Now, (5.52) and (5.53) imply

$$\varepsilon\|R_{1\varepsilon}(\cdot,t)\|_0^2 + \|R_{2\varepsilon}(\cdot,t)\|_0^2 + kR_{2\varepsilon}(1,t)^2 \le M_7\varepsilon^3 \tag{5.54}$$

(we denoted by $\|\cdot\|_0$ the usual norm of $L^2(0,1)$).

In the following we are going to derive some estimates for $(R_{1\varepsilon t}, R_{2\varepsilon t})$. First, we see that $R_\varepsilon'(t) =: Q_\varepsilon(t) = (Q_{1\varepsilon}(\cdot,t), Q_{2\varepsilon}(\cdot,t), Q_{2\varepsilon}(1,t))$ satisfies the problem

$$\begin{cases} \varepsilon Q_{1\varepsilon t} + Q_{2\varepsilon x} + rQ_{1\varepsilon} = L_{1\varepsilon t}(t), \\ Q_{2\varepsilon t} + Q_{1\varepsilon x} + gQ_{2\varepsilon} = L_{2\varepsilon t}(t), \\ Q_{2\varepsilon t}(1,t) - k^{-1}Q_{1\varepsilon t}(1,t) + \sigma_\varepsilon'(t) = L_{3\varepsilon}'(t), \\ Q_\varepsilon(x,0) = L_\varepsilon(x,0), \\ r_0Q_{1\varepsilon}(0,t) + Q_{2\varepsilon}(0,t) = 0. \end{cases} \tag{5.55}$$

Taking the inner product in $H_1$ of equations $(5.55)_{1,2,3}$ by $Q_\varepsilon(t)$ and integrating the resulting equation over $[0,t]$, we obtain

$$\begin{aligned} &\frac{1}{2}\|Q_\varepsilon(t)\|^2 + r\parallel Q_{1\varepsilon}\parallel_{L^2(D_t)}^2 \\ &\qquad + g\parallel Q_{2\varepsilon}\parallel_{L^2(D_t)}^2 + k\int_0^t \sigma_\varepsilon'(s)R_{2\varepsilon s}(1,s)ds \\ &\qquad \le \frac{1}{2}\|L_\varepsilon(0)\|^2 + \int_0^t \|L_\varepsilon'(s)\|\cdot\|Q_\varepsilon(s)\|ds \quad \text{for all } t\in[0,T], \end{aligned} \tag{5.56}$$

where $D_t = (0,1)\times(0,t)$. Obviously, $\|L_\varepsilon(0)\|^2 \le M_8\varepsilon^2$, and

$$\begin{aligned} \int_0^t \sigma_\varepsilon'(s)R_{2\varepsilon s}(1,s)ds &= \int_0^t \Big[f_0'(\theta_\varepsilon(s) + R_{2\varepsilon}(1,s)) \\ &\qquad \times \big(R_{2\varepsilon s}(1,s) + \theta_{\varepsilon s}(s)\big) - f_0'(\theta_\varepsilon(s))\theta_{\varepsilon s}(s)\Big]R_{2\varepsilon s}(1,s)ds \\ &\ge \int_0^t [f_0'(\theta_\varepsilon(s) + R_{2\varepsilon}(1,s)) - f_0'(\theta_\varepsilon(s))]\theta_{\varepsilon s}(s)R_{2\varepsilon s}(1,s)ds \\ &\ge -M_9\int_0^t |R_{2\varepsilon}(1,s)|\cdot|R_{2\varepsilon s}(1,s)|ds \\ &\ge -M_{10}\varepsilon^{3/2}\int_0^t |R_{2\varepsilon s}(1,s)|ds \end{aligned}$$

for all $t\in[0,T]$ (in the last inequality we have used (5.54)).

It follows from (5.56) that

$$\frac{1}{2}\|Q_\varepsilon(t)\|^2 + r \parallel Q_{1\varepsilon} \parallel^2_{L^2(D_t)} + g \parallel Q_{2\varepsilon} \parallel^2_{L^2(D_t)}$$
$$\leq \frac{1}{2}M_8\varepsilon^2 + M_{11}\varepsilon^{3/2}\int_0^t |R_{2\varepsilon s}(1,s)|ds + \int_0^t \|L'_\varepsilon(s)\| \cdot \|Q_\varepsilon(s)\|ds$$
$$\leq \frac{1}{2}M_8\varepsilon^2 + \int_0^t (\|L'_\varepsilon(s)\| + M_{12}\varepsilon^{3/2})\|Q_\varepsilon(s)\|ds \quad \text{for all } t \in [0,T].$$

Therefore,

$$\|Q_\varepsilon(t)\| \leq \sqrt{M_8}\varepsilon + \int_0^t \|L'_\varepsilon(s)\|ds + M_{13}\varepsilon^{3/2} \quad \text{for all } t \in [0,T], \tag{5.57}$$

and

$$r \parallel Q_{1\varepsilon} \parallel^2_{L^2(D_t)} + g \parallel Q_{2\varepsilon} \parallel^2_{L^2(D_t)} \leq \frac{1}{2}M_8\varepsilon^2 + \int_0^t (\|L'_\varepsilon(s)\| + M_{12}\varepsilon^{3/2})\|Q_\varepsilon(s)\|ds \tag{5.58}$$

for all $t \in [0,T]$. Since

$$L'_\varepsilon(t) = \begin{pmatrix} -\varepsilon X_{1tt}(\cdot,t) \\ (\beta' r + r^{-1}\alpha^{(3)} - \alpha^{(3)}\tau + g\alpha')e^{-r\tau} \\ k^{-1}\big(-r\beta(1) - \alpha''(1)r^{-1} + \tau\alpha''(1)\big)e^{-r\tau} + \varepsilon S_\varepsilon(t), \end{pmatrix}^T$$

where $S_\varepsilon(t)$ is a bounded function, i.e.,

$$\sup\,\{|\, S_\varepsilon(t)\,|;\ \ \varepsilon > 0,\ \ 0 \leq t \leq T\} < \infty,$$

it follows that

$$\int_0^T \|L'_\varepsilon(s)\|ds \leq M_{14}\varepsilon. \tag{5.59}$$

Now, from (5.57) and (5.59) we infer that

$$\|Q_\varepsilon(t)\|^2 = \varepsilon\|R_{1\varepsilon t}(\cdot,t)\|_0^2 + \|R_{2\varepsilon t}(\cdot,t)\|_0^2 + kR_{2\varepsilon t}(1,t)^2 \leq M_{15}\varepsilon^2. \tag{5.60}$$

On the other hand, combining the obvious inequality

$$\int_0^t \langle B_{0\varepsilon}R_\varepsilon(s), R_\varepsilon(s)\rangle ds \geq r\|R_{1\varepsilon}\|^2_{L^2(D_t)} + g\|R_{2\varepsilon}\|^2_{L^2(D_t)} \quad \text{for all } t \in [0,T],$$

with (5.51), (5.53) and (5.54), we infer

$$r\|R_{1\varepsilon}\|^2_{L^2(D_T)} + g\|R_{2\varepsilon}\|^2_{L^2(D_T)} = \mathcal{O}(\varepsilon^3). \tag{5.61}$$

Similarly, from (5.58), (5.59), (5.60) one gets

$$r\|R_{1\varepsilon t}\|^2_{L^2(D_T)} + g\|R_{2\varepsilon t}\|^2_{L^2(D_T)} = \mathcal{O}(\varepsilon^2). \tag{5.62}$$

As a consequence of (5.61) and of (5.62) we find

$$\begin{aligned}\|R_{1\varepsilon}(\cdot,t)\|_0^2 &= 2\int_0^t \langle R_{1\varepsilon s}(\cdot,s), R_{1\varepsilon}(\cdot,s)\rangle_0 ds \\ &\leq 2\|R_{1\varepsilon t}\|_{L^2(D_T)}\|R_{1\varepsilon}\|_{L^2(D_T)} = \mathcal{O}(\varepsilon^{5/2})\end{aligned} \tag{5.63}$$

(we have denoted by $\langle\cdot,\cdot\rangle_0$ the usual scalar product of $L^2(0,1)$).
From (5.60), (5.63) and system (5.8) it follows that

$$\|R_{2\varepsilon x}(\cdot,t)\|_0 \leq M_{16}\varepsilon^{5/4}. \tag{5.64}$$

By a similar reasoning we can show that

$$\|R_{1\varepsilon x}(\cdot,t)\|_0 \leq M_{17}\varepsilon. \tag{5.65}$$

By using the obvious formulae

$$R_{2\varepsilon}(x,t)^2 - R_{2\varepsilon}(1,t)^2 = -2\int_x^1 R_{2\varepsilon}(\xi,t)R_{2\varepsilon x}(\xi,t)d\xi,$$

$$R_{1\varepsilon}(x,t)^2 - R_{1\varepsilon}(0,t)^2 = 2\int_0^x R_{1\varepsilon}(\xi,t)R_{1\varepsilon x}(\xi,t)d\xi,$$

for all $(x,t) \in \overline{D}_T$, on account of (5.54) and of (5.63)–(5.65), we derive the following estimates

$$R_{2\varepsilon}(x,t)^2 \leq R_{2\varepsilon}(1,t)^2 + 2\|R_{2\varepsilon}(\cdot,t)\|_0 \cdot \|R_{2\varepsilon x}(\cdot,t)\|_0 \leq M_{18}\varepsilon^{11/4},$$
$$R_{1\varepsilon}(x,t)^2 \leq (1/r_0^2)R_{2\varepsilon}(0,t)^2 + 2\|R_{1\varepsilon}(\cdot,t)\|_0 \cdot \|R_{1\varepsilon x}(\cdot,t)\|_0 \leq M_{19}\varepsilon^{9/4},$$

for all $(x,t) \in \overline{D}_T$. The proof is complete. □

## 5.2 A zeroth order asymptotic expansion for the solution of problem $(NS)$, $(IC)$, $(BC.1)$

In this section we examine the case in which one of the equations of the telegraph system is nonlinear and construct a zeroth order asymptotic expansion for the corresponding boundary value problem.

One can extend this investigation to higher order asymptotic expansions (in particular to first order expansions, as in the previous section) at the expense of more assumptions on the data as well as some additional technical difficulties.

More precisely, the problem we are going to investigate consists of the system

$$\begin{cases} \varepsilon u_t + v_x + ru = f_1, \\ v_t + u_x + g(x,v) = f_2 \ \text{ in } D_T, \end{cases} \tag{$NS$}$$

the initial conditions

$$u(x,0) = u_0(x), \quad v(x,0) = v_0(x), \ 0 \le x \le 1, \tag{$IC$}$$

and boundary conditions

$$\begin{cases} r_0 u(0,t) + v(0,t) = 0, \\ u(1,t) - kv_t(1,t) = f_0\big(v(1,t)\big) + e_0(t), \ 0 \le t \le T, \end{cases} \tag{$BC.1$}$$

where $g : [0,1] \times \mathbb{R} \to \mathbb{R}$, $f_1, f_2 : \overline{D}_T \to \mathbb{R}$, $e_0 : [0,T] \to \mathbb{R}$, $u_0, v_0 : [0,1] \to \mathbb{R}$, $f_0 : \mathbb{R} \to \mathbb{R}$ are known functions, $r, r_0, k$ are constants, $r \ge 0$, $k$, $r_0 > 0$, and $0 < \varepsilon \ll 1$.

*Remark* 5.2.1. Note that all the results of this section can be extended to the more general boundary conditions

$$\begin{cases} -u(0,t) = r^*(v(0,t)), \\ u(1,t) - kv_t(1,t) = f_0\big(v(1,t)\big) + e_0(t), \ 0 \le t \le T, \end{cases}$$

where $r^*$ is a nonlinear nondecreasing function. However, we prefer $(BC.1)$ for the sake of simplicity.

We will denote again by $U_\varepsilon = (u_\varepsilon(x,t), v_\varepsilon(x,t))$ the solution of problem $P_\varepsilon$, which consists of $(NS)$, $(IC)$, $(BC.1)$. $P_0$ will designate the corresponding reduced problem. Again, $P_\varepsilon$ is a singularly perturbed problem of the boundary layer type, with respect to the sup norm of $C(\overline{D}_T)^2$ (see Chapter 1), and the boundary layer is a thin strip $[0,1] \times [0,\delta]$, $\delta > 0$.

### 5.2.1 Formal expansion

Using previous arguments one can see that a zeroth order asymptotic expansion is

$$U_\varepsilon = U_0(x,t) + V_0(x,\tau) + R_\varepsilon(x,t), \tag{5.66}$$

where:

$U_0 = (X(x,t), Y(x,t))$ is the zeroth order regular term;
$V_0 = (c_0(x,\tau), d_0(x,\tau))$, $\tau = t/\varepsilon$, is the boundary layer function;
$R_\varepsilon = (R_{1\varepsilon}(x,t), R_{2\varepsilon}(x,t))$ is the remainder of the order zero.

By the reasoning used in Subsection 5.1.1 one can determine

$$d_0 \equiv 0, \ c_0(x,\tau) = \alpha(x)e^{-r\tau}, \ 0 \le x \le 1, \ \tau \ge 0,$$

where $\alpha(x) = u_0(x) + (1/r)\left(v_0'(x) - f_1(x,0)\right)$, $0 \le x \le 1$. Moreover, $X$ and $Y$ satisfy

$$X = (1/r)(f_1 - Y_x) \text{ in } D_T, \tag{5.67}$$

$$\begin{cases} Y_t - r^{-1}Y_{xx} + g(x,Y) = f_2 - r^{-1}f_{1x} \text{ in } D_T, \\ Y(x,0) = v_0(x), \quad 0 \le x \le 1, \\ Y(0,t) - (r_0/r)Y_x(0,t) = -(r_0/r)f_1(0,t), \\ kY_t(1,t) + r^{-1}Y_x(1,t) + f_0\big(Y(1,t)\big) = r^{-1}f_1(1,t) - e_0(t), \quad 0 \le t \le T. \end{cases} \tag{5.68}$$

For the components of the remainder we derive the following boundary value problem

$$\begin{cases} \varepsilon R_{1\varepsilon t} + R_{2\varepsilon x} + rR_{1\varepsilon} = -\varepsilon X_t, \\ R_{2\varepsilon t} + R_{1\varepsilon x} + g(x, R_{2\varepsilon} + Y) - g(x,Y) = -c_{0x} \text{ in } D_T, \end{cases} \tag{5.69}$$

$$R_{1\varepsilon}(x,0) = R_{2\varepsilon}(x,0) = 0, \quad 0 \le x \le 1, \tag{5.70}$$

$$\begin{cases} r_0 R_{1\varepsilon}(0,t) + R_{2\varepsilon}(0,t) = 0, \\ R_{1\varepsilon}(1,t) - kR_{2\varepsilon t}(1,t) - f_0\left(v_\varepsilon(1,t)\right) + f_0\left(Y(1,t)\right) \\ = -c_0(1,\tau), \quad 0 \le t \le T. \end{cases} \tag{5.71}$$

By the identification procedure we also obtain

$$c_0(0,\tau) = 0 \Leftrightarrow \alpha(0) = 0.$$

This equation will appear again in the next subsection as a necessary compatibility condition for the data. So, with this condition, our correction $c_0$ does not introduce any discrepancy at the corner point $(0,0)$ of $\overline{D}_T$. Concerning the corner point $(1,0)$, here the situation is completely different. Since we have a dynamic boundary condition at $x = 1$, we may include the term $c_0(1,\tau)$ into the (dynamic) equation $(5.71)_2$. This term is sizeable at $t = 0$, but the remainder remains "small" (even if the derivative $R_{2\varepsilon t}$ is not small around $t = 0$). Thus, $c_0$ does not introduce any discrepancy at $(1,0)$ and we do not need to require that $\alpha(1) = 0$ (as needed in the case of an algebraic boundary condition at $x = 1$). This discussion shows that the case of dynamic boundary conditions makes a difference.

Summarizing, we can see that $U_0 = (X,Y)$ satisfies the reduced problem $P_0$, which is made up by the algebraic equation (5.67) and the boundary value problem (5.68), while the pair $(R_{1\varepsilon}, R_{2\varepsilon})$ satisfies the above problem (5.69), (5.70), (5.71), which is of the same type as the original problem $P_\varepsilon$.

### 5.2.2 Existence, uniqueness and regularity of the solutions of problems $P_\varepsilon$ and $P_0$

In this subsection we are going to investigate problems $P_\varepsilon$ and $P_0$, which are formulated in Subsection 5.2.1. First we consider problem $P_\varepsilon$. The presence of the nonlinear term $g$ in system $(NS)$ makes our problem more complex. We choose as in Subsection 5.1.2 the Hilbert space $H_1 = \left(L^2(0,1)\right)^2 \times \mathbb{R}$, endowed with the same scalar product and norm denoted $\|\cdot\|$. Instead of operator $B_\varepsilon$, we now define the operator $B_{1\varepsilon} : D(B_{1\varepsilon}) \subset H_1 \to H_1$,

$$\begin{aligned} D(B_{1\varepsilon}) &:= \left\{(p,q,a); p,q \in H^1(0,1), a = q(1), r_0 p(0) + q(0) = 0\right\}, \\ B_{1\varepsilon}(p,q,a) &:= \left(\varepsilon^{-1}\left(q' + rp\right),\ p' + g(\cdot,q),\ k^{-1}\left(-p(1) + f_0\left(a\right)\right)\right). \end{aligned}$$

We associate with $P_\varepsilon$ the following Cauchy problem in $H_1$:

$$\begin{cases} w_\varepsilon'(t) + B_{1\varepsilon} w_\varepsilon(t) = F_\varepsilon(t),\ 0 < t < T, \\ w_\varepsilon(0) = w_0, \end{cases} \tag{5.72}$$

where

$$\begin{aligned} w_\varepsilon(t) &= (u_\varepsilon(\cdot,t), v_\varepsilon(\cdot,t), \xi_\varepsilon(t)),\ w_0 = (u_0, v_0, \xi_0) \in H_1, \\ F_\varepsilon(t) &= \left(\varepsilon^{-1} f_1(\cdot,t), f_2(\cdot,t), -k^{-1} e_0(t)\right),\ 0 < t < T. \end{aligned}$$

For the time being, let us restate Theorem 2.0.34 for the case of our problem (which is a particular case of problem (2.14), (2.15), (2.17)):

**Theorem 5.2.2.** *Assume that*

$$r \text{ is a nonnegative constant, and } r_0, k > 0; \tag{5.73}$$

$$f_0 : \mathbb{R} \to \mathbb{R} \ \text{ is a continuous nondecreasig function}; \tag{5.74}$$

$$\begin{aligned} &x \to g(x,\xi) \in L^2(0,1) \text{ for all } \xi \in \mathbb{R} \text{ and } \xi \to g(x,\xi) \\ &\text{is continuous and nondecreasing on } \mathbb{R} \text{ for a.a. } x \in (0,1); \end{aligned} \tag{5.75}$$

$$w_0 \in D(B_{1\varepsilon}),\quad F_\varepsilon \in W^{1,1}(0,T;H_1). \tag{5.76}$$

*Then, problem* (5.72) *has a unique strong solution* $w_\varepsilon \in W^{1,\infty}(0,T;H_1)$, *with* $\xi_\varepsilon(t) = v_\varepsilon(1,t)$ *for all* $t \in [0,T]$. *In addition,* $u_\varepsilon, v_\varepsilon \in L^\infty(0,T;H^1(0,1))$.

Next, we will state and prove a result concerning higher smoothness of solution $w_\varepsilon$:

**Theorem 5.2.3.** *Assume that* (5.73), (5.76) *hold and, in addition,*

$$f_0 \in C^1(\mathbb{R}),\ f_0'(x) \ge 0 \text{ for all } x \in \mathbb{R}; \tag{5.77}$$

$$x \to g(x,\xi) \in L^2(0,1)\ \forall \xi \in \mathbb{R} \text{ and } \partial g/\partial \xi =: g_\xi \in C([0,1]\times\mathbb{R}),\ g_\xi \ge 0. \tag{5.78}$$

*Then, the strong solution* $w_\varepsilon$ *of the Cauchy problem* (5.72) *satisfies*

$$w_\varepsilon \in C^1([0,T];H_1),\quad u_\varepsilon, v_\varepsilon \in C([0,T];H^1(0,1)). \tag{5.79}$$

*Proof.* By Theorem 5.2.2 there exists a unique strong solution of problem (5.72), $w_\epsilon \in W^{1,\infty}(0,T;H_1)$, such that $u_\varepsilon, v_\varepsilon \in L^\infty(0,T;H^1(0,1))$.

It remains to show assertions (5.79). To this end, we consider the Cauchy problem in $H_1$:

$$\begin{cases} \overline{w}'_\varepsilon(t) + B_{0\varepsilon}\overline{w}_\varepsilon(t) = G_\varepsilon(t), \quad 0<t<T, \\ \overline{w}_\varepsilon(0) = F_\varepsilon(0) - B_{1\varepsilon}w_0, \end{cases} \tag{5.80}$$

where $\overline{w}_\varepsilon(t) = \left(\overline{u}_\varepsilon(\cdot,t), \overline{v}_\varepsilon(\cdot,t), \overline{\xi}_\varepsilon(t)\right), \quad G_\varepsilon : (0,T) \to H_1,$

$$G_\varepsilon(t) = \begin{pmatrix} \varepsilon^{-1} f_{1t}(\cdot,t) \\ f_{2t}(\cdot,t) - g_\xi\left(\cdot, v_\varepsilon(\cdot,t)\right) v_{\varepsilon t}(\cdot,t) \\ -k^{-1}[e'_0(t) + f'_0\left(v_\varepsilon(1,t)\right) v_{\varepsilon t}(1,t)] \end{pmatrix}^T.$$

Here $v_\varepsilon(\cdot,t),\ v_\varepsilon(1,t)$ are the last two components of our solution $w_\varepsilon$;

$$B_{0\varepsilon} : D(B_{1\varepsilon}) \subset H_1 \to H_1 \text{ is the operator } B_{1\varepsilon} \text{ with } g \equiv 0,\ f_0 \equiv 0.$$

Now, (5.77) and (5.78) imply that $G_\varepsilon \in L^1(0,T;H_1)$.
Therefore, we can apply Theorem 2.0.21 to infer that problem (5.80) has a unique weak solution $\overline{w}_\varepsilon \in C\left([0,T];H_1\right)$, which satisfies the equation

$$\overline{w}_\varepsilon(t) = S_\varepsilon(t)\overline{w}_\varepsilon(0) + \int_0^t S_\varepsilon(t-s)G_\varepsilon(s)\,ds, \quad 0 \le t \le T, \tag{5.81}$$

where $\{S_\varepsilon(t),\ t \ge 0\}$ is the contractions semigroup generated by $-B_{0\varepsilon}$. Since $w_\varepsilon$ is the strong solution of problem (5.72) and the function

$$F_{1\varepsilon}(t) := F_\varepsilon(t) - \left(0, g\left(\cdot, v_\varepsilon(\cdot,t)\right), k^{-1} f_0\left(v_\varepsilon(1,t)\right)\right), \quad 0 \le t \le T,$$

belongs to $W^{1,1}(0,T;H_1)$, we have

$$w'_\varepsilon(t) = S_\varepsilon(t)\overline{w}_\varepsilon(0) + \int_0^t S_\varepsilon(t-s)F'_{1\varepsilon}(s)ds, \quad 0 \le t \le T.$$

Obviously, $F'_{1\varepsilon} = G_\varepsilon$ and from (5.81) and the last equation it follows that $\overline{w}_\varepsilon(t) = w_\varepsilon(t)$ for all $t \in [0,T]$, so $w_\varepsilon \in C^1([0,T];H_1)$.
Finally, from $(NS)$ we infer that $u_\varepsilon, v_\varepsilon \in C([0,T];H^1(0,1))$. Thus, the theorem is completely proved. □

In the sequel, we are dealing with problem (5.68). We write it as a Cauchy problem in the Hilbert space $H_2 = L^2(0,1) \times \mathbb{R}$, defined in Subsection 5.1.2, with the same scalar product and the corresponding induced norm, denoted $\|\cdot\|_{H_2}$. To this purpose, we define the operator $A(t) : D(A(t)) \subset H_2 \to H_2$ by:

$$D(A(t)) = \left\{(p,a) \in H_2;\ p \in H^2(0,1),\ a = p(1),\ r^{-1}p'(0) + \alpha_1(t) = r_0^{-1}p(0)\right\},$$
$$A(t)(p,a) = \left(-r^{-1}p'' + g(\cdot,p),\ k^{-1}[r^{-1}p'(1) + f_0(a)]\right),\ \alpha_1(t) = -r^{-1}f_1(0,t),$$

and associate with problem (5.68) the following Cauchy problem in $H_2$ :

$$\begin{cases} Z'(t) + A(t)Z(t) = h(t), \ 0 < t < T, \\ Z(0) = Z_0, \end{cases} \tag{5.82}$$

where

$$Z(t) = (Y(\cdot,t), \varsigma(t)), \ Z_0 = (v_0, \zeta_0), \ h(t) = (h_1(\cdot,t), h_2(t)),$$
$$h_1(x,t) = (f_2 - r^{-1} f_{1x})(x,t), \ h_2(t) = k^{-1}(r^{-1} f_1(1,t) - e_0(t)).$$

Let us also define the operator $B(t) : D(B(t)) \subset H_2 \to H_2$,

$$D(B(t)) = \{(p,a) \in H_2; \ p \in H^2(0,1), \ a = p(1), \ r^{-1}p'(0) + \alpha_1'(t) = r_0^{-1}p(0)\},$$
$$B(t)(p,a) = (-r^{-1}p'', \ k^{-1}r^{-1}p'(1)),$$

which will be needed to prove the following higher regularity result:

**Theorem 5.2.4.** *Assume that* (5.74) *holds, and*

$$r, \ r_0, \ k \textit{ are given positive constants}; \tag{5.83}$$

$$\begin{aligned} &g = g(x,\xi) : [0,1] \times \mathbb{R} \to \mathbb{R} \textit{ is continuous, and} \\ &\textit{it is nondecreasing with respect to its second variable}; \end{aligned} \tag{5.84}$$

$$h \in W^{1,1}(0,T; H_2), \ \alpha_1 \in H^1(0,T); \tag{5.85}$$

$$Z_0 \in D(A(0)). \tag{5.86}$$

*Then, problem* (5.82) *has a unique strong solution* $Z \in W^{1,2}(0,T;H_2)$, *with* $Y \in L^2(0,T;H^2(0,1)) \bigcap W^{1,2}(0,T;H^1(0,1))$ *and* $\varsigma(t) = Y(1,t)$ *for all* $t \in [0,T]$.

*If, in addition,*

$$f_0 \in C^2(\mathbb{R}), \ f_0' \ge 0, \ h \in W^{2,1}(0,T;H_2), \ \alpha_1 \in H^2(0,T); \tag{5.87}$$

$$g_\xi, \ g_{\xi\xi} \textit{ exist and are continuous on } [0,1] \times \mathbb{R}, \ g_\xi \ge 0; \tag{5.88}$$

$$Z_{01} := h(0) - A(0)Z_0 \in D(B(0)), \tag{5.89}$$

*then* $Z \in W^{2,2}(0,T;H_2)$ *and* $Y \in W^{2,2}(0,T;H^1(0,1)) \bigcap W^{1,2}(0,T;H^2(0,1))$.

*Proof.* The first part of the theorem follows by an argument similar to that used in the proof of Theorem 5.1.5 (see also [24], p. 101). Indeed, it is sufficient to point out that $A(t) = \partial\phi(t,\cdot), \ t \in [0,T]$, where

$$\phi(t,\cdot) : H_2 \to (-\infty, +\infty],$$

$$\phi(t,(p,a)) = \begin{cases} \frac{1}{2r}\int_0^1 p'(x)^2 dx + \int_0^1 G(x,p(x))dx + \frac{1}{2r_0}p(0)^2 \\ -\alpha_1(t)p(0) + l_0(a), \ \text{if } p \in H^1(0,1) \text{ and } a = p(1), \\ +\infty, \ \text{otherwise,} \end{cases}$$

in which $l_0 : \mathbb{R} \to \mathbb{R}$ is a primitive of $f_0$ whilst $G(x,\xi) = \int_0^\xi g(x,\tau)d\tau$ (see also [24], p. 93). For every $t \in [0,T]$, $D(\phi(t,\cdot)) = \{(p,a);\ p \in H^1(0,1),\ a = p(1)\} =: V$, which is independent of $t$. $V$ is a Hilbert space with the the scalar product defined in Subsection 5.1.2. According to Theorem 2.0.32, problem (5.82) has a unique strong solution $Z \in W^{1,2}(0,T;H_2)$, with $Y \in W^{1,2}(0,T;H^1(0,1)) \bigcap L^2(0,T;H^2(0,1))$ and $\zeta(t) = Y(1,t)$ for all $t \in [0,T]$. Now, assume that our additional assumptions (5.87)–(5.89) hold. Then, $Z$ belongs to $W^{2,2}(0,T;V^*)$, where $V^*$ is the dual of $V$, and $Z(t) = (Y(\cdot,t), Y(1,t))$ satisfies the equation (see the proof of Theorem 5.1.5)

$$\begin{aligned} &\langle(\varphi,\varphi(1)),(Y_{tt}(\cdot,t),Y_{tt}(1,t))\rangle + r^{-1}\langle\varphi',Y_{tx}(\cdot,t)\rangle_0 \\ &\quad - \alpha_1'(t)\varphi(0) + r_0^{-1}Y_t(0,t)\varphi(0) \\ &\quad = \langle(\varphi,\varphi(1)),\overline{h}(t)\rangle_{H_2} \quad \forall\ (\varphi,\varphi(1)) \in V,\ a.a.t \in (0,T) \end{aligned} \tag{5.90}$$

and the initial condition

$$Z'(0) = Z_{01}, \tag{5.91}$$

where

$$\overline{h}(t) = h'(t) - (g_\xi(\cdot,Y(\cdot,t))Y_t(\cdot,t), k^{-1}f_0'(Y(1,t))Y_t(1,t))$$

(we have denoted by $\langle\cdot,\cdot\rangle_0$ the usual scalar product of $L^2(0,1)$ and by $\langle\cdot,\cdot\rangle$ the pairing between $V$ and $V^*$). Obviously, $Z'(t)$ is the unique solution of problem (5.90), (5.91). Now, let us consider the Cauchy problem in $H_2$

$$\begin{cases} \overline{Z}'(t) + B(t)\overline{Z}(t) = \overline{h}(t),\ 0 < t < T, \\ \overline{Z}(0) = Z_{01}, \end{cases}$$

where $\overline{Z}(t) = (\overline{Y}(\cdot,t), \overline{\zeta}(t))$. This is obtained by formal differentiation with respect to $t$ of problem (5.82). Since $B(t)$ is similar to $A(t)$, one can make use again of Theorem 2.0.32. Thus, since $Z_{01} \in D(B(0))$ and $\overline{h} \in L^2(0,T;H_2)$, it follows that

$$\overline{Z} \in W^{1,2}(0,T;H_2),\ \overline{Y}_x \in L^2(0,T;L^2(0,1)),\ \ \overline{\zeta}(t) = \overline{Y}(1,t),\ t \in [0,T].$$

Since $\overline{Z}$ also satisfies problem (5.90), (5.91) we infer that $\overline{Z} = Z'$. Therefore,

$$Y \in W^{2,2}(0,T;L^2(0,1)) \bigcap W^{1,2}(0,T;H^2(0,1)),\ Y(1,\cdot) \in W^{2,2}(0,T),$$

from which it follows that $\overline{h}$ belongs to $W^{1,1}(0,T;H_2)$ and thus, by usual arguments (see Theorem 5.1.5), one can conclude that $\overline{Y} = Y_t \in W^{1,2}(0,T;H^1(0,1))$. The proof is complete. □

*Remark* 5.2.5. From our hypotheses $w_0 \in D(B_{1\varepsilon})$ and $Z_0 \in D(A(0))$ (see Theorems 5.2.3, 5.2.4), it follows that $\alpha(0) = 0$. This condition was determined in Subsection 5.2.1 as a result of identifications.

Taking into account the three theorems above we are going to state two corollaries which show that our expansion (5.66) is well defined. Moreover, they will be used in the next section to get estimates for the remainder components in different norms, which will validate completely our asymptotic expansion.

**Corollary 5.2.6.** *Assume that* (5.74), (5.83), (5.84) *hold and, in addition, the following conditions are fulfilled*

$$\begin{cases} f_1 \in W^{1,1}(0,T;H^1(0,1)),\ f_2 \in W^{1,1}(0,T;L^2(0,1)), \\ f_1(0,\cdot) \in H^1(0,T),\ e_0 \in W^{1,1}(0,T),\ u_0 \in H^1(0,1), v_0 \in H^2(0,1), \\ r_0 u_0(0) + v_0(0) = 0,\ r_0 v_0'(0) - r v_0(0) = r_0 f_1(0,0). \end{cases}$$

*Then, for every* $\varepsilon > 0$, *problem* $P_\varepsilon$ *admits a unique strong solution*

$$(u_\varepsilon, v_\varepsilon) \in W^{1,\infty}(0,T;L^2(0,1))^2 \bigcap L^\infty(0,T;H^1(0,1))^2,$$

*and problem* $P_0$ *admits a unique solution*

$$\begin{aligned}(X,Y) \in &(L^2(0,T;H^1(0,1)) \bigcap W^{1,1}(0,T;L^2(0,1)) \\ &\times (L^2(0,T;H^2(0,1)) \bigcap W^{1,2}(0,T;H^1(0,1)).\end{aligned}$$

**Corollary 5.2.7.** *Assume that* (5.83) *holds. If* $u_0, v_0, f_1, f_2$ *and* $e_0$ *are smooth enough,* $f_0 \in C^2(\mathbb{R})$, $f_0' \geq 0$, $g$ *satisfies* (5.84), (5.88), *and the following compatibility conditions are fulfilled:*

$$\begin{cases} r_0 u_0(0) + v_0(0) = 0, \\ r_0 v_0'(0) - r v_0(0) = r_0 f_1(0,0), \\ r(r^{-1} v_0''(0) - g(0, v_0(0)) - r^{-1} f_{1x}(0,0) + f_2(0,0)) + r_0 f_{1t}(0,0) \\ = r_0(r^{-1} v_0^{(3)}(0) - g_x(0, v_0(0)) \\ - g_\xi(0, v_0(0)) v_0'(0) + f_{2x}(0,0) - r^{-1} f_{1xx}(0,0)), \\ r^{-1} v_0'(1) + f_0(v_0(1)) + e_0(0) - r^{-1} f_1(1,0) \\ = k[-r^{-1} v_0''(1) + g(1, v_0(1)) - f_2(1,0) + r^{-1} f_{1x}(1,0)], \end{cases}$$

*then, for every* $\varepsilon > 0$, *problem* $P_\varepsilon$ *admits a unique strong solution*

$$(u_\varepsilon, v_\varepsilon) \in C^1([0,T];L^2(0,1))^2 \bigcap C([0,T];H^1(0,1))^2$$

*and problem* $P_0$ *admits a unique solution*

$$\begin{aligned}(X,Y) \in &(W^{2,1}(0,T;L^2(0,1)) \bigcap W^{1,2}(0,T;H^1(0,1))) \\ &\times (W^{1,2}([0,T];H^2(0,1)) \bigcap W^{2,2}(0,T;H^1(0,1))).\end{aligned}$$

### 5.2.3 Estimates for the remainder components

We begin with some estimates concerning our expansion (5.66) with respect to the $L^p(0,T;L^2(0,1)) \times L^\infty(0,T;C[0,1])$ norm, under fairly weak assumptions on the data. We have the following result:

**Theorem 5.2.8.** *Let* $p \in [2,\infty)$. *Assume that all the assumptions of Corollary* 5.2.6 *hold. Then, for every* $\varepsilon > 0$, *the solution of problem* $P_\varepsilon$ *admits an asymptotic expansion of the form* (5.66) *and the following estimates hold:*

$$\begin{aligned} &\|u_\varepsilon - X\|_{L^p(0,T;L^2(0,1))} = \mathcal{O}(\varepsilon^{1/p}), \\ &\|v_\varepsilon - Y\|_{C([0,T];L^2(0,1))} = \mathcal{O}(\varepsilon^{1/2}), \quad \|v_\varepsilon - Y\|_{L^\infty(0,T;C[0,1])} = \mathcal{O}(\varepsilon^{1/4}). \end{aligned}$$

*Proof.* Throughout this proof we denote by $M_k$ $(k = 1,2,\dots)$ different positive constants which depend on the data, but are independent of $\varepsilon$. According to Corollary 5.2.6, problem (5.69), (5.70), (5.71) has a unique solution

$$\begin{aligned} R_\varepsilon \in &\left(W^{1,1}(0,T;L^2(0,1)) \cap L^2(0,T;H^1(0,1))\right) \\ &\times \left(W^{1,2}(0,T;L^2(0,1)) \cap L^2(0,T;H^1(0,1))\right), \end{aligned}$$

and (see Theorems 5.2.2 and 5.2.4) $R_{2\varepsilon}(1,\cdot) \in W^{1,2}(0,T)$. It is easily seen that

$$R_\varepsilon(t) := (R_{1\varepsilon}(\cdot,t), R_{2\varepsilon}(\cdot,t), R_{2\varepsilon}(1,t))$$

is the strong solution of the following Cauchy problem in $H_1$:

$$\begin{cases} R_\varepsilon'(t) + B_{0\varepsilon}R_\varepsilon(t) + \sigma_\varepsilon(t) = G_{1\varepsilon}(t), \quad 0<t<T, \\ R_\varepsilon(0) = 0, \end{cases} \tag{5.92}$$

where $H_1$ and $B_{0\varepsilon}$ were defined in the proof of Theorem 5.2.3, and

$$\begin{aligned} \sigma_\varepsilon(t) &:= (0, \sigma_{1\varepsilon}(t), \sigma_{2\varepsilon}(t)), \quad \sigma_{1\varepsilon}(t) = g(\cdot, v_\varepsilon(\cdot,t)) - g(\cdot, Y(\cdot,t)), \\ \sigma_{2\varepsilon}(t) &= k^{-1}[f_0(v_\varepsilon(1,t)) - f_0(Y(1,t))], \\ G_{1\varepsilon}(t) &= \big(-X_t(\cdot,t), -c_{0x}(\cdot,\tau),\ k^{-1}c_0(1,\tau)\big) \text{ for all } t \in (0,T). \end{aligned}$$

By a standard computation we get

$$\|R_\varepsilon(t)\|^2 + 2\int_0^t \langle B_{0\varepsilon}R_\varepsilon(s) + \sigma_\varepsilon(s), R_\varepsilon(s)\rangle ds = 2\int_0^t \langle G_{1\varepsilon}(s), R_\varepsilon(s)\rangle ds \tag{5.93}$$

for all $t \in [0,T]$. Together with

$$\langle B_{0\varepsilon}R_\varepsilon(s) + \sigma_\varepsilon(s), R_\varepsilon(s)\rangle \geq 0, \ s \in [0,T],$$

and Gronwall's lemma, (5.93) yields

$$\|R_\varepsilon(t)\| \leq \int_0^t \|G_{1\varepsilon}(s)\| ds, \ t \in [0,T]. \tag{5.94}$$

On the other hand,

$$\|G_{1\varepsilon}(s)\|^2 = \varepsilon \int_0^1 X_s(x,s)^2 dx + \left(\int_0^1 \alpha'(x)^2 dx + k^{-1}\alpha(1)^2\right) e^{-2rs/\varepsilon}$$

for all $s \in [0,T]$, which yields

$$\int_0^T \|G_{1\varepsilon}(s)\| ds = \mathcal{O}(\varepsilon^{1/2}). \tag{5.95}$$

Combining (5.94) with (5.95) one gets

$$\|R_\varepsilon(t)\|^2 = \varepsilon \|R_{1\varepsilon}(\cdot,t)\|_0^2 + \|R_{2\varepsilon}(\cdot,t)\|_0^2 + kR_{2\varepsilon}(1,t)^2 \leq M_1\varepsilon \tag{5.96}$$

for all $t \in [0,T]$, where $\|\cdot\|_0$ denotes the usual norm of $L^2(0,1)$.
As $f_0$ and $\xi \to g(x,\xi)$, $x \in [0,1]$, are nondecreasing functions, we have

$$\begin{aligned}\int_0^T &\langle B_{0\varepsilon} R_\varepsilon(s) + \sigma_\varepsilon(s), R_\varepsilon(s)\rangle ds \\ &\geq r_0 \int_0^T R_{1\varepsilon}(0,s)^2 ds + r\int_0^T \|R_{1\varepsilon}(\cdot,s)\|_0^2 ds.\end{aligned} \tag{5.97}$$

Since $r > 0$, it follows from (5.93), (5.95), (5.96) and (5.97)

$$\int_0^T \|R_{1\varepsilon}(\cdot,s)\|_0^2 ds = \|R_{1\varepsilon}\|_{L^2(D_T)}^2 = \mathcal{O}(\varepsilon). \tag{5.98}$$

Combining the obvious inequality

$$\begin{aligned}\int_0^T \|R_{1\varepsilon}(\cdot,t)\|_0^p dt &= \int_0^T \|R_{1\varepsilon}(\cdot,t)\|_0^2 \cdot \|R_{1\varepsilon}(\cdot,t)\|_0^{p-2} dt \\ &\leq \|R_{1\varepsilon}\|_{C([0,T];L^2(0,1))}^{p-2} \int_0^T \|R_{1\varepsilon}(\cdot,t)\|_0^2 dt,\end{aligned}$$

with (5.96) and (5.98), we infer that

$$\|R_{1\varepsilon}\|_{L^p(0,T;L^2(0,1))} \leq \|R_{1\varepsilon}\|_{L^2(D_T)}^{2/p} \|R_{1\varepsilon}\|_{C([0,T];L^2(0,1))}^{1-2/p} = \mathcal{O}(\varepsilon^{1/p}).$$

Since $u_\varepsilon = X + c_0 + R_{1\varepsilon}$, and $\|c_0\|_{L^p(0,T;L^2(0,1))} = \mathcal{O}(\varepsilon^{1/p})$, we find

$$\|u_\varepsilon - X\|_{L^p(0,T;L^2(0,1))} \leq \|R_{1\varepsilon}\|_{L^p(0,T;L^2(0,1))} + \|c_0\|_{L^p(0,T;L^2(0,1))} = \mathcal{O}(\varepsilon^{1/p}).$$

On the other hand, (5.96) implies that $\|R_{2\varepsilon}(\cdot,t)\|_0^2 \leq M_1\varepsilon$ for all $t \in [0,T]$. Since $w_\varepsilon(t) := \big(u_\varepsilon(\cdot,t), v_\varepsilon(\cdot,t), v_\varepsilon(1,t)\big)$ is the strong solution of problem (5.72), we have (see Theorem 2.0.20)

$$\|w_\varepsilon'(t)\| \leq \|F_\varepsilon(0) - B_{1\varepsilon}w_0\| + \int_0^t \|F_\varepsilon'(s)\| ds \leq M_2\varepsilon^{-1/2} \text{ for a.a. } t \in (0,T).$$

Therefore,

$$\varepsilon\|u_{\varepsilon t}(\cdot,t)\|_0^2 + \|v_{\varepsilon t}(\cdot,t)\|_0^2 \le M_3\varepsilon^{-1}$$

for a.a. $t \in (0,T)$. On the other hand,

$$\|u_\varepsilon(\cdot,t)\|_0 \le \|X(\cdot,t)\|_0 + \|c_0(\cdot,\tau)\|_0 + \|R_{1\varepsilon}(\cdot,t)\|_0 \le M_4 \text{ for all } t \in [0,T],$$

and from $v_{\varepsilon x} = f_1 - \varepsilon u_{\varepsilon t} - ru_\varepsilon$ in $D_T$, we derive $\|v_{\varepsilon x}(\cdot,t)\|_0 \le M_5$ for a.a. $t \in [0,T]$. Therefore,

$$\|R_{2\varepsilon x}(\cdot,t)\|_0 \le \|v_{\varepsilon x}(\cdot,t)\|_0 + \|Y_x(\cdot,t)\|_0 \le M_6 \text{ for a.a. } t \in (0,T).$$

Finally, using (5.96), we find

$$R_{2\varepsilon}(x,t)^2 \le R_{2\varepsilon}(1,t)^2 + 2\|R_{2\varepsilon}(\cdot,t)\|_0 \cdot \|R_{2\varepsilon x}(\cdot,t)\|_0 \le M_7\varepsilon^{1/2}$$

for a.a. $t \in (0,T)$ and for all $x \in [0,1]$, and thus $\|R_{2\varepsilon}\|_{L^\infty(0,T;C[0,1])} = \mathcal{O}(\varepsilon^{1/4})$. This concludes the proof. □

*Remark* 5.2.9. Note that our boundary layer function $c_0$ does not appear in the statement of Theorem 5.2.8. This is quite natural because $\| c_0 \|_{L^p(0,T;L^2(0,1))} = \mathcal{O}(\varepsilon^{1/p})$. In fact, this means that our problem is regularly perturbed with respect to the given norm. In other words, the boundary layer is not visible in this norm.

Now, we are going to establish estimates with respect to the uniform norm for the two components of the remainder. To this purpose, we need to require more smoothness of the data as well as higher compatibility conditions.

**Theorem 5.2.10.** *Assume that all the assumptions of Corollary* 5.2.7 *are fulfilled. Then, for every $\varepsilon > 0$, the solution of problem $P_\varepsilon$ admits an asymptotic expansion of the form* (5.66) *and the following estimates hold*

$$\|R_{1\varepsilon}\|_{C(\overline{D}_T)} = \mathcal{O}(\varepsilon^{1/8}), \quad \|R_{2\varepsilon}\|_{C(\overline{D}_T)} = \mathcal{O}(\varepsilon^{3/8}).$$

*Proof.* We will use the same notation as in the proof of the previous theorem. Of course, the estimates established there remain valid, since our present assumptions are stronger. Note that $X$ (the first component of the solution of $P_0$) belongs to $W^{2,1}(0,T;L^2(0,1))$ (see Corollary 5.2.7) and thus $Q_\varepsilon(t) := R_\varepsilon'(t)$ is the mild solution of the Cauchy problem in $H_1$:

$$\begin{cases} Q_\varepsilon'(t) + B_{0\varepsilon}Q_\varepsilon(t) + \sigma_\varepsilon'(t) = G_{1\varepsilon}'(t), & 0 < t < T, \\ Q_\varepsilon(0) = G_{1\varepsilon}(0), \end{cases}$$

which is obtained by differentiating the Cauchy problem (5.92) with respect to $t$.

Now, we continue with an auxiliary result:

**Lemma 5.2.11.** *If $u = (u^1, u^3, u^3) \in C([0,T]; H_1)$ is the mild solution of the following Cauchy problem in $H_1$*

$$\begin{cases} u'(t) + B_{0\varepsilon} u(t) = f\big(t, u(t)\big), & 0 < t < T, \\ u(0) = u_0, \end{cases}$$

*where $f : [0,T] \times H_1 \to H_1$, $u_0 \in H_1$ and $f\big(\cdot, u(\cdot)\big) \in L^1(0,T; H_1)$, then*

$$\|u(t)\|^2 + 2r\|u^1\|^2_{L^2(D_t)} \le \|u_0\|^2 + 2\int_0^t \langle f(s, u(s)), u(s)\rangle ds, \quad 0 \le t \le T, \tag{5.99}$$

*where $D_t := (0,1) \times (0,t)$.*

*Proof.* Obviously,

$$u(t) = S_\varepsilon(t) u_0 + \int_0^t S_\varepsilon(t-\tau) f\big(\tau, u(\tau)\big) d\tau, \quad 0 \le t \le T, \tag{5.100}$$

where $\{S_\varepsilon(t),\ t \ge 0\}$ is the continuous semigroup of contractions generated by $-B_{0\varepsilon}$. Since $D(B_{0\varepsilon})$ is dense in $H_1$ and $C^1([0,T]; H_1)$ is dense in $L^1(0,T; H_1)$, there exist sequences $\{u_{0n}\}_{n\in\mathbb{N}} \subset D(B_{0\varepsilon})$ and $\{f_n\}_{n\in\mathbb{N}} \subset C^1([0,T]; H_1)$, such that $u_{0n} \to u_0$ in $H_1$ and $f_n \to h$ in $L^1(0,T; H_1)$, as $n \to \infty$, where $h(t) = f\big(t, u(t)\big)$. On account of Theorem 2.0.27, the Cauchy problem in $H_1$

$$\begin{cases} u_n'(t) + B_{0\varepsilon} u_n(t) = f_n(t), & 0 < t < T, \\ u_n(0) = u_{0n}, \end{cases}$$

has a unique strong solution $u_n = (u_n^1, u_n^2, u_n^3) \in C^1([0,T]; H_1)$, which satisfies

$$u_n(t) = S_\varepsilon(t) u_{0n} + \int_0^t S_\varepsilon(t-\tau) f_n(\tau) d\tau, \quad 0 \le t \le T,\ \forall n \in \mathbb{N}. \tag{5.101}$$

From the obvious equality

$$\|u_n(t)\|^2 + 2\int_0^t \langle B_{0\varepsilon} u_n(s), u_n(s)\rangle ds = \|u_{0n}\| + 2\int_0^t \langle f_n(s), u_n(s)\rangle ds \tag{5.102}$$

for all $t \in [0,T]$, $n \in \mathbb{N}$, we derive

$$\|u_n(t)\|^2 + 2r\|u_n^1\|^2_{L^2(D_t)} \le \|u_{0n}\|^2 + 2\int_0^t \langle f_n(s), u_n(s)\rangle ds \tag{5.103}$$

for all $t \in [0,T]$. By (5.100) and (5.101) we obtain that $u_n \to u$ as $n \to \infty$ in $C([0,T]; H_1)$. Therefore, inequality (5.99) follows from (5.103) by letting $n \to \infty$. □

By Lemma 5.2.11 and the obvious inequality $\|G_{1\varepsilon}(0)\|^2 \le M_8$, we have

$$\|Q_\varepsilon(t)\|^2 + 2r\|R_{1\varepsilon t}\|^2_{L^2(D_t)} \le M_8 + 2\int_0^t \langle G'_{1\varepsilon}(s) - \sigma'_\varepsilon(s), Q_\varepsilon(s)\rangle ds \quad (5.104)$$

for all $t \in [0, T]$.
On the other hand,

$$\int_0^t \langle \sigma'_\varepsilon(s), Q_\varepsilon(s)\rangle ds \ge -M_9\Big(\int_0^t \mid R_{2\varepsilon s}(1,s) \mid ds + \int_0^t \|R_{2\varepsilon s}(\cdot,s)\|_0 ds\Big), \quad (5.105)$$

$$\begin{aligned}\int_0^t \langle G'_{1\varepsilon}(s), Q_\varepsilon(s)\rangle ds \le & \varepsilon\int_0^t \|X_{ss}(\cdot,s)\|_0 \cdot \|R_{1\varepsilon s}(\cdot,s)\|_0 ds \\ & + \frac{r}{\varepsilon}\int_0^t e^{-\frac{rs}{\varepsilon}} \left(\|\alpha'\|_0 \cdot \|R_{2\varepsilon s}(\cdot,s)\|_0 + \mid \alpha(1) R_{2\varepsilon s}(1,s) \mid\right) ds \\ \le & M_{10}\Big(\int_0^t \Big(\sqrt{\varepsilon}\|X_{ss}(\cdot,s)\|_0 + \frac{r}{\varepsilon}e^{-\frac{rs}{\varepsilon}}\Big) \parallel Q_\varepsilon(s) \parallel ds\Big)\end{aligned}$$

for all $t \in [0, T]$. Combining the last inequality with (5.104) and (5.105), we get

$$\|Q_\varepsilon(t)\|^2 \le M_{11}\int_0^t \Big(1 + \sqrt{\varepsilon}\|X_{ss}(\cdot,s)\|_0 + \frac{r}{\varepsilon}e^{-\frac{rs}{\varepsilon}}\Big) \parallel Q_\varepsilon(s) \parallel ds, \quad 0 \le t \le T,$$

and thus, using Gronwall's inequality, we can infer that

$$\|Q_\varepsilon(t)\|^2 = \varepsilon\|R_{1\varepsilon t}(\cdot,t)\|_0^2 + \|R_{2\varepsilon t}(\cdot,t)\|_0^2 + kR_{2\varepsilon t}(1,t)^2 \le M_{12} \quad (5.106)$$

for all $t \in [0, T]$. Therefore,

$$\varepsilon\|R_{1\varepsilon t}(\cdot,t)\|_0^2 \le M_{12}, \; \|R_{2\varepsilon t}(\cdot,t)\|_0 \le M_{12} \quad (5.107)$$

for all $t \in [0, T]$. Now, combining (5.104) with (5.106), we find

$$\|R_{1\varepsilon t}\|^2_{L^2(D_T)} = \mathcal{O}(1). \quad (5.108)$$

In addition (see (5.70), (5.98) and (5.108)), we have

$$\begin{aligned}\|R_{1\varepsilon}(\cdot,t)\|_0^2 = & 2\int_0^t \langle R_{1\varepsilon s}(\cdot,s), R_{1\varepsilon}(\cdot,s)\rangle_0 ds \\ \le & 2\int_0^t \|R_{1\varepsilon s}(\cdot,s)\|_0 \cdot \|R_{1\varepsilon}(\cdot,s)\|_0 ds \\ \le & 2\|R_{1\varepsilon t}\|_{L^2(D_T)} \, \|R_{1\varepsilon}\|_{L^2(D_T)} \le M_{13}\varepsilon^{1/2}.\end{aligned} \quad (5.109)$$

Now, by (5.69), (5.107) and (5.109), it follows that

$$\|R_{1\varepsilon x}(\cdot,t)\|_0 \le M_{14}, \; \|R_{2\varepsilon x}(\cdot,t)\|_0 \le M_{15}\varepsilon^{1/4} \; \forall\, t \in [0, T]. \quad (5.110)$$

Taking into account (5.96), (5.109) and (5.110), we obtain

$$\begin{aligned} R_{2\varepsilon}(x,t)^2 &= R_{2\varepsilon}(1,t)^2 - 2\int_x^1 R_{2\varepsilon\xi}(\xi,t)R_{2\varepsilon}(\xi,t)d\xi \\ &\le R_{2\varepsilon}(1,t)^2 + 2\|R_{2\varepsilon}(\cdot,t)\|_0 \cdot \|R_{2\varepsilon x}(\cdot,t)\|_0 \le M_{16}\varepsilon^{3/4}, \\ R_{1\varepsilon}(x,t)^2 &\le R_{1\varepsilon}(0,t)^2 + 2\|R_{1\varepsilon}(\cdot,t)\|_0 \cdot \|R_{1\varepsilon x}(\cdot,t)\|_0 \\ &= (1/r_0^2)R_{2\varepsilon}(0,t)^2 + 2\|R_{1\varepsilon}(\cdot,t)\|_0 \cdot \|R_{1\varepsilon x}(\cdot,t)\|_0 \le M_{17}\varepsilon^{1/4} \end{aligned}$$

for all $(x,t) \in \overline{D}_T$. Thus, $\|R_{1\varepsilon}\|_{C(\overline{D}_T)} = \mathcal{O}(\varepsilon^{1/8})$, $\|R_{2\varepsilon}\|_{C(\overline{D}_T)} = \mathcal{O}(\varepsilon^{3/8})$. The proof is complete. □

## 5.3 A zeroth order asymptotic expansion for the solution of problem $(NS)$, $(IC)$, $(BC.2)$

In this section we consider the problem $P_\varepsilon$ consisting of the system

$$\begin{cases} \varepsilon u_t + v_x + ru = f_1, \\ v_t + u_x + g(x,v) = f_2 \text{ in } D_T, \end{cases} \tag{NS}$$

the initial conditions

$$u(x,0) = u_0(x), \quad v(x,0) = v_0(x), \quad 0 \le x \le 1, \tag{IC}$$

and the nonlinear boundary conditions

$$\begin{cases} u(0,t) + k_0 v_t(0,t) = -r_0(v(0,t)) - l_0 \int_0^t v(0,s)ds + e_0(t), \\ u(1,t) - k_1 v_t(1,t) = \ \ r_1(v(1,t)) + l_1 \int_0^t v(1,s)ds - e_1(t), \ 0 \le t \le T, \end{cases} \tag{BC.2}$$

where $r, k_0, k_1, l_0, l_1$ are given nonnegative constants and $g : [0,1] \times \mathbb{R} \to \mathbb{R}$, $f_1, f_2 : \overline{D}_T \to \mathbb{R}$, $e_0, e_1 : [0,T] \to \mathbb{R}$, $u_0, v_0 : [0,1] \to \mathbb{R}$, $r_0, r_1 : \mathbb{R} \to \mathbb{R}$ are known functions.

Note that in this case both the boundary conditions are of the dynamic type and contain some integral terms.

Again, the present $P_\varepsilon$ is a singularly perturbed problem of the boundary layer type with respect to the $C(\overline{D}_T)^2$-norm, the boundary layer being a neighborhood of the segment $\{(x,0);\ 0 \le x \le 1\}$.

### 5.3.1 Formal expansion

In the following we will determine formally a zeroth order asymptotic expansion of the solution $U_\varepsilon = (u_\varepsilon(x,t), v_\varepsilon(x,t))$ of problem $P_\varepsilon$, which will be of the same form as in Subsection 5.2.1 (see (5.66)), i.e.,

$$U_\varepsilon = U_0(x,t) + V_0(x,\tau) + R_\varepsilon(x,t). \tag{5.111}$$

We have relabeled the expansion for convenience, but $U_0$, $V_0$, $R_\varepsilon$, $\tau$ have the same meaning. Again, we make use of (5.111) in $(NS)$ and $(IC)$ to derive by identification the same systems of equations and initial conditions, as previously derived in Subsection 5.2.1. Here, we relabel them as follows:

$$\begin{cases} X = r^{-1}(f_1 - Y_x), \\ Y_t - r^{-1}Y_{xx} + g(x,Y) = f_2 - r^{-1}f_{1x} \text{ in } D_T, \end{cases} \tag{5.112}$$

$$Y(x,0) = v_0(x),\ 0 \le x \le 1, \tag{5.113}$$

$$d_0 \equiv 0,\ c_0(x,\tau) = \alpha(x)e^{-r\tau},\ 0 \le x \le 1,\ \tau \ge 0, \tag{5.114}$$

where $\alpha(x) = u_0(x) + (1/r)(v_0'(x) - f_1(x,0)),\ 0 \le x \le 1,$

$$\begin{cases} \varepsilon R_{1\varepsilon t} + R_{2\varepsilon x} + rR_{1\varepsilon} = -\varepsilon X_t, \\ R_{2\varepsilon t} + R_{1\varepsilon x} + g(x,v_\varepsilon) - g(x,Y) = -c_{0x} \text{ in } D_T, \end{cases} \tag{5.115}$$

$$R_{1\varepsilon}(x,0) = R_{2\varepsilon}(x,0) = 0,\ 0 \le x \le 1. \tag{5.116}$$

Finally, we derive from $(BC.2)$ by the identification procedure the following boundary conditions (see also $(5.112)_1$)

$$\begin{cases} k_0Y_t(0,t) - r^{-1}Y_x(0,t) = -r_0\big(Y(0,t)\big) - l_0\int_0^t Y(0,s)ds \\ -r^{-1}f_1(0,t) + e_0(t), \\ -k_1Y_t(1,t) - r^{-1}Y_x(1,t) = r_1\big(Y(1,t)\big) + l_1\int_0^t Y(1,s)ds \\ -r^{-1}f_1(1,t) - e_1(t),\ 0 \le t \le T, \end{cases} \tag{5.117}$$

$$\begin{cases} R_{1\varepsilon}(0,t) + k_0R_{2\varepsilon t}(0,t) + r_0\big(v_\varepsilon(0,t)\big) - r_0\big(Y(0,t)\big) \\ = -l_0\int_0^t R_{2\varepsilon}(0,s)ds - c_0(0,\tau), \\ R_{1\varepsilon}(1,t) - k_1R_{2\varepsilon t}(1,t) - r_1\big(v_\varepsilon(1,t)\big) + r_1\big(Y(1,t)\big) \\ = l_1\int_0^t R_{2\varepsilon}(1,s)ds - c_0(1,\tau),\ 0 \le t \le T. \end{cases} \tag{5.118}$$

It is worth mentioning that in this case we do not have any condition for $\alpha$ at $x = 0$ or $x = 1$. This is due to the fact that both the boundary conditions of $P_\varepsilon$ are of the dynamic type (see Subsection 5.2.1 for an explanation).

### 5.3.2 Existence, uniqueness and regularity of the solutions of problems $P_\varepsilon$ and $P_0$

We start with problem $P_\varepsilon$ consisting of $(NS)$, $(IC)$, $(BC.2)$. Our basic framework for this problem will be the product space $H_3 := L^2(0,1)^2 \times \mathbb{R}^4$, which is a real Hilbert space with the scalar product defined by

$$\begin{aligned} \langle \omega_1, \omega_2 \rangle_{H_3} :=& \varepsilon\int_0^1 p_1(x)p_2(x)dx + \int_0^1 q_1(x)q_2(x)dx \\ &+ k_0y_1^1y_1^2 + k_1y_2^1y_2^2 + l_0y_3^1y_3^2 + l_1y_4^1y_4^2 \end{aligned}$$

for all $\omega_i = (p_i, q_i, y_1^i, y_2^i, y_3^i, y_4^i) \in H_3,\ i = 1, 2$. The corresponding norm will be denoted by $\|\cdot\|_{H_3}$. Next, we define the operator $J_\varepsilon : D(J_\varepsilon) \subset H_3 \to H_3$,

$$\begin{aligned} D(J_\varepsilon) :=& \{(p, q, y) \in H_3;\ \ p, q \in H^1(0,1),\ y_1 = q(0),\ y_2 = q(1)\},\\ J_\varepsilon(p, q, y) :=& \Big(\varepsilon^{-1}(q' + rp),\ p' + g(\cdot, q),\ k_0^{-1}[p(0) + r_0(y_1) + l_0 y_3],\\ & k_1^{-1}[-p(1) + r_1(y_2) + l_1 y_4], -y_1, -y_2\Big), \end{aligned}$$

where $y := (y_1, y_2, y_3, y_4)$. We associate with $P_\varepsilon$ the Cauchy problem in $H_3$:

$$\begin{cases} w_\varepsilon'(t) + J_\varepsilon w_\varepsilon(t) = F_\varepsilon(t), & 0 < t < T,\\ w_\varepsilon(0) = w_0, \end{cases} \tag{5.119}$$

where

$$\begin{aligned} w_\varepsilon(t) :=& \left(u_\varepsilon(\cdot, t), v_\varepsilon(\cdot, t), y_{1\varepsilon}(t), y_{2\varepsilon}(t), y_{3\varepsilon}(t), y_{4\varepsilon}(t)\right),\\ F_\varepsilon(t) :=& \left(\varepsilon^{-1} f_1(\cdot, t), f_2(\cdot, t), k_0^{-1} e_0(t), k_1^{-1} e_1(t), 0, 0\right),\ t \in [0, T],\\ w_0 :=& (u_0, v_0, y_0) \in H_3,\ y_0 := (y_{01}, y_{02}, y_{03}, y_{04}). \end{aligned}$$

As far as problem $P_\varepsilon$ is concerned, we have:

**Theorem 5.3.1.** *Assume that:*

$$r, l_0, l_1 \text{ are given nonnegative constants},\ k_0,\ k_1 > 0; \tag{5.120}$$

$$r_0, r_1 \in C^1(\mathbb{R}),\ r_0' \ge 0,\ r_1' \ge 0; \tag{5.121}$$

$$x \to g(x, \xi) \in L^2(0,1)\ \forall \xi \in \mathbb{R} \text{ and } \partial g/\partial \xi =: g_\xi \in C([0,1] \times \mathbb{R}),\ g_\xi \ge 0; \tag{5.122}$$

$$F_\varepsilon \in W^{1,1}(0, T; H_3); \tag{5.123}$$

$$w_0 \in D(J_\varepsilon),\ y_{03} = y_{04} = 0. \tag{5.124}$$

*Then, problem* (5.119) *has a unique strong solution* $w_\varepsilon \in C^1([0,T]; H_3)$. *In addition,* $u_\varepsilon, v_\varepsilon$ *belong to the space* $C\left([0,T]; H^1(0,1)\right)$, *and* $y_{1\varepsilon}(t) = v_\varepsilon(0,t),\ y_{2\varepsilon}(t) = v_\varepsilon(1,t)$ *for all* $t \in [0,T]$.

*Proof.* First we will prove that $J_\varepsilon$ is a maximal monotone operator. Indeed, the monotonicity of $J_\varepsilon$ follows by an elementary computation, and for its maximality we make use of the equivalent Minty condition, i.e., $R(I + J_\varepsilon) = H_3$, where $I$ is the identity operator on $H_3$ (see Theorem 2.0.6). Thus, it suffices to show that for every $(h, k, z) \in H_3$ the equation

$$(p, q, y) + J_\varepsilon\left((p, q, y)\right) = (h, k, z)$$

has a solution $(p, q, y) \in D(J_\varepsilon)$. This equation can be rewritten as

$$\begin{cases} p + \varepsilon^{-1} q' + \varepsilon^{-1} rp = h,\\ q + p' + g(\cdot, q) = k,\\ (-p(0), p(1)) = \beta\left((q(0), q(1))\right), \end{cases} \tag{5.125}$$

where $\beta : \mathbb{R}^2 \to \mathbb{R}^2$ is some maximal monotone operator which comes out from calculation. Problem (5.125) has a solution, since this problem represents the Minty maximality condition for operator $B$ given in *Example* 3, Chapter 2. Thus $J_\varepsilon$ is indeed maximal monotone. Now, by Theorem 2.0.20, it follows that problem (5.119) has a unique strong solution $w_\varepsilon \in W^{1,\infty}(0,T;H_3)$, $w_\varepsilon(t) \in D(J_\varepsilon)$ for all $t \in [0,T]$, and hence $y_{1\varepsilon}(t) = v_\varepsilon(0,t)$, $y_{2\varepsilon}(t) = v_\varepsilon(1,t)$ $\forall\, t \in [0,T]$. Moreover, $u_{\varepsilon x}, v_{\varepsilon x} \in L^\infty(0,T;L^2(0,1))$ (see $(NS)$).

It remains to prove that $w_\varepsilon \in C^1([0,T];H_3)$ and $u_{\varepsilon x}, v_{\varepsilon x} \in C([0,T];L^2(0,1))$. These properties follow by arguments we have already used before (see the proof of Theorem 5.2.3). We leave the details to the reader. □

We continue with problem $(5.112)_2$, (5.113), (5.117). In this case, we choose as a basic setup the space $H_4 := L^2(0,1) \times \mathbb{R}^4$, with the scalar product defined by

$$\langle \zeta^1, \zeta^2 \rangle_{H_4} := \int_0^1 p_1(x)p_2(x)dx + k_0 y_1^1 y_1^2 + k_1 y_2^1 y_2^2 + l_0 y_3^1 y_3^2 + l_1 y_4^1 y_4^2$$

for all $\zeta^i = (p_i, y_1^i, y_2^i, y_3^i, y_4^i) \in H_4$, $i = 1,2$. Denote the associated norm by $\|\cdot\|_{H_4}$. We associate with problem $(5.112)_2$, (5.113), (5.117) the Cauchy problem in $H_4$:

$$\begin{cases} z'(t) + Lz(t) = h(t), \ 0 < t < T, \\ z(0) = z_0, \end{cases} \tag{5.126}$$

where $L : D(L) \subset H_4 \to H_4$,

$$\begin{aligned} D(L) :=& \left\{ (p,y) \in H_4;\ p \in H^2(0,1),\ y_1 = p(0), y_2 = p(1) \right\}, \\ L(p,y) :=& \Big( -r^{-1}p'' + g(\cdot,p),\ k_0^{-1}[-r^{-1}p'(0) + r_0\big(p(0)\big) + l_0 y_3], \\ & k_1^{-1}[r^{-1}p'(1) + r_1\big(p(1)\big) + l_1 y_4], -y_1, -y_2 \Big), \end{aligned}$$

$$\begin{aligned} y =& (y_1, y_2, y_3, y_4),\ z(t) = (Y(\cdot,t), \xi_1(t), \xi_2(t), \xi_3(t), \xi_4(t)),\ z_0 \in H_4, \\ h(t) :=& \Big( f_2(\cdot,t) - r^{-1}f_{1x}(\cdot,t), k_0^{-1}\left(-r^{-1}f_1(0,t) + e_0(t)\right), \\ & k_1^{-1}\left(r^{-1}f_1(1,t) + e_1(t)\right),\ 0,\ 0 \Big),\ 0 < t < T. \end{aligned}$$

It is easy to show that operator $L$ is not cyclically monotone, i.e., $L$ is not a subdifferential. However, we are able to prove the following (fairly strong) result:

**Theorem 5.3.2.** *Assume that*

$$l_0, l_1 \text{ are given nonnegative constants, and } r,\ k_0,\ k_1 > 0; \tag{5.127}$$

$$r_0, r_1 \in C^2(\mathbb{R}),\ r_0' \geq 0,\ r_1' \geq 0; \tag{5.128}$$

$$g = g(x,\xi),\ g_\xi,\ g_{\xi\xi} : [0,1]\times\mathbb{R}\to\mathbb{R} \text{ are continuous functions, } g_\xi \geq 0; \tag{5.129}$$

$$h \in W^{2,1}(0,T;H_4); \tag{5.130}$$

$$z_0 = (v_0, z_{01}, z_{02}, 0, 0) \in D(L),\ h(0) - Lz_0 \in D(L). \tag{5.131}$$

*Then, problem (5.126) has a unique strong solution $z \in C^2([0,T];H_4)$, with $Y \in W^{2,2}(0,T;H^1(0,1)) \bigcap C^1([0,T];H^2(0,1))$.*

*Proof.* One can easily see that $L$ is a monotone operator in $H_4$. To show its maximality we need to prove that for each $(q,a) \in H_4$ there exists a $(p,y) \in D(L)$ such that

$$(p,y) + L(p,y) = (q,a).$$

This equation is equivalent to the problem of finding $(p,y) \in D(L)$ which satisfies

$$\begin{cases} p - r^{-1}p'' + g(x,p) = q, \quad 0 < x < 1, \\ (p'(0), -p'(1)) = \alpha(p(0), p(1)), \end{cases}$$

where $\alpha : D(\alpha) = \mathbb{R}^2 \to \mathbb{R}^2$ is some maximal monotone operator which comes out by computation. Therefore, the maximality of $L$ follows by the maximality of the operator discussed in *Example* 1 of Chapter 2.

Since $z_0 \in D(L)$ and $h \in W^{2,1}(0,T;H_4)$, one can apply Theorem 2.0.20 to derive that problem (5.126) has a unique strong solution $z \in W^{1,\infty}(0,T;H_4)$ and $z(t) \in D(L)$ for all $t \in [0,T]$, so $\xi_1(t) = Y(0,t)$, $\xi_2(t) = Y(1,t)$ for all $t \in [0,T]$.

Now, let us show that $Y \in W^{1,2}(0,T;H^1(0,1))$. We can use our standard reasoning. Indeed, starting with the obvious inequality

$$\frac{1}{2}\int_0^{T-\delta} \frac{d}{dt}\|z(t+\delta) - z(t)\|_{H_4}^2 + \frac{1}{r}\int_0^{T-\delta} \|Y_x(\cdot,t+\delta) - Y_x(\cdot,t)\|_{L^2(0,1)}^2$$
$$\leq \int_0^{T-\delta} \|h(t+\delta) - h(t)\|_{H_4}\|z(t+\delta) - z(t)\|_{H_4} \ \forall\, t \in [0, T-\delta],$$

and taking into account that $h \in W^{2,1}(0,T;H_4)$ and $z \in W^{1,\infty}(0,T;H_4)$, we infer that

$$\int_0^{T-\delta} \|Y_x(\cdot,t+\delta) - Y_x(\cdot,t)\|_{L^2(0,1)}^2 \leq C\delta^2,$$

where $C$ is some positive constant, which implies that (see Theorem 2.0.3) $Y \in W^{1,2}(0,T;H^1(0,1))$.

Next, we will show that $Y \in C^2([0,T];L^2(0,1))$. For this we will rewrite problem $(5.112)_2$, (5.113), (5.117) as another Cauchy problem in the Hilbert space $H_5 := L^2(0,1)\times\mathbb{R}^2$, endowed with the scalar product

$$\langle\theta_1,\theta_2\rangle_{H_5} := \int_0^1 p_1p_2dx + k_0y_1^1y_1^2 + k_1y_2^1y_2^2 \ \ \forall\, \theta_i = (p_i, y_1^i, y_2^i) \in H_5,\ i = 1,2,$$

using the operator $L_0 : D(L_0) \subset H_5 \to H_5$,

$$D(L_0) := \left\{(p, y_1, y_2);\ p \in H^2(0,1),\ y_1 = p(0),\ y_2 = p(1)\right\},$$
$$L_0(p, y_1, y_2) := \left(-r^{-1}p'', -(rk_0)^{-1}p'(0), (rk_1)^{-1}p'(1)\right),$$

which is the subdifferential of the function $\Phi_0 : H_5 \to (-\infty, +\infty]$,

$$\Phi_0\left((p, y_1, y_2)\right) = \begin{cases} \frac{1}{2r}\int_0^1 p'(x)^2 dx, & \text{if } p \in H^1(0,1), y_1 = p(0), y_2 = p(1), \\ +\infty, & \text{otherwise.} \end{cases}$$

Thus we have a new Cauchy problem in the new space $H_5$:

$$\begin{cases} w'(t) + L_0 w(t) = h_1(t),\ 0 < t < T, \\ w(0) = \zeta_0, \end{cases} \tag{5.132}$$

where $w(t) = (Y(\cdot,t), Y(0,t), Y(1,t))$, $\zeta_0 = (v_0, v_0(0), v_0(1)) \in D(L_0)$,

$$h_1(t) := \begin{pmatrix} f_2(\cdot,t) - r^{-1}f_{1x}(\cdot,t) - g\left(\cdot, Y(\cdot,t)\right) \\ -k_0{}^{-1}[r^{-1}f_1(0,t) - e_0(t) + r_0\left(Y(0,t)\right) + l_0\int_0^t Y(0,s)ds] \\ k_1{}^{-1}[r^{-1}f_1(1,t) + e_1(t) - r_1\left(Y(1,t)\right) - l_1\int_0^t Y(1,s)ds] \end{pmatrix}^T,$$

$0 < t < T$. This new Cauchy problem is associated with a subdifferential, $L_0$, and we are going to take advantage of this fact. Since $z \in W^{1,\infty}(0,T;H_4)$ and $Y \in W^{1,2}(0,T;H^1(0,1))$, we have $h_1 \in W^{1,2}(0,T;H_5)$. On the other hand, $\xi_0 \in D(L_0)$, and applying Theorem 2.0.27 we obtain that

$$w \in C^1([0,T];H_5),\ L_0 w \in C([0,T];H_5) \Rightarrow,$$
$$Y \in C([0,T];H^2(0,1)) \bigcap C^1([0,T];L^2(0,1)),\ Y(0,\cdot),\ Y(1,\cdot) \in C^1[0,T].$$

Using (5.131) we can see that $h_1(0) - L_0\zeta_0 \in D(L_0)$. Thus, according to Theorem 2.0.24 we obtain that $\overline{w} := w'$ is the strong solution of the problem

$$\begin{cases} \overline{w}'(t) + L_0\overline{w}(t) = h_1'(t),\ 0 < t < T, \\ \overline{w}(0) = h_1(0) - L_0\zeta_0. \end{cases} \tag{5.133}$$

Therefore,

$$\overline{w} \in W^{1,2}(0,T;H_5),\ \overline{w}(t) \in D(L_0) \text{ for a.a. } t \in (0,T),$$

so $h_1 \in W^{2,1}(0,T;H_5)$ and using Theorem 2.0.27 we derive that $\overline{w} \in C^1([0,T];H_5)$, $L_0\overline{w} \in C([0,T];H_5)$ and $\overline{w}(t) \in D(L_0)$ for all $t \in [0,T]$. Thus we have proved that

$$Y \in C^2([0,T];L^2(0,1)) \bigcap C^1([0,T];H^2(0,1)),\ Y(0,\cdot),\ Y(1,\cdot) \in C^2[0,T].$$

Finally, using a standard reasoning one can show that $Y_t \in W^{1,2}\left(0,T;H^1(0,1)\right)$. This completes the proof. □

**Corollary 5.3.3.** *Assume that* (5.127) *holds. If* $u_0$, $v_0$, $f_1$, $f_2$, $e_0$, $e_1$ *are smooth enough,* $r_0$, $r_1 \in C^2(\mathbb{R})$, $r_0'$, $r_1' \geq 0$, $g$ *satisfies* (5.129) *and the following compatibility conditions are fulfilled*

$$\begin{cases} (k_0 r)^{-1} v_0'(0) - {k_0}^{-1} \left(r_0 \left(v_0(0)\right) + r^{-1} f_1(0,0) - e_0(0)\right) \\ = r^{-1} v_0''(0) - g\left(0, v_0(0)\right) + f_2(0,0) - r^{-1} f_{1x}(0,0), \\ -(k_1 r)^{-1} v_0'(1) - {k_1}^{-1} \left(r_1 \left(v_0(1)\right) - r^{-1} f_1(1,0) + e_1(0)\right) \\ = r^{-1} v_0''(1) - g\left(1, v_0(1)\right) + f_2(1,0) - r^{-1} f_{1x}(1,0), \end{cases} \tag{5.134}$$

*then, for every* $\varepsilon > 0$, *problem* $P_\varepsilon$ *has a unique strong solution*

$$(u_\varepsilon, v_\varepsilon) \in C^1([0,T]; L^2(0,1))^2 \bigcap C([0,T]; H^1(0,1))^2$$

*and problem* $P_0$ *admits a unique solution*

$$\begin{aligned} (X,Y) \in & (W^{2,1}(0,T; L^2(0,1)) \bigcap C^1([0,T]; H^1(0,1))) \\ & \times (C^1([0,T]; H^2(0,1)) \bigcap W^{2,2}(0,T; H^1(0,1))). \end{aligned}$$

### 5.3.3 Estimates for the remainder components

In this subsection we will prove some estimates for the components of the remainder term with respect to the uniform convergence norm, which validate our asymptotic expansion.

**Theorem 5.3.4.** *Assume that all the assumptions of Corollary* 5.3.3 *are fulfilled. Then, for every* $\varepsilon > 0$, *the solution of problem* $P_\varepsilon$ *admits an asymptotic expansion of the form* (5.111) *and the following estimates hold:*

$$\|R_{1\varepsilon}\|_{C(\overline{D}_T)} = \mathcal{O}(\varepsilon^{1/8}), \quad \|R_{2\varepsilon}\|_{C(\overline{D}_T)} = \mathcal{O}(\varepsilon^{3/8}).$$

*Proof.* In the sequel we will denote by $M_k$, $k = 1, 2, \ldots$, different positive constants which do not depend on $\varepsilon$. Since all the assumptions of Corollary 5.3.3 are fulfilled, we obtain by virtue of Theorems 5.3.1, 5.3.2 and (5.111) that

$$(R_{1\varepsilon}, R_{2\varepsilon}) \in \left(C^1\left([0,T]; L^2(0,1)\right) \bigcap C\left([0,T]; H^1(0,1)\right)\right)^2,$$

with $R_{2\varepsilon}(0,\cdot)$, $R_{2\varepsilon}(1,\cdot) \in C^1[0,T]$, is the unique solution of problem (5.115), (5.116), (5.118).

In the sequel we will denote

$$R_\varepsilon(t) := \Big(R_{1\varepsilon}(\cdot,t), R_{2\varepsilon}(\cdot,t), R_{2\varepsilon}(0,t), R_{2\varepsilon}(1,t), \int_0^t R_{1\varepsilon}(0,s)ds, \int_0^t R_{2\varepsilon}(0,s)ds\Big).$$

By a reasoning similar to that we have already used in the proofs of Theorems 5.1.10, 5.2.8 and 5.2.10 we derive the estimates

$$\varepsilon\|R_{1\varepsilon}(\cdot,t)\|_0^2 + \|R_{2\varepsilon}(\cdot,t)\|_0^2 + k_0 R_{2\varepsilon}(0,t)^2 + k_1 R_{2\varepsilon}(1,t)^2 \leq M_1 \varepsilon \tag{5.135}$$

for all $t \in [0,T]$ (we denote by $\|\cdot\|_0$ the usual norm of $L^2(0,1)$), and

$$\|R_{1\varepsilon}\|^2_{L^2(D_T)} = \mathcal{O}(\varepsilon). \tag{5.136}$$

On the other hand, $Q_\varepsilon(t) := R'_\varepsilon(t)$, $t \in [0,T]$, is the mild solution of the following Cauchy problem in $H_3$ :

$$\begin{cases} Q'_\varepsilon(t) + J_{0\varepsilon} Q_\varepsilon(t) + \overline{J}_\varepsilon(t) = N'_\varepsilon(t), \ 0 < t < T, \\ Q_\varepsilon(0) = N_\varepsilon(0), \end{cases}$$

where

$$N_\varepsilon(t) := -\Big(X_t(\cdot,t), c_{0x}(\cdot,\tau), {k_0}^{-1} c_0(0,\tau), -{k_1}^{-1} c_0(1,\tau), 0, 0\Big),\ t \in [0,T],$$

$J_{0\varepsilon}$ is equal to a particular $J_\varepsilon$, which was defined in the previous subsection, where $g$, $r_0$, $r_1$ are null functions, and

$$\overline{J}_\varepsilon(t) := \begin{pmatrix} 0 \\ g_\xi\big(\cdot, v_\varepsilon(\cdot,t)\big) R_{2\varepsilon t}(\cdot,t) + \Big(g_\xi\big(\cdot, v_\varepsilon(\cdot,t)\big) - g_\xi\big(\cdot, Y(\cdot,t)\big)\Big) Y_t(\cdot,t) \\ k_0^{-1}\Big(r'_0\big(v_\varepsilon(0,t)\big) R_{2\varepsilon t}(0,t) + \Big(r'_0\big(v_\varepsilon(0,t)\big) - r'_0(Y(0,t)\Big) Y_t(0,t)\Big) \\ k_1^{-1}\Big(\big(r'_1\big(v_\varepsilon(1,t)\big) R_{2\varepsilon t}(1,t) + \Big(r'_1\big(v_\varepsilon(1,t)\big) - r'_1(Y(1,t)\Big) Y_t(1,t)\Big) \\ 0 \\ 0 \end{pmatrix}^T$$

for all $t \in [0,T]$.

Now, by arguments similar to those used in the proof of Theorem 5.2.10 (see Lemma 5.2.11) and by the obvious estimate $\|N_\varepsilon(0)\| \le M_2$, we obtain

$$\begin{aligned} \|Q_\varepsilon(t)\|^2_{H_3} &+ 2r\|R_{1\varepsilon t}\|^2_{L^2(D_t)} \\ &\le M_2^2 + 2\int_0^t \langle N'_\varepsilon(s) - \overline{J}_\varepsilon(s), Q_\varepsilon(s)\rangle_{H_3} ds. \end{aligned} \tag{5.137}$$

On the other hand, $\| v_\varepsilon \|_{C(\overline{D}_T)} = \mathcal{O}(1)$ (see the proof of Theorem 5.2.8) and thus, since $g_{\xi\xi} \in C([0,1]\times\mathbb{R})$ and $r_0$, $r_1 \in C^2(\mathbb{R})$, we have

$$\begin{aligned} \int_0^t \langle \overline{J}_\varepsilon(s), Q_\varepsilon(s)\rangle_{H_3} ds \ge & \int_0^t \langle (g_\xi(\cdot, v_\varepsilon(\cdot,s)) - g_\xi(\cdot, Y(\cdot,s))) Y_s(\cdot,s), R_{2\varepsilon s}(\cdot,s)\rangle_0 ds \\ & + \int_0^t \Big(r'_0\big(v_\varepsilon(0,s)\big) - r'_0\big(Y(0,s)\big)\Big) Y_s(0,s) R_{2\varepsilon s}(0,s) ds \\ & + \int_0^t \Big(r'_1\big(v_\varepsilon(1,s)\big) - r'_1\big(Y(1,s)\big)\Big) Y_s(1,s) R_{2\varepsilon s}(1,s) ds \\ \ge & - M_3\Big(\int_0^t \|R_{2\varepsilon s}(\cdot,s)\|_0 ds + \int_0^t \mid R_{2\varepsilon s}(0,s) \mid ds + \int_0^t \mid R_{2\varepsilon s}(1,s) \mid ds\Big) \end{aligned}$$

$\forall\, t \in [0,T]$ (we have denoted by $\langle\cdot,\cdot\rangle_0$ the usual scalar product of $L^2(0,1)$). Note also that

$$\int_0^t \langle N_\varepsilon'(s), Q_\varepsilon(s)\rangle_{H_3} ds \le \varepsilon \int_0^t \|X_{ss}(\cdot,s)\|_0 \cdot \|R_{1\varepsilon s}(\cdot,s)\|_0 ds$$
$$+\frac{r}{\varepsilon}\|\alpha'\|_0 \int_0^t e^{-\frac{rs}{\varepsilon}} \|R_{2\varepsilon s}(\cdot,s)\|_0 ds$$

$$+\frac{r}{\varepsilon}\alpha(0)\int_0^t e^{-\frac{rs}{\varepsilon}} R_{2\varepsilon s}(0,s)ds + \frac{r}{\varepsilon}\alpha(1)\int_0^t e^{-\frac{rs}{\varepsilon}} R_{2\varepsilon s}(1,s)ds$$
$$\le M_4 \int_0^t \left(\sqrt{\varepsilon}\|X_{ss}(\cdot,s)\|_0 + \frac{r}{\varepsilon}e^{-\frac{rs}{\varepsilon}}\right) \| Q_\varepsilon(s) \|_{H_3}\, ds$$

for all $t \in [0,T]$.
By (5.137) and the last two inequalities we infer

$$\|Q_\varepsilon(t)\|_{H_3}^2 \le M_5 \int_0^t \left(1+\sqrt{\varepsilon}\|X_{ss}(\cdot,s)\|_0 + \frac{r}{\varepsilon}e^{-\frac{rs}{\varepsilon}}\right) \| Q_\varepsilon(s) \|_{H_3}\, ds, \quad 0 \le t \le T$$

and, using Gronwall's lemma, we arrive at $\|Q_\varepsilon(t)\|_{H_3} \le M_6$ for all $t \in [0,T]$. Therefore, by (5.137), we have

$$\|R_{1\varepsilon t}\|_{L^2(D_T)} = \mathcal{O}(1) \tag{5.138}$$

and obviously

$$\varepsilon\|R_{1\varepsilon t}(\cdot,t)\|_0^2 + \|R_{2\varepsilon t}(\cdot,t)\|_0^2 + k_0 R_{2\varepsilon t}(0,t)^2 + k_1 R_{2\varepsilon t}(1,t)^2 \le M_7 \tag{5.139}$$

for all $t \in [0,T]$. Now, using the inequality

$$\|R_{1\varepsilon}(\cdot,t)\|_0^2 \le 2\|R_{1\varepsilon t}\|_{L^2(D_T)}\, \|R_{1\varepsilon}\|_{L^2(D_T)},$$

together with (5.135) and (5.138), we can easily obtain that

$$\|R_{1\varepsilon}(\cdot,t)\|_0 \le M_8\varepsilon^{1/4} \text{ for all } t \in [0,T]. \tag{5.140}$$

On the other hand, (5.139) implies

$$\sqrt{\varepsilon}\|R_{1\varepsilon t}(\cdot,t)\|_0 \le \sqrt{M_7},\ \|R_{2\varepsilon t}(\cdot,t)\|_0 \le \sqrt{M_7} \text{ for all } t \in [0,T].$$

Taking into account these estimates, as well as (5.115), (5.135) and (5.140), we find that

$$\|R_{1\varepsilon x}(\cdot,t)\|_0 \le M_9, \|R_{2\varepsilon x}(\cdot,t)\|_0 \le M_{10}\varepsilon^{1/4}\ \forall\, t \in [0,T]. \tag{5.141}$$

From (5.135) and (5.141) we derive

$$\begin{aligned} R_{2\varepsilon}(x,t)^2 =& R_{2\varepsilon}(0,t)^2 + 2\int_0^x R_{2\varepsilon\xi}(\xi,t)R_{2\varepsilon}(\xi,t)d\xi \\ \leq& R_{2\varepsilon}(0,t)^2 + 2\|R_{2\varepsilon}(\cdot,t)\|_0 \cdot \|R_{2\varepsilon x}(\cdot,t\|)_0 \leq M_{11}\varepsilon^{3/4} \end{aligned}$$

for all $(x,t) \in \overline{D}_T$. Therefore, $\|R_{2\varepsilon}\|_{C(\overline{D}_T)} = \mathcal{O}(\varepsilon^{3/8})$.
Now we are going to prove the estimate for $R_{1\varepsilon}$. For each $t\in[0,T]$ and $\varepsilon > 0$ there exists an $x_{t\varepsilon}\in[0,1]$, such that $R_{1\varepsilon}(x_{t\varepsilon},t)^2 \leq M_8^2\varepsilon^{1/2}$. Therefore, from

$$\begin{aligned} R_{1\varepsilon}(x,t)^2 =& R_{1\varepsilon}(x_{t\varepsilon},t)^2 + 2\int_{x_{t\varepsilon}}^x R_{1\varepsilon\xi}(\xi,t)R_{1\varepsilon}(\xi,t)d\xi \\ \leq& R_{1\varepsilon}(x_{t\varepsilon},t)^2 + 2\|R_{1\varepsilon}(\cdot,t)\|_0 \cdot \|R_{1\varepsilon x}(\cdot,t)\|_0, \end{aligned}$$

we get $\|R_{1\varepsilon}\|_{C(\overline{D}_T)} = \mathcal{O}(\varepsilon^{1/8})$ (see also (5.140) and (5.141)). This concludes the proof. □

## 5.4 A zeroth order asymptotic expansion for the solution of problem $(LS)'$, $(IC)$, $(BC.1)$

In what follows we will be interested in the case of a small capacitance. Therefore, the small parameter $\varepsilon$ will appear in the second equation of the system, multiplying the term $v_t$. More precisely, let us consider in $D_T = [0,1] \times [0,T]$ the following problem, denoted again $P_\varepsilon$, which consists of the system

$$\begin{cases} u_t + v_x + ru = f_1, \\ \varepsilon v_t + u_x + gv = f_2, \end{cases} \qquad (LS)'$$

the initial conditions

$$u(x,0) = u_0(x), \;\; v(x,0) = v_0(x), \quad 0 \leq x \leq 1, \qquad (IC)$$

and the nonlinear boundary conditions:

$$\begin{cases} r_0 u(0,t) + v(0,t) = 0, \\ u(1,t) - kv_t(1,t) = f_0\left(v(1,t)\right) + e(t), \quad 0 \leq t \leq T, \end{cases} \qquad (BC.1)$$

where $f_1, f_2 : \overline{D}_T \to \mathbb{R}$, $u_0, v_0 : [0,1] \to \mathbb{R}$, $f_0 : \mathbb{R} \to \mathbb{R}$, $e : [0,T] \to \mathbb{R}$ are given functions, while $r, g, r_0, k$ are given nonnegative constants.

Again, by a reasoning similar to that used in Chapter 1 we can see that problem $P_\varepsilon$ is singularly perturbed, of the boundary layer type, with respect to the uniform norm (i.e., the $C(\overline{D}_T)^2$-norm). The boundary layer is located near the line segment $\{(x,0);\ \ 0 \le x \le 1\}$.

Obviously, we may study the case in which the term $ru$ of system $(LS)'$ is replaced by a nonlinear term of the form $r^*(\cdot, u)$. We prefer the linear case for simplicity and clarity.
Moreover, condition $(BC.1)_1$ may be replaced by a nonlinear one. We also note that our approach may be used to investigate either pure algebraic boundary conditions, or pure dynamic boundary conditions (i.e., both conditions involve the time derivative $v_t$).

### 5.4.1 Formal expansion

We look for a zeroth order asymptotic expansion for the solution $U_\varepsilon$ of problem $P_\varepsilon$ formulated above,

$$U_\varepsilon = U_0(x,t) + V_0(x,\tau) + R_\varepsilon(x,t) \tag{5.142}$$

(see (5.66)), where $U_0,\ V_0,\ R_\varepsilon,\ \tau$ were defined in Subsection 5.2.1. By the standard identification procedure, we are led to the following reduced problem:

$$Y = g^{-1}(f_2 - X_x) \text{ in } D_T, \tag{5.143}$$

$$\begin{cases} X_t - g^{-1}X_{xx} + rX = f_1 - g^{-1}f_{2x} \text{ in } D_T, \\ X(x,0) = u_0(x),\ 0 \le x \le 1, \\ r_0X(0,t) - g^{-1}X_x(0,t) = -g^{-1}f_2(0,t), \\ kX_{tx}(1,t) + gX(1,t) - gf_0\left(g^{-1}(f_2(1,t) - X_x(1,t))\right) \\ = ge(t) + kf_{2t}(1,t),\ 0 \le t \le T. \end{cases} \tag{5.144}$$

Our boundary layer functions (correctors) are

$$c_0 \equiv 0,\ d_0(x,\tau) = \beta(x)e^{-g\tau},\ 0 \le x \le 1,\ \tau \ge 0,$$

where $\beta(x) = v_0(x) + g^{-1}(u_0'(x) - f_2(x,0))$.

The remainder $R_\varepsilon$ satisfies the boundary value problem

$$\begin{cases} R_{1\varepsilon t} + R_{2\varepsilon x} + rR_{1\varepsilon} = -d_{0x}, \\ \varepsilon R_{2\varepsilon t} + R_{1\varepsilon x} + gR_{2\varepsilon} = -\varepsilon Y_t \text{ in } \ D_T, \\ R_{1\varepsilon}(x,0) = R_{2\varepsilon}(x,0) = 0,\ 0 \le x \le 1, \\ r_0R_{1\varepsilon}(0,t) + R_{2\varepsilon}(0,t) = 0, \\ kR_{2\varepsilon t}(1,t) - R_{1\varepsilon}(1,t) + f_0\left(v_\varepsilon(1,t)\right) - f_0\left(Y(1,t)\right) = 0,\ 0 \le t \le T. \end{cases} \tag{5.145}$$

In addition, we derive from $(BC.1)$ the following two conditions

$$\beta(0) = \beta(1) = 0. \tag{5.146}$$

It is worth pointing out that in this case we do not need any correction for the first component of the solution. This is quite natural since the small parameter multiplies $v_t$. Unlike all the previous cases, at the corner $(1,0)$ of $D_T$ we have the condition $\beta(1)=0$. To explain this, note that the term $v_t(1,t)$ of $(BC.1)_2$ includes $-g\beta(1)\varepsilon^{-1}e^{-g\tau}$, hence $\beta(1)$ must be zero. Otherwise, a discrepancy would appear at the corner $(1,0)$.

### 5.4.2 Existence, uniqueness and regularity of the solutions of problems $P_\varepsilon$ and $P_0$

We start with problem $P_\varepsilon$. This is similar to problem $P_\varepsilon$ studied in Subsection 5.2.2. Our basic framework in this case will be the product space $\mathcal{H}_1 = \left(L^2(0,1)\right)^2 \times \mathbb{R}$, endowed with the scalar product

$$\langle \zeta_1, \zeta_2\rangle_{\mathcal{H}_1} := \int_0^1 p_1(x)p_2(x)dx + \varepsilon \int_0^1 q_1(x)q_2(x)dx + ka_1a_2,$$

for all $\zeta_i = (p_i, q_i, a_i) \in \mathcal{H}_1,\ i = 1,2$ and the corresponding norm denoted by $\|\cdot\|_{\mathcal{H}_1}$. We associate with our problem the following Cauchy problem in $\mathcal{H}_1$:

$$\begin{cases} w_\varepsilon'(t) + \mathcal{B}_\varepsilon w_\varepsilon(t) = F_\varepsilon(t),\ 0<t<T, \\ w_\varepsilon(0) = w_0, \end{cases} \tag{5.147}$$

where $\mathcal{B}_\varepsilon : D(\mathcal{B}_\varepsilon) \subset \mathcal{H}_1 \to \mathcal{H}_1$,

$$\begin{aligned} D(\mathcal{B}_\varepsilon) &:= \left\{(p,q,a) \in \mathcal{H}_1;\ p,q \in H^1(0,1),\ r_0p(0)+q(0)=0,\ a=q(1)\right\}, \\ \mathcal{B}_\varepsilon\left((p,q,a)\right) &:= \left(q'+rp, \varepsilon^{-1}(p'+gq), k^{-1}\left(-p(1)+f_0(a)\right)\right), \\ w_\varepsilon(t) &:= \left(u_\varepsilon(\cdot,t), v_\varepsilon(\cdot,t), \xi_\varepsilon(t)\right),\ w_0 \in \mathcal{H}_1, \\ F_\varepsilon(t) &:= \left(f_1(\cdot,t), \varepsilon^{-1}f_2(\cdot,t), -k^{-1}e(t)\right),\ 0<t<T. \end{aligned}$$

One can prove the following result, which is similar to Theorem 5.2.3

**Theorem 5.4.1.** *Assume that*

$$r, g \text{ are nonnegative constants, } r_0, k > 0; \tag{5.148}$$

$$f_0 \in C^1(\mathbb{R}),\quad f_0' \geq 0,\ F_\varepsilon \in W^{1,1}(0,T;\mathcal{H}_1); \tag{5.149}$$

$$w_0 \in D(\mathcal{B}_\varepsilon). \tag{5.150}$$

*Then, problem* (5.147) *has a unique strong solution* $w_\varepsilon \in C^1([0,T];\mathcal{H}_1)$ *with* $\xi_\varepsilon(t) = v_\varepsilon(1,t)$ *for all* $t \in [0,T]$. *In addition,* $u_\varepsilon, v_\varepsilon \in C([0,T];H^1(0,1))$.

In the sequel we deal with the reduced problem $P_0$, which consists of the algebraic equation (5.143) and the boundary value problem (5.144). In order to homogenize the boundary conditions $(5.144)_{3,4}$, we set

$$\overline{X}(x,t) = X(x,t) - \gamma(t)x - \delta(t),$$

where $\gamma$ and $\delta$ will be determined from the system:

$$\begin{cases} r_0\overline{X}(0,t) - g^{-1}\overline{X}_x(0,t) = 0, \\ f_2(1,t) - X_x(1,t) = -\overline{X}_x(1,t), \ 0 \le t \le T. \end{cases}$$

One gets

$$\gamma(t) = f_2(1,t), \ \delta(t) = (r_0 g)^{-1}(f_2(1,t) - f_2(0,t)), t \in [0,T].$$

Obviously, $\overline{X}$ satisfies the following problem

$$\begin{cases} \overline{X}_t - g^{-1}\overline{X}_{xx} + r\overline{X} = h_1 \text{ in } D_T, \\ \overline{X}(x,0) = \overline{u}_0(x), \ 0 < x < 1, \\ r_0\overline{X}(0,t) - g^{-1}\overline{X}_x(0,t) = 0, \\ kg^{-1}\overline{X}_{tx}(1,t) + \overline{X}(1,t) - f_0\left(-g^{-1}\overline{X}_x(1,t)\right) = h_2(t), \ 0 < t < T, \end{cases} \tag{5.151}$$

where

$$\begin{cases} h_1(x,t) = f_1(x,t) - g^{-1}f_{2x}(x,t) - \gamma'(t)x - \delta'(t) - r\gamma(t)x - r\delta(t), \\ \overline{u}_0(x) = u_0(x) - \gamma(0)x - \delta(0), \\ h_2(t) = e(t) + kg^{-1}f_{2t}(1,t) - kg^{-1}\gamma'(t) - \gamma(t) - \delta(t). \end{cases}$$

Next, we choose as a basic setup the space $\mathcal{H}_2 := L^2(0,1) \times \mathbb{R}$, which is a Hilbert space with respect to the scalar product

$$\langle \omega_1, \omega_2 \rangle_{\mathcal{H}_2} := g\int_0^1 p_1(x)p_2(x)dx + kg^{-1}a_1a_2, \ \omega_i = (p_i, a_i) \in \mathcal{H}_2, \ i = 1,2,$$

and associate with problem (5.151) the following Cauchy problem in $\mathcal{H}_2$:

$$\begin{cases} z'(t) + \mathcal{A}z(t) = h(t), \ 0 < t < T, \\ z(0) = z_0, \end{cases} \tag{5.152}$$

where $\mathcal{A} : D(\mathcal{A}) \subset \mathcal{H}_2 \to \mathcal{H}_2$,

$$\begin{aligned} D(\mathcal{A}) &= \left\{(p,a) \in \mathcal{H}_2; \ p \in H^2(0,1), \ a = p'(1), \ gr_0p(0) = p'(0)\right\}, \\ \mathcal{A}(p,a) &= \left(-g^{-1}p'' + rp, \ k^{-1}g\left(p(1) - f_0(-g^{-1}p'(1))\right)\right), \\ z(t) &= \left(\overline{X}(\cdot,t), \overline{\xi}(t)\right), \ \ z_0 \in \mathcal{H}_2, \ h(t) = \left(h_1(\cdot,t), gk^{-1}h_2(t)\right). \end{aligned}$$

**Theorem 5.4.2.** *Assume that*

$$r \geq 0; \ k, \ r_0, \ g > 0; \tag{5.153}$$

$$f_0 \in C^2(\mathbb{R}), \ f_0' \geq 0, \ h \in W^{2,\infty}(0,T;\mathcal{H}_2); \tag{5.154}$$

$$z_0 \in D(\mathcal{A}), \ h(0) - \mathcal{A}z_0 \in D(\mathcal{A}). \tag{5.155}$$

*Then, problem* (5.152) *has a unique strong solution* $z \in C^2([0,T];\mathcal{H}_2)$ *and* $\overline{\xi}(t) = \overline{X}_x(1,t) \ \forall \ t \in [0,T]$. *In addition,* $\overline{X}_{ttx} \in L^2(D_T)$.

*Proof.* It follows by a standard reasoning that $\mathcal{A}$ is maximal monotone (but not cyclically monotone). The rest of the proof of the present theorem is also based on previously used ideas. Let us just describe the main steps to be followed. According to Theorem 2.0.20, problem (5.152) has a strong solution $z \in W^{1,\infty}(0,T;\mathcal{H}_2)$, and $z(t) \in D(\mathcal{A})$, which implies that $\overline{\xi}(t) = \overline{X}_x(1,t) \ \forall \ t \in [0,T]$. In addition, $\overline{X} \in W^{1,2}(0,T;H^1(0,1))$. Then, we differentiate problem (5.152) with respect to $t$. In view of Theorem 2.0.31, the resulting problem has a solution belonging to $W^{1,\infty}(0,T;\mathcal{H}_2)$. Since this solution is equal to $z'$, we conclude that $z \in W^{2,\infty}(0,T;\mathcal{H}_2)$. In addition, by standard arguments, it follows that $\overline{X} \in W^{2,2}(0,T;H^1(0,1))$. In fact, since $z_{tt}$ is the mild solution for a linear inhomogeneous equation associated with the linear part of operator $\mathcal{A}$, we conclude that $z \in C^2([0,T];\mathcal{H})$. □

*Remark* 5.4.3. Obviously, one can formulate sufficient conditions involving regularity assumptions on $u_0$, $v_0$, $f_1$, $f_2$, $e$, such that the hypotheses of both Theorems 5.4.1, 5.4.2 are fulfilled. Also, conditions (5.150), (5.155) can be expressed explicitly in terms of the data. The reader may state a common existence and smoothness result for the solutions of problems $P_\varepsilon$ and $P_0$ using explicit assumptions on the data.

On the other hand, it is worth noting that from the zeroth order compatibility condition (5.150) we necessarily obtain that $\beta(0) = 0$ (see (5.146)). But, the other condition on $\beta$, $\beta(1) = 0$, cannot be derived from the two previously quoted compatibility conditions. It would follow if we required first order compatibility conditions for the data of $P_\varepsilon$. However, we will require this additional condition on $\beta$ to obtain estimates for the remainder components. In fact, as pointed out before, if this condition is not satisfied, then a discrepancy occurs at $(x,t) = (1,0)$.

### 5.4.3 Estimates for the remainder components

We will prove some estimates with respect to the uniform norm for the remainder components of the asymptotic expansion determined in Subsection 5.4.1. They validate our asymptotic expansion completely.

**Theorem 5.4.4.** *Assume that all the assumptions of Theorems* 5.4.1, 5.4.2 *are fulfilled and, in addition,* $gv_0(1) = f_2(1,0) - u_0'(1)$ *(i.e.,* $\beta(1) = 0$*). Then, for* *every $\varepsilon > 0$, the solution of problem $P_\varepsilon$ admits an asymptotic expansion of the form* (5.142) *and the following estimates hold*

$$\|R_{1\varepsilon}\|_{C(\overline{D}_T)} = \mathcal{O}(\varepsilon^{3/8}), \ \|R_{2\varepsilon}\|_{C(\overline{D}_T)} = \mathcal{O}(\varepsilon^{1/8}).$$

*Proof.* One can use previous ideas (see Theorems 5.1.10, 5.2.8, 5.2.10). We will just describe the basic steps to be followed. First of all, by Theorems 5.4.1, 5.4.2 we can see that

$$R_\varepsilon(t) := \Big(R_{1\varepsilon}(\cdot,t), R_{2\varepsilon}(\cdot,t), R_{2\varepsilon}(1,t)\Big)$$

is the unique strong solution of the following Cauchy problem in $\mathcal{H}_2$

$$\begin{cases} R_\varepsilon'(t) + \mathcal{B}_{0\varepsilon}R_\varepsilon(t) = S_\varepsilon(t), \ 0 < t < T, \\ R_\varepsilon(0) = 0, \end{cases}$$

where $\mathcal{B}_{0\varepsilon}$ is equal to $\mathcal{B}_\varepsilon$ with $f_0 \equiv 0$ ( $\mathcal{B}_\varepsilon$ and $\mathcal{H}_2$ have been defined in the previous subsection), and

$$\mathcal{S}_\varepsilon(t) := -\big(d_{0x}(\cdot,\tau), Y_t(\cdot,t), k^{-1}(f_0(R_{2\varepsilon}(1,t) + Y(1,t)) - f_0(Y(1,t)))\big).$$

By performing some computations that are comparable with those in the proofs of Theorems 5.1.10, 5.2.8, and 5.2.10, we derive the estimates

$$\|R_{1\varepsilon}(\cdot,t)\|_0^2 + \varepsilon\|R_{2\varepsilon}(\cdot,t)\|_0^2 + kR_{2\varepsilon}(1,t)^2 \le M_1\varepsilon \text{ for all } t \in [0,T],$$

$$\|R_{2\varepsilon}\|^2_{L^2(D_T)} \le M_2\varepsilon,$$

(we denote by $\|\cdot\|_0$ the usual norm of $L^2(0,1)$).

On the other hand, $Q_\varepsilon(t) := R_\varepsilon'(t), \ 0 \le t \le T$, is the mild solution of the Cauchy problem in $\mathcal{H}_2$

$$\begin{cases} Q_\varepsilon'(t) + \mathcal{B}_{0\varepsilon}Q_\varepsilon(t) = S_\varepsilon'(t), \ 0 < t < T, \\ Q_\varepsilon(0) = S_\varepsilon(0). \end{cases}$$

By standard computations, we arrive at the estimate

$$\|R_{1\varepsilon t}(\cdot,t)\|_0^2 + \varepsilon\|R_{2\varepsilon t}(\cdot,t)\|_0^2 + kR_{2\varepsilon t}(1,t)^2 \le M_3 \text{ for all } t \in [0,T].$$

Employing again our standard device (that has been used, for example, in the last part of the proof of Theorem 5.2.10), we can show that

$$\|R_{1\varepsilon}\|_{C(\overline{D}_T)} = \mathcal{O}(\varepsilon^{3/8}), \ \ \|R_{2\varepsilon}\|_{C(\overline{D}_T)} = \mathcal{O}(\varepsilon^{1/8}). \qquad \square$$

# Part III

# Singularly Perturbed Coupled Boundary Value Problems

# Chapter 6

# Presentation of the Problems

In this chapter we introduce the problems we are going to investigate in the next chapters of this part. They are mathematical models for diffusion-convection-reaction processes. We are particularly interested in coupled problems in which a small parameter is present. More precisely, let us consider in the rectangle $Q_T = (a,c) \times (0,T)$, $-\infty < a < c < \infty$, $0 < T < \infty$, the following system of parabolic equations

$$\begin{cases} u_t + (-\varepsilon u_x + \alpha_1(x)u)_x + \beta_1(x)u = f(x,t) \text{ in } Q_{1T}, \\ v_t + (-\mu(x)v_x + \alpha_2(x)v)_x + \beta_2(x)v = g(x,t) \text{ in } Q_{2T}, \end{cases} \tag{$S$}$$

with which we associate initial conditions

$$u(x,0) = u_0(x),\ a \le x \le b;\ \ v(x,0) = v_0(x),\ b \le x \le c, \tag{$IC$}$$

transmissions conditions at $x = b$:

$$\begin{cases} u(b,t) = v(b,t), \\ -\varepsilon u_x(b,t) + \alpha_1(b)u(b,t) = -\mu(b)v_x(b,t) + \alpha_2(b)v(b,t), \quad 0 \le t \le T, \end{cases} \tag{$TC$}$$

as well as one of the following types of boundary conditions:

$$u(a,t) = v(c,t) = 0, \quad 0 \le t \le T; \tag{$BC.1$}$$

$$u_x(a,t) = v(c,t) = 0, \quad 0 \le t \le T; \tag{$BC.2$}$$

$$u(a,t) = 0, \quad -v_x(c,t) = \gamma(v(c,t)), \quad 0 \le t \le T, \tag{$BC.3$}$$

where $Q_{1T} = (a,b) \times (0,T)$, $Q_{2T} = (b,c) \times (0,T)$, $b \in \mathbb{R}$, $a < b < c$, $\gamma$ is a given nonlinear function and $\varepsilon$ is a small parameter, $0 < \varepsilon \ll 1$.

Denote by $(P.k)_\varepsilon$ the problem which consists of $(S)$, $(IC)$, $(TC)$, $(BC.k)$, for $k = 1,2,3$.

It is well known that transport (transfer) of energy or mass is fundamental to many biological, chemical, environmental and industrial processes. The basic transport mechanisms of such processes are diffusion (or dispersion) and bulk flow. Therefore, the corresponding flux has two components: a diffusive one and a convective one. Here we pay attention to the case in which the spatial domain is one-dimensional. It is represented by the interval $[a, c]$.

Let us recall some things about heat transfer. We know that conductive heat transfer is nothing else but the movement of thermal energy through the corresponding medium (material) from the more energetic particles to the others. Of course, the local temperature is given by the energy of the molecules situated at that place. So, thermal energy is transferred from points with higher temperature to other points. If the temperature of some area of the medium increases, then the random molecular motion becomes more intense in that area. Thus a transfer of thermal energy is produced, which is called heat diffusion. The corresponding conductive heat flux, denoted $q_1(x,t)$, is given by the Fourier's rate law

$$q_1 = -kT_x,$$

where $k$ is the thermal conductivity of the medium, and $T = T(x,t)$ is the temperature at point $x$, at time instant $t$. We will assume that $k$ depends on $x$ only.

On the other hand, convective heat transfer is a result of the bulk flow (or net motion) in the medium, if present. The heat flux due to convection, say $q_2(x,t)$, is given by

$$q_2 = \alpha\rho c_p(T - T_0),$$

where $\alpha$ is the velocity of the medium in the $x$ direction, $\rho$ is the density, $c_p$ is the specific heat of the material, and $T_0$ is a reference temperature.
Therefore, the total flux is

$$q = q_1 + q_2 = -kT_x + \alpha\rho c_p(T - T_0).$$

We will assume that $\alpha$ depends on $x$ only.
The first law of thermodynamics (conservation of energy) gives in a standard manner (see, e.g., [16], pp. 29–31) the well-known heat equation

$$(kT_x)_x - \rho c_p(\alpha T)_x + Q = \rho c_p T_t,$$

where $Q$ is the rate of generated heat per unit volume. By generation we mean the transformation of energy from one form (e.g., mechanical, electrical, etc.) into heat. The right-hand side of the above equation represents the stored heat. In other words, the heat equation looks like

$$T_t + (-\mu T_x + \alpha T)_x = S(x,t),$$

where

$$\mu = \frac{k}{\rho c_p}, \quad S = \frac{Q}{\rho c_p}.$$

Recall that $\mu$ is called thermal diffusivity. In our model we do not take into account radiation which is due to the presence of electromagnetic waves. Note that the same partial differential equation is a model for mass transfer, but in this case $T(x,t)$ represents the mass density of the material at point $x$, at time $t$. Instead of the Fourier law, a similar law is available in this case, which is called Fick's law of mass diffusion. An additional term depending on the density may appear in the equation due to reaction. Again, both the diffusion coefficient $\mu$ and the velocity field $\alpha$ are assumed to depend on $x$ only.

Now, let us assume that the diffusion in the subinterval $[a,b]$ is negligible. Thus, as an approximation of the physical process, we set in our model $\mu(x)=\varepsilon$ for $x\in[a,b]$. On the other hand, we assume that $\mu$ is sizeable in $[b,c]$. Denoting by $u$ and $v$ the restrictions of $T$ (which means temperature or mass density) to $[a,b]$ and $[b,c]$, we get the above system $(S)$. Of course, a complete mathematical model should include initial conditions, boundary conditions, as well as transmission conditions at $x=b$, as formulated above. Concerning our transmission conditions $(TC)$, they are naturally associated with system $(S)$. Indeed, they express the continuity of the solution as well as the conservation of the total flux at $x=b$. The most important boundary conditions which may occur in applications are the following:

a) The temperature (or mass density) is known on a part of the boundary of the domain or on the whole boundary, in our case at $x=a$ and/or $x=c$. For example, let us suppose that $u(a,t)=s_1(t)$, $u(c,t)=s_2(t)$, $0\le t\le T$, where $s_1$, $s_2$ are two given smooth functions. In this case, one can reduce the problem to another similar problem in which we have homogeneous Dirichlet boundary conditions at $x=a$ and $x=c$. Indeed, we may use the simple transformations

$$\begin{cases} \tilde{u}(x,t)=u(x,t)+h_1(t)x^2+h_2(t)x+h_3(t), \\ \tilde{v}(x,t)=v(x,t)+h_4(t)x^2+h_5(t)x+h_6(t), \end{cases}$$

where $h_i(t)$, $i=1,\dots,6$, are determined from the following algebraic systems

$$\begin{cases} a^2h_1(t)+ah_2(t)+h_3(t)=-s_1(t), \\ b^2h_1(t)+bh_2v_t+h_3(t)=0, \\ 2bh_1(t)+h_2(t)=0, \end{cases}$$

$$\begin{cases} c^2h_4(t)+ch_5(t)+h_6(t)=-s_2(t), \\ b^2h_4(t)+bh_5v_t+h_6(t)=0, \\ 2bh_4(t)+h_5(t)=0. \end{cases}$$

So the pair $(\tilde{u},\tilde{v})$ satisfies a coupled problem of the form $(P.1)_\varepsilon$. Note however that the right-hand side of the first partial differential equation of the new system includes an $\mathcal{O}(\varepsilon)$-term. But the treatment is basically the same as for the case in which $\varepsilon$ is not present there.

b) Sometimes the heat flux is specified on a part of the boundary. In this case, we have a boundary condition of the following form, say at $x = a$,

$$-k(a)u_x(a,t) = s_3(t).$$

Let us also assume that a (possibly nonhomogeneous) Dirichlet boundary condition is satisfied at $x = c$. Then, as before, the corresponding coupled problem may be transformed into a problem of the type $(P.2)_\varepsilon$. Again, a term depending on $\varepsilon$ occurs in the right-hand side of the first equation of system $(S)$, but this does not essentially change the treatment.
If the boundary point $x = a$ is highly insulated, then $s_3(t) = 0$, and so we do not need to change $u$.

c) The most usual situation is that in which heat (or mass) conducted out of the boundary is convected away by the fluid. Let us suppose this happens at $x = c$. The balance equation will be of the following Robin's type

$$-k(c)v_x(c) = h[v(c,t) - v^*],$$

where $h$ is the transfer coefficient, and $v^*$ is a constant representing the temperature (or mass density) of the environment. Thus, if we assume a Dirichlet condition at $x = a$, we have a coupled problem of the form $(P.3)_\varepsilon$. Sometimes, the above boundary balance equation is nonlinear.

For specific problems describing heat or mass transfer, we refer the reader, e.g., to A.K. Datta. [16] (see also the references therein).
In the case of stationary (steady state) processes, we have coupled problems in which $u$ and $v$ depend on $x$ only (so initial conditions are no longer necessary). See the next chapter.

# Chapter 7

# The Stationary Case

In this chapter we consider the following coupled boundary value problems of the elliptic-elliptic type

$$\begin{cases} \big( - \varepsilon u'(x) + \alpha_1(x)u(x)\big)' + \beta_1(x)u(x) = f(x), \ x \in (a,b), \\ \big( - \mu(x)v'(x) + \alpha_2(x)v(x)\big)' + \beta_2(x)v(x) = g(x), \ x \in (b,c), \end{cases} \tag{$S$}$$

with the following natural transmission conditions at $x = b$

$$u(b) = v(b), \quad - \varepsilon u'(b) + \alpha_1(b)u(b) = -\mu(b)v'(b) + \alpha_2(b)v(b), \tag{$TC$}$$

and one of the following types of boundary conditions

$$u(a) = v(c) = 0, \tag{$BC.1$}$$

$$u'(a) = v(c) = 0, \tag{$BC.2$}$$

$$u(a) = 0, \quad - v'(c) = \gamma_0(v(c)), \tag{$BC.3$}$$

where $a,\ b,\ c \in \mathbb{R}$, $a < b < c$ and $\varepsilon$ is a small parameter, $0 < \varepsilon \ll 1$. In this chapter, we denote again by $(P.k)_\varepsilon$ the problem which comprises $(S)$, $(TC)$, and $(BC.k)$, $k = 1, 2, 3$, just formulated above. We will make use of the following assumptions on the data:

$(h_1)$ $\alpha_1 \in H^1(a,b),\ \beta_1 \in L^2(a,b),\ (1/2)\alpha_1' + \beta_1 \geq 0,$ a.e. in $(a,b)$;
$(h_2)$ $\alpha_2 \in H^1(b,c),\ \beta_2 \in L^2(b,c),\ (1/2)\alpha_2' + \beta_2 \geq 0,$ a.e. in $(b,c)$;
$(h_3)$ $\mu \in H^1(b,c),\ \mu(x) > 0$ for all $x \in [b,c]$ (equivalently, there is a constant $\mu_0 > 0$ such that $\mu(x) \geq \mu_0$ for all $x \in [b,c]$);
$(h_4)$ $f \in L^2(a,b),\ g \in L^2(b,c)$;
$(h_5)$ $\alpha_1 > 0$ in $[a,b]$ or
$(h_5)'$ $\alpha_1 < 0$ in $[a,b]$;
$(h_6)$ $\gamma_0 : D(\gamma_0) = \mathbb{R} \to \mathbb{R}$ is a continuous nondecreasing function.

Note that the sign of $\alpha_1$ is essential for the asymptotic analysis we are going to perform. That is why we have introduced two separate conditions, $(h_5)$ and $(h_5)'$. We will omit the case in which $\alpha_1$ possesses zeros in $[a,b]$. This case is more complicated and leads to a turning-point problem, which requires separate analysis, based on techniques different from those used in the present monograph (see, e.g., R.E. O'Malley [37] and A.M. Watts [51]).

This chapter comprises three sections.

In the first section we investigate problem $(P.1)_\varepsilon$, under assumptions $(h_1)$–$(h_5)$. We show that it is singularly perturbed of the boundary layer type with respect to the uniform convergence norm. In fact we have an internal boundary layer (or transition layer), which is a left vicinity of the coupling point $x = b$. An asymptotic $N$th order expansion is determined formally, by using the method presented in Chapter 1. Then some existence and regularity results for the problems satisfied by the regular terms are formulated and proved. The main tool we make use of is the Lax-Milgram lemma. Note that classical ODE tools could also be employed to obtain existence for such problems and the reader is encouraged to employ this alternative approach. Here we prefer such an approach since this might also be used for similar problems in more dimensions. In the last part of this section we prove some estimates for the remainder components with respect to the sup-norm. These estimates together with our existence and regularity results validate completely our asymptotic expansion. Of course, appropriate assumptions on smoothness of the data and compatibility conditions are required to achieve our asymptotic analysis.

Note that, if assumption $(h_5)'$ is required instead of $(h_5)$, then problem $(P.1)_\varepsilon$ is also singularly perturbed with respect to the sup-norm, but in this case the boundary layer is a right vicinity of the endpoint $x = a$. So, this case is very close to a classic one and we will omit its investigation.

In Section 7.2 we study problem $(P.2)_\varepsilon$, under assumptions $(h_1)$–$(h_4)$ and $(h_5)'$. Unlike $(P.1)_\varepsilon$, this problem is regularly perturbed of the order zero, but singularly perturbed of higher orders (of the boundary layer type) with respect to the sup-norm. The corresponding boundary layers are right vicinities of the point $x = a$. For the sake of simplicity, we restrict our investigation to an asymptotic expansion of the first order. Again, after the formal derivation of such an expansion, we prove that it is well defined. More precisely, we show that $(P.2)_\varepsilon$ as well as the problems satisfied by the zeroth and first order terms of the expansion have unique solutions which are sufficiently smooth. We also prove some estimates for the remainder components in the sup-norm, which show that our expansion is a real first order expansion (which means that the remainder divided by $\varepsilon$ goes to zero uniformly).

In the last section we analyze problem $(P.3)_\varepsilon$, under assumptions $(h_1)$–$(h_5)$ and $(h_6)$. This is a nonlinear problem due to the nonlinear boundary condition at $x = c$. That is why we construct only a zeroth order asymptotic expansion of the solution. Of course, higher order asymptotic expansions could be constructed as well, at the expense of some more complicated calculations. As in the case of $(P.1)_\varepsilon$, the present problem is singularly perturbed with respect to the sup-norm, with a transition layer on the left-hand side of $x = b$. We perform a similar asymptotic analysis. Of course, the nonlinear character of the problems involved in this analysis creates some difficulties. However, this is a good exercise which shows again how to handle nonlinear problems. This section ends with some estimates which validate our asymptotic expansion.

Mention should be made of similar results for problem $(P.1)_\varepsilon$ obtained in [13] and [14].

Also, we recall that problems similar to $(P.1)_\varepsilon$ and $(P.2)_\varepsilon$ have been studied by F. Gastaldi and A. Quarteroni in [20], in an attempt to explain things related to parallel coupled Navier-Stokes/Euler problems, but they did not develop a complete asymptotic analysis.

## 7.1 Asymptotic analysis of problem $(P.1)_\varepsilon$

Here we are concerned with problem $(P.1)_\varepsilon$ formulated at the very beginning of this chapter, under our assumptions $(h_1)$–$(h_5)$. This problem is singularly perturbed with respect to the sup-norm. In order to justify this assertion, let us suppose that the solution $(u_\varepsilon, v_\varepsilon)$ of problem $(P.1)_\varepsilon$ would converge in $C[a,b] \times C[b,c]$ to some pair $(X, Y)$. It is easily seen that, under the assumptions mentioned above, $(u_\varepsilon, v_\varepsilon) \in H^2(a,b) \times H^2(b,c)$ (see also Theorem 7.1.1 below). Now, if we multiply the equations of system $(S)$ by $u_\varepsilon$, $v_\varepsilon$, and integrate over $[a,b]$ and $[b,c]$, respectively, we get

$$\varepsilon \parallel u_\varepsilon' \parallel_1^2 + \mu_0 \parallel v_\varepsilon' \parallel_2^2 + \int_a^b \left(\beta_1 + \frac{\alpha_1'}{2}\right) u_\varepsilon^2 dx + \int_b^c \left(\beta_2 + \frac{\alpha_2'}{2}\right) v_\varepsilon^2 dx$$
$$+ \frac{\alpha_2(b) - \alpha_1(b)}{2} v_\varepsilon(b)^2 \leq \int_a^b f u_\varepsilon dx + \int_b^c g v_\varepsilon dx.$$

Note that in this chapter we will use the symbols $\parallel \cdot \parallel_1$ and $\parallel \cdot \parallel_2$ to denote the usual norms of $L^2(a,b)$ and $L^2(b,c)$, respectively. The above inequality shows that sequences $\{\sqrt{\varepsilon} u_\varepsilon'\}_{\varepsilon>0}$, $\{v_\varepsilon'\}_{\varepsilon>0}$ are bounded in $L^2(a,b)$ and $L^2(b,c)$, respectively. So, $\varepsilon u_\varepsilon'$ converges to zero in $L^2(a,b)$ as $\varepsilon \to 0$. In addition, $Y \in H^1(b,c)$ and $v_\varepsilon$ converges to $Y$ weakly in $H^1(b,c)$ as $\varepsilon \to 0$, at least on a subsequence. On the other hand, one can use $(S)$ to derive the boundedness of $\{-\varepsilon u_\varepsilon' + \alpha_1 u_\varepsilon\}_{\varepsilon>0}$ and $\{-\mu v_\varepsilon' + \alpha_2 v_\varepsilon\}_{\varepsilon>0}$ in $H^1(a,b)$ and $H^1(b,c)$, respectively. Since $H^1(b,c)$ is compactly embedded in $C[b,c]$, we have $Y \in H^2(b,c)$,

$$-\mu v_\varepsilon' + \alpha_2 v_\varepsilon \to -\mu Y' + \alpha_2 Y \text{ in } C[b,c],$$

and

$$(-\mu v_\varepsilon' + \alpha_2 v_\varepsilon)' \to (-\mu Y' + \alpha_2 Y)' = g - \beta_2 Y, \text{ weakly in } L^2(b,c),$$

as $\varepsilon \to 0$, on a subsequence. Similarly,

$$-\varepsilon u_\varepsilon' + \alpha_1 u_\varepsilon \to \alpha_1 X \text{ in } C[a,b],$$

and

$$(-\varepsilon u_\varepsilon' + \alpha_1 u_\varepsilon)' \to (\alpha_1 X)' = f - \beta_1 X, \text{ weakly in } L^2(a,b).$$

Therefore, $(X,Y)$ satisfies the system

$$\begin{cases} (\alpha_1 X)' + \beta_1 X = f(x) \text{ in } (a,b), \\ \big(-\mu Y' + \alpha_2 Y\big)' + \beta_2 Y = g(x) \text{ in } (b,c), \end{cases}$$

as well as the following four conditions

$$X(a) = Y(c) = 0, \quad X(b) = Y(b), \quad -\mu(b)Y'(b) + \alpha_2(b)Y(b) = \alpha_1(b)X(b),$$

which follow by passing to the limit in $(TC)$ and $(BC.1)$, as $\varepsilon \to 0$. Obviously, there are too many conditions for the transmission problem satisfied by $(X,Y)$. Thus, problem $(P.1)_\varepsilon$ is singularly perturbed with respect to the uniform convergence topology. Moreover, it is of the boundary layer type, as we will show later. Here, in order to figure out the nature of the phenomenon, let us consider a particular case of problem $(P.1)_\varepsilon$: $\mu = 1$, $\alpha_1 = \alpha_2 := \alpha$, $\alpha$ a real nonzero constant, $\beta_1 = \beta_2 = 0$, $f = f_0$ a constant function. Let $\tilde{u}$ be the function defined by

$$\tilde{u}(x) = \frac{f_0}{\alpha}(x-a), \quad x \in [a,b],$$

which clearly satisfies equation $(S)_1$. Let $\tilde{v}$ be the solution of $(S)_2$ which satisfies in addition $\tilde{v}(b) = 0 = \tilde{v}'(b)$, i.e.,

$$\tilde{v}(x) = -e^{\alpha x}\int_b^x e^{-\alpha y}\int_b^y g(s)\, ds dy.$$

Now, wee seek the solution of $(P.1)_\varepsilon$ in the form

$$\begin{cases} u_\varepsilon(x) = T_1(\varepsilon) + T_2(\varepsilon)e^{\frac{\alpha x}{\varepsilon}} + \tilde{u}(x), & x \in [a,b], \\ v_\varepsilon(x) = T_3(\varepsilon) + T_4(\varepsilon)e^{\alpha x} + \tilde{v}(x), & x \in [b,c]. \end{cases}$$

A straightforward computation gives

$$\begin{cases} T_1(\varepsilon) = \dfrac{\varepsilon\alpha^{-1}\tilde{u}'(b)\big(1-e^{\alpha(b-c)}\big)+(\tilde{u}(b)+\tilde{v}(c))e^{\alpha(b-c)}}{e^{\alpha\frac{b-a}{\varepsilon}}-e^{\alpha(b-c)}}, \\[2ex] T_2(\varepsilon) = -e^{-\frac{\alpha a}{\varepsilon}}T_1(\varepsilon), \\[2ex] T_4(\varepsilon) = e^{-\alpha b}\big(-e^{\frac{\alpha(b-a)}{\varepsilon}}T_1(\varepsilon) + \frac{\varepsilon}{\alpha}\tilde{u}'(b)\big), \\[2ex] T_3(\varepsilon) = -\tilde{v}(c) - e^{\alpha c}T_4(\varepsilon). \end{cases}$$

Now, if $\alpha > 0$, it is easily seen that

$$T_1(\varepsilon) = e^{-\alpha\frac{b-a}{\varepsilon}}\big((\tilde{u}(b) + \tilde{v}(c))e^{\alpha(b-c)} + \mathcal{O}(\varepsilon)\big).$$

Thus

$$T_2(\varepsilon)e^{\frac{\alpha x}{\varepsilon}} = -e^{-\alpha\frac{b-x}{\varepsilon}}\big((\tilde{u}(b) + \tilde{v}(c))e^{\alpha(b-c)} + \mathcal{O}(\varepsilon)\big).$$

Therefore, the following decomposition holds

$$u_\varepsilon(x) = X(x) + [Y(b) - X(b)]e^{-\alpha\frac{b-x}{\varepsilon}} + r_{1\varepsilon}(x),\ x \in [a,b],$$

and, similarly,

$$v_\varepsilon(x) = Y(x) + r_{2\varepsilon}(x),\ x \in [b,c],$$

where

$$X(x) := \tilde{u}(x),\quad Y(x) := \tilde{v}(x) + \tilde{u}(b) - [\tilde{u}(b) + \tilde{v}(c)]e^{\alpha(x-c)},$$

while $r_{1\varepsilon}$, $r_{2\varepsilon}$ are some functions which approach zero, uniformly in $[a,b]$, $[b,c]$, respectively. In addition, $(X,Y)$ satisfies the reduced transmission problem we found before, except the transmission condition $X(b) = Y(b)$. It is clear that $u_\varepsilon$ converges uniformly to $X$ in any subinterval $[a,d]$, $a < d < b$, but not in the whole interval $[a,b]$. In other words, $u_\varepsilon$ has a singular behavior in a left neighborhood of the coupling point $x = b$, which is an internal boundary layer (or transition layer). On the other hand, $v_\varepsilon$ converges uniformly to $Y$ in $[b,c]$. This is exactly what we expected since the small parameter is present (hence strongly active) in $[a,b]$.

Note that $\varepsilon u'_\varepsilon(b)$ converges to $\alpha\big(Y(b) - X(b)\big)$, which is different from zero. However, after passing to the limit in $(TC)_2$, we obtain the transmission condition $\alpha X(b) = (-\mu Y' + \alpha Y)(b)$, which says that the total flux is conserved, even if a discontinuity at $x = b$ occurs in the limit solution $(X,Y)$.

The above decomposition is actually a zeroth order expansion for $(u_\varepsilon, v_\varepsilon)$, where the term $[Y(b)-X(b)]e^{-\alpha\frac{b-x}{\varepsilon}}$ represents the corresponding boundary (transition) layer function, which compensates for the discrepancy at $x = b$ in the limit solution $(X,Y)$. We have also derived our fast variable, $\xi = (b-x)/\varepsilon$. We guess that the above asymptotic analysis fits in well the general problem $(P.1)_\varepsilon$, including the form of the fast variable. In fact, the form of the fast variable could also be derived by using some arguments in W. Eckhaus [18], J. Kervorkian and J.D. Cole [29].

Using the above computations, one can show that the same particular case of $(P.1)_\varepsilon$, but with $\alpha < 0$, is also a singular perturbation problem with respect to the uniform convergence topology, with a boundary layer at the end point $x = a$. The fast variable will be of a similar form, $\xi = (x-a)/\varepsilon$. This time the condition $X(a) = 0$ is no longer satisfied. This situation is the same for the general case of $(P.1)_\varepsilon$, under assumptions $(h_1)$–$(h_4)$ and $(h_5)'$. The reader is encouraged to check this assertion by using arguments similar to ours.

Note that we could go further in investigating the above particular case of problem $(P.1)_\varepsilon$, by constructing higher order asymptotic expansions. This would illustrate some issues which will be tackled in the very next subsection. However, we leave it to the reader as an exercise, since the corresponding computations are easy.

### 7.1.1 Higher order asymptotic expansion

The classical perturbation theory (see Chapter 1) can be adapted to our specific singular perturbation problem. Following this theory and taking into account the above discussion on a particular case, we seek an expansion of the solution $(u_\varepsilon, v_\varepsilon)$ of problem $(P.1)_\varepsilon$ in the form

$$\begin{cases} u_\varepsilon(x) = \sum_{k=0}^N \varepsilon^k X_k(x) + \sum_{k=0}^N \varepsilon^k i_k(\xi) + R_{1\varepsilon}(x), \ x \in [a,b], \\ v_\varepsilon(x) = \sum_{k=0}^N \varepsilon^k Y_k(x) + R_{2\varepsilon}(x), \ x \in [b,c], \end{cases} \tag{7.1}$$

where:

$\xi := \varepsilon^{-1}(b-x)$ is the stretched (fast) variable;

$X_k,\ Y_k,\ k = 0, 1, \ldots, N$, are the first $(N+1)$ regular terms;

$i_k,\ k = 0, 1, \ldots, N$, are the corresponding boundary layer functions;

$(R_{1\varepsilon},\ R_{2\varepsilon})$ denotes the $N$th order remainder.

In fact, one may start with three more "fast" variables, corresponding to the right sides of points $x = a$, $x = b$ and to the left side of $x = c$, respectively. However, by the identification procedure one can see that there are no boundary layers there. The identification procedure is based on the assumption that $(u_\varepsilon, v_\varepsilon)$ given by (7.1) satisfies problem $(P.1)_\varepsilon$ formally. Indeed, if we introduce (7.1) in $(P.1)_\varepsilon$ and equate the coefficients of $\varepsilon^k$, $k = -1, 0, \ldots, (N-1)$, separately those depending on $x$ from those depending on $\xi$, we can determine all the terms involved in (7.1). Thus, from $(S)$ we derive:

$$\begin{cases} (\alpha_1(x)X_k(x))' + \beta_1(x)X_k(x) = f_k(x), \ a < x < b, \\ \big(-\mu(x)Y_k'(x) + \alpha_2(x)Y_k(x)\big)' + \beta_2(x)Y_k(x) = g_k(x), \ b < x < c, \end{cases} \tag{7.2}$$

for $k = 0, \ldots, N$, where

$$f_k(x) = \begin{cases} f(x), \ k = 0, \\ X_{k-1}''(x), \ k = 1, \ldots, N, \end{cases} \tag{7.3}$$

$$g_k(x) = \begin{cases} g(x), \ k = 0, \\ 0, \ k = 1, \ldots, N, \end{cases} \tag{7.4}$$

$$i_0''(\xi) + \alpha_1(b)i_0'(\xi) = 0, \tag{7.5}$$

$$i_k''(\xi) + \alpha_1(b)i_k'(\xi) = I_k(\xi), \ k = 1, \ldots, N, \tag{7.6}$$

where $I_k$ are functions which are to be determined recursively and depend on $i_s, i'_s, i''_s$, with $s < k$, e.g.,

$$I_1(\xi) = \xi\alpha'_1(b)i'_0(\xi) + (\alpha'_1(b) + \beta_1(b))i_0(\xi).$$

Since all the boundary layer functions should be negligible far from the boundary layer, i.e.,

$$i_k(\xi) \to 0 \text{ as } \xi \to \infty,$$

from (7.5) and (7.6) we infer that

$$\begin{cases} i_0(\xi) = A_{00}e^{-\alpha_1(b)\xi}, \\ i_k(\xi) = (A_{k+1,k}\xi^{k+1} + \cdots + A_{1k}\xi + A_{0k})e^{-\alpha_1(b)\xi}, \ k = 1, \ldots, N, \end{cases} \tag{7.7}$$

where $A_{lk}$, $l = 0, \ldots, k+1$, are constants which can be determined recursively and depend on $\alpha_1^{(m)}(b)$, $m = 0, \ldots, k$, $\beta_1^{(q)}(b)$, $q = 0, \ldots, k-1$, $Y_0(b) - X_0(b)$, $\ldots$, $Y_k(b) - X_k(b)$, $A_{ls}$, $s = 0, \ldots, k$ for all $k = 1, \ldots, N$. Constants $A_{0k}$, $k = 0, \ldots, N$ will be determined later from $(TC)_1$.

The boundary layer is located on the left side of point $x = b$, as seen in the previously investigated particular case. This fact will be proved rigorously later on. The remainder satisfies the system

$$\begin{cases} \big(-\varepsilon(R'_{1\varepsilon} + \varepsilon^N X'_N) + \alpha_1(x)R_{1\varepsilon}\big)' + \beta_1(x)R_{1\varepsilon} = h_{N\varepsilon}(x) \text{ in } [a,b], \\ \big(-\mu R'_{2\varepsilon} + \alpha_2(x)R_{2\varepsilon}\big)' + \beta_2(x)R_{2\varepsilon} = 0 \text{ in } [b,c], \end{cases} \tag{7.8}$$

where $h_{N\varepsilon}(x)$ is a function depending on $\alpha_1^{(m)}(b)$, $m = 0, \ldots, N$, $\beta_1^{(q)}(b)$, $q = 0, \ldots, (N-1)$, $\alpha_1$, $\alpha'_1$, $\beta_1$, $i_k$, $i'_k$, $k = 0, \ldots, N$, and $\varepsilon$. For example,

$$h_{0\varepsilon}(x) = -(\beta_1(x) + \alpha'_1(x))i_0(\xi) + (\alpha_1(x) - \alpha_1(b))\varepsilon^{-1}i'_0(\xi), \tag{7.9}$$

$$\begin{aligned} h_{1\varepsilon}(x) =& [\varepsilon^{-1}(\alpha_1(x) - \alpha_1(b)) + \xi\alpha'_1(b)]i'_0(\xi) \\ &+ (\alpha_1(x) - \alpha_1(b))i'_1(\xi) \\ &- (\alpha'_1(x) - \alpha'_1(b))i_0(\xi) - (\beta_1(x) - \beta_1(b))i_0(\xi) \\ &- \varepsilon(\beta_1(x) + \alpha'_1(x))i_1(\xi), \ x \in [a,b]. \end{aligned} \tag{7.10}$$

From $(BC.1)$ it follows

$$\begin{cases} X_k(a) = 0, \\ Y_k(c) = 0, \ k = 0, \ldots, N, \end{cases} \tag{7.11}$$

$$\begin{cases} R_{1\varepsilon}(a) = \Gamma_{N\varepsilon}, \\ R_{2\varepsilon}(c) = 0, \end{cases} \tag{7.12}$$

where $\Gamma_{N\varepsilon} = -\big(i_0(\xi(a)) + \cdots + \varepsilon^N i_N(\xi(a))\big)$, $\xi(a) = \varepsilon^{-1}(b-a)$.

Finally, making use of $(TC)$, we get

$$Y_k(b) - X_k(b) = i_k(0), \text{ hence } A_{0k} = Y_k(b) - X_k(b), \ k = 0, \dots, N, \tag{7.13}$$

$$\begin{cases} \alpha_1(b)X_0(b) - \alpha_2(b)Y_0(b) = -(\mu Y_0')(b), \\ \alpha_1(b)X_k(b) - \alpha_2(b)Y_k(b) = \\ -(\mu Y_k')(b) - A_{1k} + X_{k-1}'(b), \ k = 1, \dots, N, \end{cases} \tag{7.14}$$

$$\begin{cases} R_{1\varepsilon}(b) = R_{2\varepsilon}(b), \\ -\varepsilon(R_{1\varepsilon}'(b) + \varepsilon^N X_N'(b)) + \alpha_1(b)R_{1\varepsilon}(b) = -\mu(b)R_{2\varepsilon}'(b) + \alpha_2(b)R_{2\varepsilon}(b). \end{cases} \tag{7.15}$$

Note that all the terms of the asymptotic expansion (7.1) are now determined. The regular terms $(X_k, Y_k)$ satisfy problems $(7.2)_k$, $(7.11)_k$, $(7.14)_k$, which will be denoted by $(P.1)_k$, for $k = 0, \dots, N$. Obviously, the reduced problem is $(P.1)_0$. The boundary layer functions are explicitly given by formulas (7.7), where the constants $A_{0k}$ are given in (7.13), while the $N$th order remainder satisfies problem (7.8) , (7.12) , (7.15).

### 7.1.2 Existence, uniqueness and regularity of the solutions of problems $(P.1)_\varepsilon$ and $(P.1)_k$

In this subsection we will prove some results concerning the existence, uniqueness and regularity of the solution $(u_\varepsilon, v_\varepsilon)$ and of the regular terms $(X_k, Y_k)$ as well. We start with problem $(P.1)_\varepsilon$ for which we are able to prove the following result:

**Theorem 7.1.1.** *Assume that $(h_1)$–$(h_5)$ are satisfied and*

$$[\alpha(b)] := \alpha_2(b) - \alpha_1(b) \geq 0. \tag{7.16}$$

*Then, problem $(P.1)_\varepsilon$ has a unique solution $(u_\varepsilon, v_\varepsilon) \in H^2(a,b) \times H^2(b,c)$.*

*Proof.* It is obvious that our problem can be written in the following variational form: find a function $w_\varepsilon \in H_0^1(a,c)$, such that

$$a_\varepsilon(w_\varepsilon, \varphi) = \int_a^c h\varphi dx \text{ for all } \varphi \in H_0^1(a,c), \tag{7.17}$$

where $h|_{(a,b)} = f$, $h|_{(b,c)} = g$, $a_\varepsilon : H_0^1(a,c) \times H_0^1(a,c) \to \mathbb{R}$,

$$a_\varepsilon(w, \varphi) := \int_a^c \mu_\varepsilon w'\varphi' dx - \int_a^c \alpha w\varphi' dx + \int_a^c \beta w\varphi dx,$$

$$\mu_\varepsilon(x) := \begin{cases} \varepsilon, \ x \in (a,b), \\ \mu(x), \ x \in (b,c), \end{cases}$$

$$\alpha(x) := \begin{cases} \alpha_1(x), \ x \in (a,b), \\ \alpha_2(x), \ x \in (b,c), \end{cases} \quad \beta(x) := \begin{cases} \beta_1(x), \ x \in (a,b), \\ \beta_2(x), \ x \in (b,c). \end{cases}$$

Next, we will show that $w_\varepsilon \in H_0^1(a,c)$ satisfies (7.17) and $u_\varepsilon := w_\varepsilon\big|_{(a,b)}$, $v_\varepsilon := w_\varepsilon\big|_{(b,c)}$ is the unique solution of $(P.1)_\varepsilon$.
First, since

$$\begin{aligned} a_\varepsilon(\varphi,\varphi) =& \varepsilon \parallel \varphi' \parallel_1^2 + \int_b^c \mu(x)\varphi'(x)^2 dx + \int_a^b \Big(\beta_1 + \frac{\alpha_1'}{2}\Big)\varphi^2 dx \\ &+ \int_b^c \Big(\beta_2 + \frac{\alpha_2'}{2}\Big)\varphi^2 dx + \frac{[\alpha(b)]}{2}\varphi(b)^2 \ \forall \ \varphi \in H_0^1(a,c), \end{aligned} \tag{7.18}$$

by using $(h_1)$-$(h_5)$, (7.16) and the Poincaré inequality, one can see that the bilinear form $a_\varepsilon$ is continuous and coercive on $H_0^1(a,c)\times H_0^1(a,c)$.
Therefore, by the Lax-Milgram lemma (see, e.g., [11], p. 84), there exists a unique solution $w_\varepsilon \in H_0^1(a,c)$ of problem (7.17). Hence, we have

$$(-\mu_\varepsilon w_\varepsilon' + \alpha w_\varepsilon)' + \beta w_\varepsilon = h \text{ in } (a,c),$$

in the sense of distributions. This implies that $-\mu_\varepsilon w_\varepsilon' + \alpha w_\varepsilon \in H^1(a,c)$, and therefore

$$u_\varepsilon \in H^2(a,b), \ \mu v_\varepsilon' \in H^1(b,c) \Rightarrow v_\varepsilon \in H^2(b,c).$$

Thus $u_\varepsilon$, $v_\varepsilon$ satisfy $(S)$ a.e. on $(a,b)$, and $(b,c)$, respectively. Since $w_\varepsilon \in H_0^1(a,c)$ we infer that the solution satisfies $(BC.1)$ and $u_\varepsilon(b) = v_\varepsilon(b)$. Finally, making use of (7.17), we obtain that $u_\varepsilon$, $v_\varepsilon$ satisfy $(TC)_2$. □

The sum of the first two terms in the right-hand side of (7.18) defines a norm in $H_0^1(a,c)$. In fact, we may formulate alternative assumptions on $\alpha_1$, $\alpha_2$, $\beta_1$, $\beta_2$ which assure coerciveness for $a_\varepsilon$. To show this let us state the following auxiliary result which will also be useful later on.

**Lemma 7.1.2.** *For all $\varphi \in H^1(b,c)$, $\varphi(c) = 0$ and $\delta > 0$ the following inequalities hold*

$$\varphi(x)^2 \le \delta\|\varphi\|_2^2 + \frac{1}{\delta}\|\varphi'\|_2^2, \ \varphi(x)^2 \le (c-b)\|\varphi'\|_2^2 \ \forall \ x \in [b,c]. \tag{7.19}$$

*Proof.* One has just to use the obvious formulas

$$\begin{aligned} \varphi(x)^2 =& -2\int_x^c \sqrt{\delta}\varphi(t)\frac{1}{\sqrt{\delta}}\varphi'(t)dt \le \delta\|\varphi\|_2^2 + \frac{1}{\delta}\|\varphi'\|_2^2, \\ \varphi(x) =& -\int_x^c \varphi'(t)dt \end{aligned}$$

for all $x \in [b,c]$. □

*Remark* 7.1.3. It follows from the above lemma that the bilinear form $a_\varepsilon$ remains coercive (hence Theorem 7.1.1 still holds) if $[\alpha(b)] < 0$, but either

$$\mu_0 + \frac{[\alpha(b)]}{2}(c-b) > 0, \tag{7.20}$$

or

$$\begin{cases} \text{there exists a positive constant } \delta_0, \text{ such that} \\ \mu_0 + \frac{[\alpha(b)]}{2\delta_0} > 0, \; \frac{[\alpha(b)]\delta_0}{2} + \frac{\alpha_2'}{2} + \beta_2 \geq 0 \text{ a.e. on } (b,c). \end{cases} \tag{7.21}$$

Moreover, any of the two inequalities in $(h_1)$ and $(h_2)$ may be relaxed if the contribution of other terms of the right-hand side of (7.18) is sufficient to preserve the coerciveness of $a_\varepsilon$.

Now, we are going to investigate problems $(P.1)_k$, $k = 0, \ldots, N$. Our purpose is to prove existence, uniqueness, and higher regularity of all the regular terms, $(X_k, Y_k)$, involved in our asymptotic expansion in order for this expansion to be well defined. We will also need higher regularity for proving estimates for the remainder components.

Let us first discuss the assumptions we should require to achieve our goal. Thus, from (7.8) we can see that we should have $X_N \in H^1(a,b)$. To obtain this, taking into account $(7.2)_{1k}$, it follows that $X_k \in H^{N-k+1}(a,b)$, $k = 0, \ldots, N$. It is quite easy to show by induction that $(7.2)_{1k}$, with $X_k(a) = 0$, has a unique solution $X_k \in H^{N-k+1}(a,b)$ for all $k = 0, \ldots, N$, if we assume that $f \in H^N(a,b)$, $\alpha_1 \in H^{N+1}(a,b)$, $\beta_1 \in H^N(a,b)$.

On the other hand, $Y_k$ is a solution of $(7.2)_{2k}$ with the boundary conditions

$$Y_k(c) = 0, \quad -(\mu Y_k')(b) + \alpha_2(b)Y_k(b) = \rho_k,$$

where

$$\rho_k := \begin{cases} \alpha_1(b)X_0(b), \; k = 0, \\ \alpha_1(b)X_k(b) + A_{1k} - X_{k-1}'(b), \; k = 1, \ldots, N. \end{cases}$$

As far as problems $(P.1)_k$ are concerned we can prove the following result:

**Theorem 7.1.4.** *Assume that $(h_1)$–$(h_5)$ hold, $\alpha_2(b) \geq 0$, and*

$$\alpha_1 \in H^{N+1}(a,b), \; \beta_1 \in H^N(a,b), \; f \in H^N(a,b). \tag{7.22}$$

*Then, problem $(P.1)_k$ has a unique solution*

$$(X_k, Y_k) \in H^{N+1-k}(a,b) \times H^2(b,c) \; \forall \; k = 0, \ldots, N.$$

*Proof.* In order to prove the theorem it is sufficient to show that problems $(7.2)_2$, $(7.11)_2$ and (7.14) admit unique solutions $Y_k \in H^2(b,c)$ for all $k = 0, 1, \ldots, N$. These problems can be written in the following variational form: find $Y_k \in V$ such that

$$a(Y_k, \varphi) = \int_b^c g_k \varphi dx + \rho_k \varphi(b) \; \forall \; \varphi \in V,$$

where $V = \{\varphi \in H^1(b,c);\ \varphi(c) = 0\}$, $a : V \times V \to \mathbb{R}$,

$$a(w,\varphi) = \int_b^c \mu w'\varphi' dx - \int_b^c \alpha_2 w\varphi' dx + \int_b^c \beta_2 w\varphi dx \ \forall\ w,\varphi \in V.$$

Clearly, $a$ is a continuous bilinear form which satisfies

$$a(\varphi,\varphi) \geq \mu_0 \parallel \varphi' \parallel_2^2 + \int_b^c \left(\beta_2 + \frac{\alpha_2'}{2}\right)\varphi^2 dx + \frac{\alpha_2(b)}{2}\varphi(b)^2 \ \forall\ \varphi \in V.$$

Thus, according to our assumptions, $a$ is coercive. The conclusions of the theorem follow from the Lax-Milgram lemma. □

*Remark* 7.1.5. It is worth pointing out that Theorem 7.1.4 remains valid under alternative assumptions which preserve the coerciveness of functional $a$. In particular, the case $\alpha_2(b) < 0$ (hence $[\alpha(b)] < 0$) is allowed and both Theorems 7.1.1 and 7.1.4 are still valid if we require that one of the assumptions (7.20), (7.21) holds. Indeed, according to Lemma 7.1.2, both the forms $a_\varepsilon$ and $a$ remain coercive.

### 7.1.3 Estimates for the remainder components

In this subsection we will establish some estimates for the two components of the $N$th order remainder of the asymptotic expansion (7.1).

**Theorem 7.1.6.** *Suppose that all the assumptions of Theorem* 7.1.4 *as well as* (7.16) *are fulfilled, and that* $\alpha_1^{(N)}$, $\beta_1^{(N-1)}$ *are Lipschitz functions on* $[a,b]$ *(for* $N=0$, $\alpha_1$ *is Lipschitz on* $[a,b]$ *and* $\beta_1 \in L^\infty(a,b)$*). Then, for every* $\varepsilon > 0$, *the solution of problem* $(P.1)_\varepsilon$ *admits an asymptotic expansion of the form* (7.1) *and the following estimates hold*

$$\|R_{1\varepsilon}\|_{C[a,b]} = \mathcal{O}(\varepsilon^{N+1/2}),\ \|R_{2\varepsilon}\|_{C[b,c]} = \mathcal{O}(\varepsilon^{N+1/2}).$$

*Proof.* By Theorems 7.1.1 and 7.1.4, $(R_{1\varepsilon}, R_{2\varepsilon})$ belongs to $H^1(a,b) \times H^2(b,c)$ (see 7.1.1). If we denote

$$\tilde{R}_{1\varepsilon}(x) := R_{1\varepsilon}(x) + (Ax+B)\Gamma_{N\varepsilon},\ A := (b-a)^{-1},\ B := -b(b-a)^{-1},$$

it is not difficult to see that $\tilde{R}_{1\varepsilon}(a) = 0,\ \tilde{R}_{1\varepsilon}(b) = R_{1\varepsilon}(b) = R_{2\varepsilon}(b)$.
Now, taking into account (7.8), (7.12), and (7.15), we obtain

$$\begin{cases} \big(-\varepsilon S_{1\varepsilon}' + \alpha_1(x)\tilde{R}_{1\varepsilon}\big)' + \beta_1(x)\tilde{R}_{1\varepsilon} = H_{N\varepsilon}(x) \text{ in } [a,b], \\ \big(-\mu R_{2\varepsilon}' + \alpha_2(x)R_{2\varepsilon}\big)' + \beta_2(x)R_{2\varepsilon} = 0 \text{ in } [b,c], \\ \tilde{R}_{1\varepsilon}(a) = 0,\ R_{2\varepsilon}(c) = 0,\ \tilde{R}_{1\varepsilon}(b) = R_{2\varepsilon}(b), \\ -\varepsilon S_{1\varepsilon}'(b) + \alpha_1(b)\tilde{R}_{1\varepsilon}(b) + \varepsilon A\Gamma_{N\varepsilon} = -\mu(b)R_{2\varepsilon}'(b) + \alpha_2(b)R_{2\varepsilon}(b), \end{cases} \tag{7.23}$$

where $S_{1\varepsilon} = \tilde{R}_{1\varepsilon} + \varepsilon^N X_N \in H^2(a,b)$, and

$$H_{N\varepsilon}(x) := h_{N\varepsilon}(x) + \Big[A\alpha_1(x) + (\beta_1(x) + \alpha_1'(x))(Ax+B)\Big]\Gamma_{N\varepsilon}\ \forall\, x \in [a,b]. \quad (7.24)$$

By a straightforward computation we infer from (7.23) that

$$R_\varepsilon := \begin{cases} \tilde{R}_{1\varepsilon} \text{ on } [a,b], \\ R_{2\varepsilon} \text{ on } (b,c], \end{cases}$$

satisfies $R_\varepsilon \in H_0^1(a,c)$, and for all $\varphi \in H_0^1(a,c)$,

$$a_\varepsilon(R_\varepsilon, \varphi) = -\varepsilon^{N+1}\int_a^b X_N'\varphi' dx + \int_a^b H_{N\varepsilon}\varphi dx + \varepsilon A\Gamma_{N\varepsilon}\varphi(b). \quad (7.25)$$

For the sake of simplicity, we assume that

$$\beta_1 + \alpha_1'/2 \geq \omega_0 > 0 \text{ a.e. on } (a,b),\ \beta_2 + \alpha_2'/2 \geq \omega_0 > 0 \text{ a.e. on } (b,c).$$

If we choose in (7.25) $\varphi = R_\varepsilon$, we can see that

$$\begin{aligned} \varepsilon\int_a^b (\tilde{R}_{1\varepsilon}')^2 dx &+ \int_b^c \mu(R_{2\varepsilon}')^2 dx + \frac{[\alpha(b)]}{2}R_{2\varepsilon}(b)^2 \\ &+ \int_a^b \left(\beta_1 + \frac{\alpha_1'}{2}\right)\tilde{R}_{1\varepsilon}^2 dx + \int_b^c \left(\beta_2 + \frac{\alpha_2'}{2}\right)R_{2\varepsilon}^2 dx \\ &= -\,\varepsilon^{N+1}\int_a^b X_N'\tilde{R}_{1\varepsilon}' dx + \int_a^b H_{N\varepsilon}\tilde{R}_{1\varepsilon} dx + \varepsilon A\Gamma_{N\varepsilon}R_{2\varepsilon}(b). \end{aligned} \quad (7.26)$$

In the general case $\beta_1 + \alpha_1'/2 \geq 0$ a.e. on $(a,b)$, $\beta_2 + \alpha_2'/2 \geq 0$ a.e. on $(b,c)$, we choose in (7.25)

$$\varphi_1(x) := \begin{cases} e^{-x}\tilde{R}_{1\varepsilon}(x) \text{ in } [a,b], \\ e^{-b}R_{2\varepsilon}(x) \text{ in } (b,c], \end{cases}$$

and we may use a slight modification of our reasoning below.
In a first stage, we suppose that $[\alpha(b)] > 0$. Using (7.26), we can easily derive

$$\begin{aligned} \varepsilon\|\tilde{R}_{1\varepsilon}'\|_1^2 &+ \mu_0\|R_{2\varepsilon}'\|_2^2 + \omega_0\big(\|\tilde{R}_{1\varepsilon}\|_1^2 + \|R_{2\varepsilon}\|_2^2\big) + \frac{[\alpha(b)]}{2}R_{2\varepsilon}(b)^2 \\ &\leq \frac{1}{2}\Big[\varepsilon^{2N+1}\|X_N'\|_1^2 + \varepsilon\|\tilde{R}_{1\varepsilon}'\|_1^2 \\ &\quad + \omega_0^{-1}\|H_{N\varepsilon}\|_1^2 + \omega_0\|\tilde{R}_{1\varepsilon}\|_1^2 + \frac{[\alpha(b)]}{2}R_{2\varepsilon}(b)^2 + \frac{2\varepsilon A \mid \Gamma_{N\varepsilon}\mid}{[\alpha(b)]}\Big]. \end{aligned} \quad (7.27)$$

On the other hand, since $\alpha_1^{(k)}$ and $\beta_1^{(k-1)}$, $k = 1, \dots, N$, are all Lipschitz functions on $[a,b]$, one can see by an easy computations that $\|H_{N\varepsilon}\|_1 = \mathcal{O}(\varepsilon^{N+1/2})$ and $\Gamma_{N\varepsilon} = \mathcal{O}(\varepsilon^k)\ \forall k \geq 0$. As a consequence of these estimates and of (7.27), we derive

$$\|\tilde{R}_{1\varepsilon}'\|_1 = \mathcal{O}(\varepsilon^N),\ \|\tilde{R}_{1\varepsilon}\|_1 = \mathcal{O}(\varepsilon^{N+1/2}),\ \|R_{2\varepsilon}'\|_2 = \mathcal{O}(\varepsilon^{N+1/2}). \quad (7.28)$$

Since

$$R_{2\varepsilon}(x) = -\int_x^c R'_{2\varepsilon}(\zeta)d\zeta \ \forall\, x \in [b,c],$$

it follows that

$$\|R_{2\varepsilon}\|_{C[b,c]} = \mathcal{O}(\varepsilon^{N+1/2}). \tag{7.29}$$

It should be noted that $S_{1\varepsilon} = \tilde{R}_{1\varepsilon} + \varepsilon^N X_N$ belongs to $H^2(a,b)$, $S_{1\varepsilon}(a) = 0$ and, making use of $(7.23)_1$ and (7.28), we obtain that

$$\varepsilon\| \ S''_{1\varepsilon} \ \|_1 = \mathcal{O}(\varepsilon^N) \Rightarrow \| \ S''_{1\varepsilon} \ \|_1 = \mathcal{O}(\varepsilon^{N-1}).$$

But $S'_{1\varepsilon}(x)^2 \leq 2\| \ S''_{1\varepsilon} \ \|_1 \cdot \| \ S'_{1\varepsilon} \ \|_1$ for all $x \in [a,b]$, therefore

$$\| \ S'_{1\varepsilon} \ \|_{C[a,b]} = \mathcal{O}(\varepsilon^{N-1/2}) \tag{7.30}$$

(we have also used the estimate $\| \ S'_{1\varepsilon} \ \|_1 \leq \varepsilon^N \| \ X_N \ \|_1 + \| \ \widetilde{R}'_{1\varepsilon} \ \|_1 = \mathcal{O}(\varepsilon^N)$).
If we integrate $(7.23)_1$ on $[a,x]$, we get

$$-\varepsilon(S'_{1\varepsilon}(x) - S'_{1\varepsilon}(a)) + \alpha_1(x)\widetilde{R}_{1\varepsilon}(x) + \int_a^x \beta_1(y)\widetilde{R}_{1\varepsilon}(y)dy = \int_a^x H_{N\varepsilon}(y)dy.$$

Now, this equation, together with $(h_5)$, (7.28) and (7.30), implies

$$| \ \alpha_1(x)\widetilde{R}_{1\varepsilon}(x) \ | \leq M\varepsilon^{N+1/2} \ \forall\, x \in [a,b] \Rightarrow \| \ \widetilde{R}_{1\varepsilon} \ \|_{C[a,b]} = \mathcal{O}(\varepsilon^{N+1/2}).$$

As

$$\|R_{1\varepsilon}\|_{C[a,b]} \leq \|\tilde{R}_{1\varepsilon}\|_{C[a,b]} + \|(Ax+B)\Gamma_{N\varepsilon}\|_{C[a,b]},$$

we infer that $\|R_{1\varepsilon}\|_{C[a,b]} = \mathcal{O}(\varepsilon^{N+1/2})$.
If $[\alpha(b)] = 0$, we take $\varphi = (u_\varepsilon, v_\varepsilon)$ in (7.17) to get

$$\begin{aligned}\varepsilon\|u'_\varepsilon\|_1^2 + &\mu_0\|v'_\varepsilon\|_2^2 + \omega_0\big(\|u_\varepsilon\|_1^2 + \|v_\varepsilon\|_2^2\big)\\ &\leq \frac{1}{2}\Big[\omega_0^{-1}\big(\|f\|_1^2 + \|g\|_2^2\big) + \omega_0\big(\|u_\varepsilon\|_1^2 + \|v_\varepsilon\|_2^2\big)\Big].\end{aligned}$$

Therefore, $\|v_\varepsilon\|_2 = \mathcal{O}(1)$ and $\|v'_\varepsilon\|_2 = \mathcal{O}(1)$. Since $v_\varepsilon(c) = 0$, $\|v_\varepsilon\|_{C[b,c]} = \mathcal{O}(1)$. In view of (7.1), $\|R_{2\varepsilon}\|_{C[b,c]} = \mathcal{O}(1)$. Therefore $| \ R_{2\varepsilon}(b) \ | = \mathcal{O}(1)$. This allows us to continue the proof as in the previous case. □

*Remark* 7.1.7. The above estimates remain valid under alternative assumptions, as described in Remarks 7.1.3 and 7.1.5 above. The proof goes similarly, with slight modifications.

## 7.2 Asymptotic analysis of problem $(P.2)_\varepsilon$

In this section we deal with problem $(P.2)_\varepsilon$ under assumptions $(h_1)$–$(h_4)$ and $(h_5)'$. For reader's convenience, we state it in detail:

$$\begin{cases} \big(-\varepsilon u'(x) + \alpha_1(x)u(x)\big)' + \beta_1(x)u(x) = f(x), \ x \in (a,b), \\ \big(-\mu(x)v'(x) + \alpha_2(x)v(x)\big)' + \beta_2(x)v(x) = g(x), \ x \in (b,c), \end{cases} \tag{$S$}$$

with transmissions conditions at $x = b$

$$u(b) = v(b), \quad -\varepsilon u'(b) + \alpha_1(b)u(b) = -\mu(b)v'(b) + \alpha_2(b)v(b), \tag{$TC$}$$

and the boundary conditions

$$u'(a) = v(c) = 0. \tag{$BC.2$}$$

As in Section 7.1, one may consider a particular case of problem $(P.2)_\varepsilon$ which satisfies the above assumptions, including $(h_5)'$. For such an example an explicit solution is available, which shows the fact that under assumption $(h_5)'$ there is a boundary layer on the right side of point $x = a$ (and no other boundary layer). The analysis of such an example is left to the reader.

### 7.2.1 First order asymptotic expansion

Taking into account the general theory of singularly perturbed problems of the boundary layer type as well as the above comments, we will construct a first order asymptotic expansion for the solution $(u_\varepsilon, v_\varepsilon)$ of $(P.2)_\varepsilon$ in the form

$$\begin{cases} u_\varepsilon(x) = X_0(x) + \varepsilon X_1(x) + i_0(\zeta) + \varepsilon i_1(\zeta) + R_{1\varepsilon}(x), \ x \in [a,b], \\ v_\varepsilon(x) = Y_0(x) + \varepsilon Y_1(x) + R_{2\varepsilon}(x), \ x \in [b,c], \end{cases} \tag{7.31}$$

where:

$\zeta := \varepsilon^{-1}(x-a)$, is the stretched (fast) variable;

$(X_k,\ Y_k),\ k = 0,1$, are the first two regular terms;

$i_k,\ k = 0,1$, are the corresponding boundary layer functions (corrections);

$(R_{1\varepsilon}, R_{2\varepsilon})$ denotes the remainder of the first order.

We have considered only one fast variable, corresponding to point $x = a$. We will see that this is indeed the case.

If we require that $(u_\varepsilon, v_\varepsilon)$ given by (7.31) satisfy formally $(P.2)_\varepsilon$, we obtain

$$\begin{cases} (\alpha_1(x)X_k(x))' + \beta_1(x)X_k(x) = f_k(x), \ a < x < b, \\ \big(-\mu(x)Y_k'(x) + \alpha_2(x)Y_k(x)\big)' + \beta_2(x)Y_k(x) = g_k(x), \ b < x < c, \end{cases} \tag{7.32}$$

$k = 0, 1$, where

$$f_k(x) = \begin{cases} f(x), \; k = 0, \\ X_0''(x), \; k = 1, \end{cases} \quad g_k(x) = \begin{cases} g(x), \; k = 0, \\ 0, \; k = 1, \end{cases}$$

$$\begin{cases} -i_0''(\zeta) + \alpha_1(a) i_0'(\zeta) = 0, \\ -i_1''(\zeta) + \alpha_1(a) i_1'(\zeta) = -\alpha_1'(a)(i_0(\zeta) + \zeta i_0'(\zeta)) - \beta_1(a) i_0(\zeta). \end{cases} \tag{7.33}$$

On the other hand, from the boundary conditions we obtain

$$Y_0(c) = 0, \; Y_1(c) = 0, \tag{7.34}$$

$$R_{1\varepsilon}'(a) = -\varepsilon X_1'(a), \; R_{2\varepsilon}(c) = 0, \tag{7.35}$$

$i_0'(0) = 0, \; i_1'(0) = -X_0'(a)$. Therefore, making use of (7.33), we find

$$i_0(\zeta) = 0, \; i_1(\zeta) = -\frac{X_0'(a)}{\alpha_1(a)} e^{\alpha_1(a)\zeta}. \tag{7.36}$$

As $i_0 \equiv 0$, our problem $(P.2)_\varepsilon$ is regularly perturbed of order zero. The remainder components satisfy the system

$$\begin{cases} \big(-\varepsilon(R_{1\varepsilon}' + \varepsilon X_1') + \alpha_1(x) R_{1\varepsilon}\big)' + \beta_1(x) R_{1\varepsilon} = h_\varepsilon(x) \text{ in } [a, b], \\ \big(-\mu(x) R_{2\varepsilon}' + \alpha_2(x) R_{2\varepsilon}\big)' + \beta_2(x) R_{2\varepsilon} = 0 \text{ in } [b, c], \end{cases} \tag{7.37}$$

where $h_\varepsilon(x) = (\alpha_1(a) - \alpha_1(x)) i_1'(\zeta) - \varepsilon(\beta_1(x) + \alpha_1'(x)) i_1(\zeta), \; x \in [a, b]$. Finally, from $(TC)$ we derive

$$\begin{cases} Y_0(b) = X_0(b), \\ \alpha_1(b) X_0(b) - \alpha_2(b) Y_0(b) = -(\mu Y_0')(b), \end{cases} \tag{7.38}$$

$$\begin{cases} Y_1(b) = X_1(b), \\ \alpha_1(b) X_1(b) - \alpha_2(b) Y_1(b) - X_0'(b) = -(\mu Y_1')(b), \end{cases} \tag{7.39}$$

$$\begin{cases} R_{1\varepsilon}(b) + \varepsilon i_1(\zeta(b)) = R_{2\varepsilon}(b), \\ -\varepsilon(R_{1\varepsilon}'(b) + \varepsilon X_1'(b)) + \alpha_1(b) R_{1\varepsilon}(b) \\ = -\mu(b) R_{2\varepsilon}'(b) + \alpha_2(b) R_{2\varepsilon}(b) + Q_\varepsilon, \end{cases} \tag{7.40}$$

where $Q_\varepsilon = \varepsilon\left(i_1'(\zeta(b)) - \alpha_1(b) i_1(\zeta(b))\right), \; \zeta(b) = \varepsilon^{-1}(b - a)$.

Thus, all the terms of expansion (7.31) have already been determined: $(X_0, Y_0)$ satisfies the reduce problem $(P.2)_0$, i.e., $(7.32)_{k=0}$, $(7.34)_1$ and (7.38), while $(X_1, Y_1)$ satisfies problem $(7.32)_{k=1}$, $(7.34)_2$ and (7.39), denoted by $(P.2)_1$.

Unlike the case studied in the preceding section, the present reduced problem $(P.2)_0$ inherits both the transmission conditions of $(P.2)_\varepsilon$.

One may also consider higher order expansions. The whole construction goes analogously to that presented above, without any essential complications if all the data are sufficiently smooth. The reader is encouraged to analyze this case.

Note again that $(P.2)_\varepsilon$ is regularly perturbed of order zero, and singularly perturbed of higher orders. This is due to the Neumann condition at $x = a$. If we would replace it by the Dirichlet condition $u(a) = 0$, we could see that the new problem is singularly perturbed of order zero, with a boundary layer located in the vicinity of the same point $x = a$. The reader may try to construct an $N$th order asymptotic expansion for this new problem and prove estimates similar to those of Theorem 7.1.6.

### 7.2.2 Existence, uniqueness and regularity of the solutions of problems $(P.2)_\varepsilon$, $(P.2)_0$ and $(P.2)_1$

Here we will state and prove some existence and regularity results for the solutions of problems $(P.2)_\varepsilon$, $(P.2)_0$ and $(P.2)_1$. We are looking for regularity properties which are minimal for our treatment. We begin with the perturbed problem, for which we have the following result

**Theorem 7.2.1.** *Assume that* $(h_1)$–$(h_4)$, $(h_5)'$ *are satisfied and*

$$[\alpha(b)] := \alpha_2(b) - \alpha_1(b) \geq 0. \tag{7.41}$$

*Then, problem* $(P.2)_\varepsilon$ *has a unique solution* $(u_\varepsilon, v_\varepsilon) \in H^2(a,b) \times H^2(b,c)$.

*Proof.* It is similar to the proof of Theorem 7.1.1, so we will just outline it. Problem $(P.2)_\varepsilon$ has the variational formulation: find a function $w_\varepsilon \in W$, such that

$$a_{1\varepsilon}(w_\varepsilon, \varphi) = \int_a^c h\varphi dx \text{ for all } \varphi \in W,$$

where $W = \{\varphi \in H^1(a,c);\ \varphi(c) = 0\}$, $h|_{(a,b)} = f$, $h|_{(b,c)} = g$, $a_{1\varepsilon} : W \times W \to \mathbb{R}$,

$$a_{1\varepsilon}(w,\varphi) = \int_a^c \mu_\varepsilon w'\varphi' dx - \int_a^c \alpha w \varphi' dx + \int_a^c \beta w \varphi dx - \alpha_1(a) w(a) \varphi(a)$$

$\forall\ \varphi, w \in W$ (for the definitions of $\mu_\varepsilon$, $\alpha$, $\beta$ see the proof of Theorem 7.1.1). Next, the obvious inequality

$$\begin{aligned} a_{1\varepsilon}(\varphi,\varphi) \geq & \varepsilon \parallel \varphi' \parallel_1^2 + \mu_0 \parallel \varphi' \parallel_2^2 + \int_a^b \left(\beta_1 + \frac{\alpha_1'}{2}\right) \varphi^2 dx \\ & + \int_b^c \left(\beta_2 + \frac{\alpha_2'}{2}\right) \varphi^2 dx + \frac{[\alpha(b)]}{2}\varphi^2(b) - \frac{\alpha_1(a)}{2}\varphi^2(a)\ \forall\ \varphi \in W, \end{aligned}$$

combined with our assumptions implies that $a_\varepsilon$ is continuous and coercive on $W \times W$. Now, using the Lax-Milgram lemma and arguing as in the proof of Theorem 7.1.1, we will derive our conclusions. □

Now, we come to problem $(P.2)_0$. First, it should be remarked that (7.38) can be written equivalently as:

$$\begin{cases} Y_0(b) = X_0(b), \\ (\alpha_1(b) - \alpha_2(b))Y_0(b) = -(\mu Y_0')(b), \end{cases} \tag{7.42}$$

therefore $Y_0$ satisfies the problem

$$\begin{cases} \big(-\mu(x)Y_0'(x) + \alpha_2(x)Y_0(x)\big)' + \beta_2(x)Y_0(x) = g(x), \ b < x < c, \\ (\alpha_1(b) - \alpha_2(b))Y_0(b) = -(\mu Y_0')(b), \ Y_0(c) = 0. \end{cases}$$

We are going to show that this problem has a unique solution $Y_0 \in H^2(b,c)$. We define the bilinear form $a_1 : V \times V \to \mathbb{R}$, $V = \{\varphi \in H^1(a,c);\ \varphi(c) = 0\}$,

$$a_1(w,\varphi) = \int_b^c \mu w'\varphi' dx - \int_b^c \alpha_2 w\varphi' dx + \int_b^c \beta_2 w\varphi dx - \alpha_1(b)w(b)\varphi(b) \ \forall\ w,\varphi \in V.$$

An easy computation leads to

$$a_1(\varphi,\varphi) = \int_b^c \mu(\varphi')^2 dx + \int_b^c \left(\frac{\alpha_2'}{2} + \beta_2\right)\varphi^2 dx + \frac{1}{2}\left([\alpha(b)] - \alpha_1(b)\right)\varphi(b)^2 \ \forall\ \varphi \in V.$$

Therefore, taking into account $(h_2)$, $(h_3)$, $(h_5)'$, (7.41) and using the Poincaré inequality, it follows that $a_1$ is continuous and coercive on $V \times V$. The Lax-Milgram lemma applied to

$$a_1(w,\varphi) = \int_b^c g\varphi dx \ \forall\ \varphi \in V \tag{7.43}$$

yields that there exists a unique function $Y_0 \in V$ which satisfies (7.43). Hence

$$(-\mu(x)Y_0' + \alpha_2(x)Y_0)' + \beta_2(x)Y_0 = g(x) \text{ on } (b,c),$$

in the sense of distributions, so $\mu Y_0' + \alpha_2 Y \in H^1(b,c) \Rightarrow Y_0 \in H^2(b,c)$. Since $Y_0 \in V$ we have $Y_0(c) = 0$ and the boundary condition in $b$ follows from (7.43). Now, $X_0$ satisfies the problem

$$(\alpha_1(x)X_0)' + \beta_1(x)X_0 = f(x) \text{ in } (a,b),\ X_0(b) = Y_0(b),$$

which under assumptions $(h_1), (h_4), (h_5)'$ has a unique solution $X_0 \in H^1(a,b)$. If in addition we suppose that

$$\alpha_1 \in H^2(a,b),\ \beta_1 \in H^1(a,b),\ f \in H^1(a,b),$$

it is quite easy to show that $X_0 \in H^2(a,b)$.

By similar arguments, one can show that problem $(P.2)_1$ has a unique solution $(X_1, Y_1) \in H^1(a,b) \times H^2(b,c)$, if assumptions $(h_1)$–$(h_4)$, $(h_5)'$, (7.41) hold, and $X_0 \in H^2(a,b)$.

Summarizing, we have the following result

**Theorem 7.2.2.** *Assume that* $(h_1)$*–*$(h_4)$*,* $(h_5)'$*,* (7.41) *hold and, in addition,*

$$\alpha_1 \in H^2(a,b), \ \beta_1 \in H^1(a,b), \ f \in H^1(a,b). \tag{7.44}$$

*Then, problems* $(P.2)_0$ *and* $(P.2)_1$ *have unique solutions* $(X_0, Y_0) \in H^2(a,b) \times H^2(b,c)$*,* $(X_1, Y_1) \in H^1(a,b) \times H^2(b,c)$*.*

Note that the conclusions of the above two theorems remain valid under alternative assumptions which preserve the coerciveness of the bilinear forms involved in the corresponding variational formulations. For example, we may assume that $[\alpha(b)] < 0$, if either (7.20) or (7.21) holds.

### 7.2.3 Estimates for the remainder components

In order to validate completely our first order expansion, as derived before (see (7.31)), it is sufficient to prove the following result:

**Theorem 7.2.3.** *Assume that* $(h_1)$*–*$(h_4)$*,* $(h_5)'$*,* (7.41) *are fulfilled. Then, problems* $(P.2)_\varepsilon$ *and* $P_0$ *have unique solutions,* $(u_\varepsilon, v_\varepsilon) \in H^2(a,b) \times H^2(b,c)$*,* $(X_0, Y_0) \in H^1(a,b) \times H^2(b,c)$*, and*

$$\|u_\varepsilon - X\|_{C[a,b]} = \mathcal{O}(\varepsilon^{1/2}), \ \|v_\varepsilon - Y\|_{C[b,c]} = \mathcal{O}(\varepsilon^{1/2}). \tag{7.45}$$

*If, in addition,* (7.44) *are satisfied, then for every* $\varepsilon > 0$*, the solution of problem* $(P.2)_\varepsilon$ *admits an asymptotic expansion of the form* (7.31) *and the following estimates are valid:*

$$\|R_{1\varepsilon}\|_{C[a,b]} = \mathcal{O}(\varepsilon^{3/2}), \ \|R_{2\varepsilon}\|_{C[b,c]} = \mathcal{O}(\varepsilon^{3/2}). \tag{7.46}$$

*Proof.* Denote $S_\varepsilon = (S_{1\varepsilon}, S_{2\varepsilon}) := (u_\varepsilon - X_0, v_\varepsilon - Y_0)$. By Theorems 7.2.1 and 7.2.2 we have $S_\varepsilon \in H^1(a,b) \times H^2(b,c)$ and, under assumptions (7.44), $R_\varepsilon = (R_{1\varepsilon}, R_{2\varepsilon}) \in H^1(a,b) \times H^2(b,c)$. On the other hand, using $(P.2)_\varepsilon$, $(P.2)_0$ and $(P.2)_1$, it is quite easy to show that

$$\begin{cases} -\varepsilon u_\varepsilon'' + (\alpha_1(x)S_{1\varepsilon})' + \beta_1(x)S_{1\varepsilon} = 0 \text{ in } [a,b], \\ \big(-\mu S_{2\varepsilon}' + \alpha_2(x)S_{2\varepsilon}\big)' + \beta_2(x)S_{2\varepsilon} = 0 \text{ in } [b,c], \\ u_\varepsilon'(a) = 0, \ S_{2\varepsilon}(c) = 0, \ S_{1\varepsilon}(b) = S_{2\varepsilon}(b), \\ -\varepsilon u_\varepsilon'(b) + \alpha_1(b)S_{1\varepsilon}(b) = -\mu(b)S_{2\varepsilon}'(b) + \alpha_2(b)S_{2\varepsilon}(b), \end{cases} \tag{7.47}$$

and

$$\begin{cases} -\varepsilon T_\varepsilon'' + (\alpha_1(x)R_{1\varepsilon})' + \beta_1(x)R_{1\varepsilon} = h_\varepsilon \text{ in } [a,b], \\ \big(-\mu R_{2\varepsilon}' + \alpha_2(x)R_{2\varepsilon}\big)' + \beta_2(x)R_{2\varepsilon} = 0 \text{ in } [b,c], \\ T_\varepsilon'(a) = 0, \ R_{2\varepsilon}(c) = 0, \ R_{1\varepsilon}(b) + \varepsilon i_1(\zeta(b)) = R_{2\varepsilon}(b), \\ -\varepsilon T_\varepsilon'(b) + \alpha_1(b)R_{1\varepsilon}(b) = -\mu(b)R_{2\varepsilon}'(b) + \alpha_2(b)R_{2\varepsilon}(b) + Q_\varepsilon, \end{cases} \tag{7.48}$$

respectively, where $T_\varepsilon(x) := R_{1\varepsilon}(x) + \varepsilon X_1(x)$ belongs to $H^2(a,b)$, hence $T'_\varepsilon(a)$ and $T'_\varepsilon(b)$ make sense. From (7.47) we obtain

$$a_{1\varepsilon}(S_\varepsilon, \varphi) = -\varepsilon \int_a^b X_0' \varphi' dx \ \forall \ \varphi \in W. \tag{7.49}$$

Arguing as in the proof of Theorem 7.1.6, we can suppose for the sake of simplicity that

$$\beta_1 + \alpha_1'/2 \geq \omega_0 > 0 \text{ a.e. on } (a,b), \ \beta_2 + \alpha_2'/2 \geq \omega_0 > 0 \text{ a.e. on } (b,c).$$

We take $\varphi = S_\varepsilon$ in (7.49) to derive

$$\begin{aligned} \varepsilon \int_a^b (S_{1\varepsilon}')^2 dx + & \int_b^c \mu (S_{2\varepsilon}')^2 dx + \int_a^b \left(\beta_1 + \frac{\alpha_1'}{2}\right) S_{1\varepsilon}^2 dx \\ & + \int_b^c \left(\beta_2 + \frac{\alpha_2'}{2}\right) S_{2\varepsilon}^2 dx + \frac{[\alpha(b)]}{2} S_{2\varepsilon}(b)^2 \leq -\varepsilon \int_a^b X_0' S_{1\varepsilon}' dx. \end{aligned} \tag{7.50}$$

Therefore,

$$\varepsilon \| S_{1\varepsilon}' \|_1^2 + \mu_0 \| S_{2\varepsilon}' \|_2^2 + \omega_0 \left( \| S_{1\varepsilon} \|_1^2 + \| S_{2\varepsilon} \|_2^2 \right) \leq \frac{1}{2} \left[ \varepsilon \| X_0' \|_1^2 + \varepsilon \| S_{1\varepsilon}' \|_1^2 \right].$$

From the above inequality it follows

$$\| S_{1\varepsilon}' \|_1 = \mathcal{O}(1), \ \| S_{1\varepsilon} \|_1 = \mathcal{O}(\varepsilon^{1/2}), \ \| S_{2\varepsilon}' \|_2 = \mathcal{O}(\varepsilon^{1/2}). \tag{7.51}$$

Now, using the formula

$$S_{2\varepsilon}(x) = -\int_x^c S_{2\varepsilon}'(s) ds \ \forall \ x \in [b,c],$$

one can see that (see (7.51))

$$\| S_{2\varepsilon} \|_{C[b,c]} = \mathcal{O}(\varepsilon^{1/2}). \tag{7.52}$$

Next, making use of $(7.47)_1$ and (7.51), we infer

$$\varepsilon \parallel u_\varepsilon'' \parallel_1 = \mathcal{O}(1) \Rightarrow \parallel u_\varepsilon'' \parallel_1 = \mathcal{O}(\varepsilon^{-1}),$$

and hence $\parallel u_\varepsilon' \parallel_{C[a,b]} = \mathcal{O}(\varepsilon^{-\frac{1}{2}})$ (we have used that $\parallel u_\varepsilon' \parallel_1 = \mathcal{O}(1)$ in view of $(7.51)_1$). Now, integrating $(7.47)_1$ over $[x,b]$, we derive

$$-\varepsilon(u_\varepsilon'(b) - u_\varepsilon'(x)) + \alpha_1(b) S_{1\varepsilon}(b) - \alpha_1(x) S_{1\varepsilon}(x) + \int_x^b \beta_1(y) S_{1\varepsilon}(y) dy = 0.$$

Since $S_{1\varepsilon}(b) = S_{2\varepsilon}(b)$, the last equality leads to (see also (7.52))

$$\| S_{1\varepsilon} \|_{C[a,b]} = \mathcal{O}(\varepsilon^{1/2}). \tag{7.53}$$

Thus, estimates (7.45) are proved. It remains to show (7.46). To this end, multiply the two equations of (7.48) by $R_{1\varepsilon}$, $R_{2\varepsilon}$, respectively, then integrate the resulting equations over $[a,b]$ and $[b,c]$, respectively, and finally add the two equations. It follows that

$$\begin{aligned}
\varepsilon \int_a^b (R_{1\varepsilon}')^2 dx &+ \int_b^c \mu (R_{1\varepsilon}')^2 dx + \int_a^b \left(\frac{\alpha_1'}{2} + \beta_1\right) R_{1\varepsilon}^2 dx \\
&+ \int_b^c \left(\frac{\alpha_2'}{2} + \beta_2\right) R_{2\varepsilon}^2 dx - \frac{\alpha_1(a)}{2} R_{1\varepsilon}(a)^2 + \frac{[\alpha(b)]}{2} R_{2\varepsilon}(b)^2 \\
&- \frac{\alpha_1(b)}{2}\varepsilon^2 i_1(\zeta(b))^2 + \varepsilon i_1(\zeta(b))\left(\mu(b)R_{2\varepsilon}'(b) - \alpha_2(b)R_{2\varepsilon}(b)\right) \\
&+ \varepsilon\alpha_1(b) i_1(\zeta(b)) R_{2\varepsilon}(b) + Q_\varepsilon(R_{2\varepsilon}(b) - \varepsilon i_1(\zeta(b))) \\
&= \int_a^b h_\varepsilon R_{1\varepsilon} dx + \varepsilon^2 \int_a^b X_1' R_{1\varepsilon}' dx.
\end{aligned}$$

An easy calculation which involves the above equality shows that

$$\begin{aligned}
\frac{\varepsilon}{2} \parallel R_{1\varepsilon}' \parallel_1^2 &+ \mu_0 \parallel R_{2\varepsilon}' \parallel_2^2 + \frac{\omega_0}{2} \parallel R_{1\varepsilon} \parallel_1^2 + \omega_0 \parallel R_{2\varepsilon} \parallel_2^2 - \frac{\alpha_1(a)}{2} R_{1\varepsilon}(a)^2 \\
&\le \frac{2}{\omega_0} \parallel h_\varepsilon \parallel_1^2 + \frac{\varepsilon^3}{2} \parallel X_1' \parallel_1^2 + M_0 \varepsilon (\mid R_{2\varepsilon}(b) \mid + \mid R_{2\varepsilon}'(b) \mid) \qquad (7.54) \\
&\times (\mid Q_\varepsilon \mid + \mid i_1(\zeta(b)) \mid) + \varepsilon \mid i_1(\zeta(b)) Q_\varepsilon \mid,
\end{aligned}$$

where $M_0$ is a positive constant independent of $\varepsilon$.
If we analyze the structure of $h_\varepsilon$, $Q_\varepsilon$ and $i_1(\zeta(b))$, we can easily see that

$$\parallel h_\varepsilon \parallel_1 = \mathcal{O}(\varepsilon^{3/2}), \quad \mid i_1(\zeta(b)) \mid = \mathcal{O}(\varepsilon^k), \quad \mid Q_\varepsilon \mid = \mathcal{O}(\varepsilon^k)\ \forall\, k \ge 0.$$

Now, we are going to derive some estimates for $R_{2\varepsilon}(b)$ and $R_{2\varepsilon}'(b)$.
Since $\|S_{2\varepsilon}\|_{C[b,c]} = \mathcal{O}(\varepsilon^{1/2})$, taking into account $(7.31)_2$, we get $\mid R_{2\varepsilon}(b) \mid = \mathcal{O}(\varepsilon^{1/2})$.
On the other hand, using $\|S_{2\varepsilon}'\|_2 = \mathcal{O}(\varepsilon^{1/2})$, we derive $\|v_\varepsilon'\|_2 = \mathcal{O}(1)$ and this together with $(S)_2$ implies that $\|v_\varepsilon''\|_2 = \mathcal{O}(1)$. Since $H^1(b,c)$ is continuously embedded into $C[b,c]$, the preceding estimates lead to $\|v_\varepsilon'\|_{C[b,c]} = \mathcal{O}(1)$, therefore $\mid R_{2\varepsilon}'(b) \mid = \mathcal{O}(1)$. Now, by virtue of these estimates and (7.54), we derive

$$\|R_{1\varepsilon}'\|_1 = \mathcal{O}(\varepsilon),\ \|R_{1\varepsilon}\|_1 = \mathcal{O}(\varepsilon^{3/2}),\ \|R_{2\varepsilon}'\|_2 = \mathcal{O}(\varepsilon^{3/2}). \qquad (7.55)$$

Since $R_{2\varepsilon}(c) = 0$, it follows from (7.55) that $\|R_{2\varepsilon}\|_{C[b,c]} = \mathcal{O}(\varepsilon^{3/2})$.
To show that $\|R_{1\varepsilon}\|_{C[a,b]} = \mathcal{O}(\varepsilon^{3/2})$, one can argue as before. Indeed, from $(7.48)_1$ and (7.55) it follows that $\parallel T_\varepsilon'' \parallel_1 = \mathcal{O}(1)$. Then, by the definition of $T_\varepsilon$ and $(7.55)_1$, we get $\parallel T_\varepsilon' \parallel_{C[a,b]} = \mathcal{O}(\varepsilon^{\frac{1}{2}})$. Now, we integrate $(7.48)_1$ over $[a,x]$. Since $\parallel T_\varepsilon' \parallel_{C[a,b]} = \mathcal{O}(\varepsilon^{\frac{1}{2}})$, $R_{1\varepsilon}(a) = \mathcal{O}(\varepsilon^{3/2})$ (cf. (7.54)), and $(7.55)_2$, we conclude by the resulting equation that $\|R_{1\varepsilon}\|_{C[a,b]} = \mathcal{O}(\varepsilon^{3/2})$. □

*Remark* 7.2.4. The above result still holds under alternative assumptions. For example, we may allow $[\alpha(b)] < 0$, if either (7.20) or (7.21)) holds. This claim can be proved by slightly modified computations.

## 7.3 Asymptotic analysis of problem $(P.3)_\varepsilon$

In this section we examine problem $(P.3)_\varepsilon$. We suppose that assumptions $(h_1)$–$(h_6)$ are fulfilled. One can justify as in Section 7.1 that our problem is singularly perturbed, with an internal transition layer in a neighborhood of point $x = b$.

It should be pointed out that $\gamma_0$ may be assumed to be multivalued and all the results which follow remain valid. However, to obtain higher order asymptotic expansions, we need to assume that $\gamma_0$ is single valued. The case $\gamma_0 = 0$ corresponds to a homogeneous Neumann boundary condition at $x = c$.

### 7.3.1 Formal expansion

Here we construct a formal zeroth order asymptotic expansion for the solution $(u_\varepsilon, v_\varepsilon)$ of problem $(P.3)_\varepsilon$ which will be of the following form:

$$\begin{cases} u_\varepsilon(x) = X_0(x) + i_0(\xi) + R_{1\varepsilon}(x), \ x \in [a, b], \\ v_\varepsilon(x) = Y_0(x) + R_{2\varepsilon}(x), \ x \in [b, c], \end{cases} \tag{7.56}$$

where: $\xi = \varepsilon^{-1}(b - x)$ is the stretched variable; $(X_0, \ Y_0)$ is the first term of the regular series; $i_0$ is the boundary layer function corresponding to the first component of the solution; $(R_{1\varepsilon}, \ R_{2\varepsilon})$ is the remainder of order zero.

Since we can use arguments similar to those we have used for the previous cases, we will not go into many details. We will determine only the problems satisfied by the terms specified before.

Thus, $(X_0, Y_0)$ satisfies the reduced problem, say $(P.3)_0$:

$$\begin{cases} (\alpha_1(x)X_0(x))' + \beta_1(x)X_0(x) = f(x), \ a < x < b, \\ \big(-\mu(x)Y_0'(x) + \alpha_2(x)Y_0(x)\big)' + \beta_2(x)Y_0(x) = g(x), \ b < x < c, \\ X_0(a) = 0, \ \alpha_1(b)X_0(b) - \alpha_2(b)Y_0(b) = -(\mu Y_0')(b), \\ -Y_0'(c) = \gamma_0(Y_0(c)). \end{cases} \tag{7.57}$$

The boundary (transition) layer function is given by

$$i_0(\xi) = (Y_0(b) - X_0(b))e^{-\alpha_1(b)\xi},$$

while the pair $(R_{1\varepsilon}, R_{2\varepsilon})$ satisfies the problem

$$\begin{cases} \big( - \varepsilon(R'_{1\varepsilon} + X'_0) + \alpha_1(x)R_{1\varepsilon}\big)' + \beta_1(x)R_{1\varepsilon} = h_\varepsilon(x) \text{ in } [a,b], \\ \big( - \mu R'_{2\varepsilon} + \alpha_2(x)R_{2\varepsilon}\big)' + \beta_2(x)R_{2\varepsilon} = 0 \text{ in } [b,c], \\ R_{1\varepsilon}(a) = -i_0(\xi(a)),\ \xi(a) = \varepsilon^{-1}(b-a), \\ R_{1\varepsilon}(b) = R_{2\varepsilon}(b), \\ -\varepsilon(R'_{1\varepsilon}(b) + X'_0(b)) + \alpha_1(b)R_{1\varepsilon}(b) = -\mu(b)R'_{2\varepsilon}(b) + \alpha_2(b)R_{2\varepsilon}(b), \\ -R'_{2\varepsilon}(c) = \gamma_0(Y_0(c) + R_{2\varepsilon}(c)) - \gamma_0(Y_0(c)), \end{cases} \tag{7.58}$$

where $h_{0\varepsilon}(x) = -(\beta_1(x) + \alpha'_1(x))i_0(\xi) + (\alpha_1(x) - \alpha_1(b))\varepsilon^{-1}i'_0(\xi),\ x \in [a,b]$.

### 7.3.2 Existence, uniqueness and regularity of the solutions of problems $(P.3)_\varepsilon$ and $(P.3)_0$

The main difficulty in studying problems $(P.3)_\varepsilon$ and $(P.3)_0$ comes up from the fact that the boundary conditions at $c$ are nonlinear. Concerning problem $(P.3)_\varepsilon$ we can state and prove the following result:

**Theorem 7.3.1.** *Assume that* $(h_1)$–$(h_6)$ *are satisfied and*

$$[\alpha(b)] := \alpha_2(b) - \alpha_1(b) \geq 0,\ \alpha_2(c) \geq 0. \tag{7.59}$$

*Then, problem* $(P.3)_\varepsilon$ *has a unique solution* $(u_\varepsilon, v_\varepsilon) \in H^2(a,b) \times H^2(b,c)$.

*Proof.* Since all the assumptions of Theorem 7.1.1 are satisfied, there exists a unique solution $(\widetilde{u}_\varepsilon, \widetilde{v}_\varepsilon) \in H^2(a,b) \times H^2(b,c)$, which satisfies system $(S)$ a.e. in $(a,b) \times (b,c)$, $(TC)$, as well as the homogeneous Dirichlet boundary conditions $\widetilde{u}_\varepsilon(a) = \widetilde{v}_\varepsilon(c) = 0$.
Notice also that $(u,v) = (u_\varepsilon, v_\varepsilon) - (\widetilde{u}_\varepsilon, \widetilde{v}_\varepsilon)$ satisfies the homogeneous system

$$\begin{cases} \big( - \varepsilon u'(x) + \alpha_1(x)u(x)\big)' + \beta_1(x)u(x) = 0,\ x \in (a,b), \\ \big( - \mu(x)v'(x) + \alpha_2(x)v(x)\big)' + \beta_2(x)v(x) = 0,\ x \in (b,c), \end{cases} \tag{7.60}$$

as well as the boundary and transmission conditions

$$\begin{cases} u(a) = 0,\ u(b) = v(b), \\ -\varepsilon u'(b) + \alpha_1(b)u(b) = -\mu(b)v'(b) + \alpha_2(b)v(b), \\ v'(c) + \gamma_0(v(c)) = -\widetilde{v}'_\varepsilon(c). \end{cases} \tag{7.61}$$

Now, let us point out that all the assertions of our theorem are satisfied if and only if there exists a unique solution $(u,v) \in H^2(a,b) \times H^2(b,c)$ of problems (7.60), (7.61). To check this, let $\{u_1,u_2\} \subset H^2(a,b)$, $\{v_1,v_2\} \subset H^2(b,c)$ be the fundamental systems of solutions for the two equations of (7.60), which satisfy the following conditions: $u_1(a) = 0$, $u_2(a) = 1$, $v_1(c) = 0$, $v'_1(c) = 1$, $v_2(c) = 1$,

$v_2'(c) = 0$. We have chosen these conditions to make our next calculations simpler. The general solution of system (7.60) will be

$$u = d_1 u_1 + d_2 u_2, \quad v = k_1 v_1 + k_2 v_2, \tag{7.62}$$

$d_1, d_2, k_1, k_2 \in \mathbb{R}$. The problem of the existence of a unique solution to (7.60), (7.61) reduces to the problem of the existence of a unique solution of an algebraic system with unknowns $d_1, d_2, k_1, k_2$. Indeed, if we use (7.62) in (7.61), we get:

$$\begin{cases} u_1(b)d_1 - v_1(b)k_1 - v_2(b)k_2 = 0, \\ d_1(-\varepsilon u_1'(b) + \alpha_1(b)u_1(b)) + k_1(\mu(b)v_1'(b) - \alpha_2(b)v_1(b)) \\ +k_2(\mu(b)v_2'(b) - \alpha_2(b)v_2(b)) = 0, \\ k_1 + \gamma_0(k_2) = -\widetilde{v}_\varepsilon(c) \end{cases} \tag{7.63}$$

(the condition $u(a) = 0$ implies $d_2 = 0$). Note that the first two equations of system (7.63) form a linear system with unknowns $d_1, k_1$, which has a unique solution for each $k_2$. To show this, it is enough to prove that the corresponding homogeneous system (which corresponds to $k_2 = 0$) has only the null solution. Indeed, if we take into account how these constants were introduced, it is necessary and sufficient to prove that functions $u = d_1 u_1, v = k_1 v_1$, where $d_1, k_1$ satisfy the homogeneous system, are identically zero. In fact, these functions satisfy (7.60) with homogeneous Dirichlet boundary conditions at $a$, $c$, and transmission conditions at $b$, as considered in (7.61). Therefore, as the solution of this problem is unique (see Theorem 7.1.1), we have $u = 0$, $v = 0$, i.e., $d_1 = k_1 = 0$. Now, we determine uniquely $d_1, k_1$ from $(7.63)_{1,2}$, as functions of $k_2$, and plug them into $(7.63)_3$. Thus the following nonlinear algebraic equation in $k_2$ is obtained:

$$\lambda k_2 + \gamma_0(k_2) = -\widetilde{v}_\varepsilon(c). \tag{7.64}$$

We have not computed the exact value of $\lambda$ since only its sign is relevant in the reasoning to follow. Indeed, by $(h_6)$, $\gamma_0$ is a continuous nondecreasing function, so (7.64) has a unique solution if $\lambda > 0$ or, equivalently, $\lambda k_2^2 = v'(c)k_2 > 0$ $\forall k_2 \neq 0$. Since $u = d_1 u_1$, $v = k_1 v_1 + k_2 v_2$, with $d_1, k_1$ solutions of equations $(7.63)_{1,2}$, satisfy

$$\begin{aligned} \varepsilon \int_a^b (u')^2 dx + \int_b^c \mu(v')^2 dx + \int_a^b \left(\frac{\alpha_1'}{2} + \beta_1\right) u^2 dx + \int_b^c \left(\frac{\alpha_2'}{2} + \beta_2\right) v^2 dx \\ + \frac{[\alpha(b)]}{2} v(b)^2 + \frac{\alpha_2(c)}{2} k_2^2 - \mu(c)v'(c)k_2 = 0, \end{aligned} \tag{7.65}$$

we infer that $\mu(c)v'(c)k_2 > 0$ $\forall k_2 \neq 0$.
Therefore, $(u_\varepsilon, v_\varepsilon) = (\widetilde{u}_\varepsilon, \widetilde{v}_\varepsilon) + (u, v)$ is the unique solution of problem $(P.3)_\varepsilon$. □

*Remark* 7.3.2. The above theorem remains valid under alternative assumptions. For example, this is the case if (7.59) does not hold, but one of the following two conditions holds:

$$\mu_0 + \frac{\Pi}{2}(c - b) > 0, \tag{7.66}$$

or

$$\begin{cases} \text{there exists a positive constant } \delta_0 \text{ such that,} \\ \mu_0 + \frac{\Pi}{2\delta_0} > 0, \ \frac{\Pi\delta_0}{2} + \frac{\alpha_2'}{2} + \beta_2 \geq 0 \text{ a.e. on } (b,c), \end{cases} \tag{7.67}$$

where

$$\Pi = \begin{cases} \alpha_2(c), \ \text{if } \alpha_2(c) < 0 \text{ and } [\alpha(b)] \geq 0, \\ [\alpha(b)], \ \text{if } \alpha_2(c) \geq 0 \text{ and } [\alpha(b)] < 0, \\ [\alpha(b)] + \alpha_2(c), \ \text{if } \alpha_2(c) < 0 \text{ and } [\alpha(b)] < 0. \end{cases}$$

Now, we are going to investigate problem $(P.3)_0$. Obviously, under our assumptions, $(7.57)_1$, with $X_0(a) = 0$, has a unique solution $X_0 \in H^1(a,b)$. Note also that $Y_0$ is a solution of $(7.57)_2$ with the boundary conditions

$$-(\mu Y_0')(b) + \alpha_2(b)Y_0(b) = \alpha_1(b)X_0(b), \ \ -Y_0'(c) = \gamma_0(Y_0(c)). \tag{7.68}$$

The next result is concerned with the existence and uniqueness of the solution of problem $(P.3)_0$:

**Theorem 7.3.3.** *Assume that* $(h_1)$*–*$(h_6)$ *are satisfied and either* $\alpha_2(b) \geq 0, \ \alpha_2(c) > 0$ *or* $\alpha_2(b) > 0, \ \alpha_2(c) \geq 0$. *Then, problem* $(P.3)_0$ *has a unique solution* $(X_0, Y_0) \in H^1(a,b) \times H^2(b,c)$.

*Proof.* Taking into account the above comments, in order to prove our theorem it is sufficient to show that problem $(7.57)_2$, $(7.68)$ has a unique solution $Y_0 \in H^2(b,c)$. Let $\tilde{y} \in H^2(b,c)$ be the unique solution of equation $(7.57)_2$ which satisfies the following boundary conditions

$$\alpha_1(b)X_0(b) = -\mu(b)\tilde{y}'(b) + \alpha_2(b)\tilde{y}(b), \ \tilde{y}(c) = 0$$

(such a solution exists and is unique on account of Theorem 7.1.4).
Since $y := Y_0 - \tilde{y}$ satisfies the homogeneous equation

$$-\left(\mu y' + \alpha_2 y\right)' + \beta_2 y = 0 \text{ on } (b,c), \tag{7.69}$$

together with the boundary conditions

$$(\mu y')(b) = \alpha_2(b)y(b), \ -\tilde{y}'(c) - y'(c) = \gamma_0(y(c)), \tag{7.70}$$

it is enough to show that $y$ is unique with this property. As in the proof of the previous theorem, let $\{y_1, y_2\} \subset H^2(b,c)$ be the fundamental system consisting of the solutions of the homogeneous equation (7.69) which satisfy $y_1(c) = 1, \ y_1'(c) = 0, \ y_2(c) = 0, \ y_2'(c) = 1$. Obviously, the general solution of equation (7.69) has the form $y = d_1y_1 + d_2y_2$, $d_1, d_2 \in \mathbb{R}$. Thus, the problem of finding $y$ reduces to that of finding constants $d_1, d_2$ which satisfy the boundary conditions (7.70). Hence, we arrive at the following algebraic system

$$\begin{cases} (\alpha_2(b)y_2(b) - \mu(b)y_2'(b))d_2 = (\mu(b)y_1'(b) - \alpha_2(b)y_1(b))d_1, \\ d_2 + \gamma_0(d_1) = -\tilde{y}'(c). \end{cases} \tag{7.71}$$

We need to show that this system has a unique solution. Let us first note that $\alpha_2(b)y_2(b)-\mu(b)y_2'(b)\neq 0$; otherwise $y:=y_2$ would be the solution of the equation (7.69) with the boundary conditions $\alpha_2(b)y(b)=\mu(b)y'(b),\ y(c)=0$. By Theorem 7.1.4, $y_2$ would be the null solution, which is impossible by the construction of $y_2$. Thus, one can calculate $d_2$ as a function of $d_1$ and plug it into $(7.71)_2$. We arrive at the the following nonlinear algebraic equation:

$$\lambda d_1+\gamma(d_1)=-\tilde{y}'(c), \tag{7.72}$$

which has a unique solution if $\lambda>0$. Let us show that $\lambda>0$. Take $y=d_1y_1+d_2y_2$, which satisfies $(7.71)_1$ and (7.69). By a simple computation, we get

$$\begin{aligned}\int_b^c \mu(y')^2dx+\int_b^c\left(\frac{\alpha_2'}{2}+\beta_2\right)y^2dx&\\ +\frac{\alpha_2(b)}{2}y(b)^2+\frac{\alpha_2(c)}{2}y(c)^2-\mu(c)y'(c)y(c)=0,&\end{aligned} \tag{7.73}$$

and hence

$$\mu(c)\lambda d_1^2=\mu(c)y'(c)y(c)\geq\frac{\alpha_2(c)}{2}d_1^2+\frac{\alpha_2(b)}{2}y(b)^2+\mu_0\parallel y'\parallel_2^2>0\ \forall d_1\neq 0.$$

Therefore, $\lambda>0$ and obviously $Y_0=y+\tilde{y}$ is the unique solution of problem $(7.57)_2$, (7.68). □

*Remark* 7.3.4. In fact, Theorem 7.3.3 remains true under alternative assumptions on the data.

### 7.3.3 Estimates for the remainder components

In this subsection we will establish some estimates for the two components of the zeroth order remainder of the asymptotic expansion (7.56).

**Theorem 7.3.5.** *Assume that $(h_1)$–$(h_6)$, (7.59) are fulfilled and $\alpha_1$ is a Lipschitz function on $[a,b]$, $\beta_1\in L^\infty(a,b)$. Then, for every $\varepsilon>0$, the solution of problem $(P.3)_\varepsilon$ admits an asymptotic expansion of the form (7.56) and the following estimates hold:*

$$\|R_{1\varepsilon}\|_{C[a,b]}=\mathcal{O}(\varepsilon^{1/2}),\ \|R_{2\varepsilon}\|_{C[b,c]}=\mathcal{O}(\varepsilon^{1/2}).$$

*Proof.* One can use arguments similar to those in the proof of Theorem 7.1.6, for $N=0$. Thus, we are not going into many details. We will use the same notation as in the proof of Theorem 7.1.6. Note that

$$\tilde{R}_{1\varepsilon}(x):=R_{1\varepsilon}(x)-(Ax+B)i_0(\xi(a)),\ A:=(b-a)^{-1},\ B:=-b(b-a)^{-1}$$

satisfy

$$\tilde{R}_{1\varepsilon}(a)=0,\ \tilde{R}_{1\varepsilon}(b)=R_{1\varepsilon}(b)=R_{2\varepsilon}(b),$$

and

$$\begin{cases} (-\varepsilon S_{1\varepsilon}' + \alpha_1(x)\tilde{R}_{1\varepsilon})' + \beta_1(x)\tilde{R}_{1\varepsilon} = H_\varepsilon(x) \text{ in } [a,b], \\ (-\mu R_{2\varepsilon}' + \alpha_2(x)R_{2\varepsilon})' + \beta_2(x)R_{2\varepsilon} = 0 \text{ in } [b,c], \\ \tilde{R}_{1\varepsilon}(a) = 0, \ \tilde{R}_{1\varepsilon}(b) = R_{2\varepsilon}(b), \\ -\varepsilon S_{1\varepsilon}'(b) + \alpha_1(b)\tilde{R}_{1\varepsilon}(b) - \varepsilon A i_0(\xi(a)) = \\ = -\mu(b)R_{2\varepsilon}'(b) + \alpha_2(b)R_{2\varepsilon}(b), \\ -R_{2\varepsilon}'(c) = \gamma_0(Y_0(c) + R_{2\varepsilon}(c)) - \gamma_0(Y_0(c)), \end{cases}$$

where $S_{1\varepsilon} = \tilde{R}_{1\varepsilon} + X_0$,

$$H_\varepsilon(x) := h_\varepsilon(x) - \Big[A\alpha_1(x) + (\beta_1(x) + \alpha_1'(x))(Ax+B)\Big] i_0(\xi(a)) \ \forall\, x \in [a,b].$$

Therefore, we have the following equation:

$$\begin{aligned} \varepsilon \int_a^b (\tilde{R}_{1\varepsilon}')^2 dx + & \int_b^c \mu (R_{2\varepsilon}')^2 dx + \int_a^b \left(\beta_1 + \frac{\alpha_1'}{2}\right) \tilde{R}_{1\varepsilon}^2 dx \\ & + \int_b^c \left(\beta_2 + \frac{\alpha_2'}{2}\right) R_{2\varepsilon}^2 dx + \frac{[\alpha(b)]}{2} R_{2\varepsilon}(b)^2 + \frac{\alpha_2(c)}{2} R_{2\varepsilon}(c)^2 \\ & + \mu(c)(\gamma_0(Y_0(c) + R_{2\varepsilon}(c)) - \gamma_0(Y_0(c)))R_{2\varepsilon}(c) \\ & = -\varepsilon \int_a^b X_0' \tilde{R}_{1\varepsilon}' dx + \int_a^b H_\varepsilon \tilde{R}_{1\varepsilon} dx - \varepsilon A i_0(\xi(a)) R_{2\varepsilon}(b). \end{aligned}$$

By assumption $(h_6)$ the term of the above equation which contains $\mu(c)$ is positive. The rest of the proof goes as in the proof of Theorem 7.1.6. □

*Remark* 7.3.6. The estimates of the above theorem remain valid under some alternative assumptions on the data.

# Chapter 8

# The Evolutionary Case

In the rectangle $Q_T = (a, c) \times (0, T)$ we consider the following partial differential system

$$\begin{cases} u_t - (\varepsilon u_x - \alpha_1(x)u)_x + \beta_1(x)u = f(x,t) \text{ in } Q_{1T}, \\ v_t - (\mu(x)v_x - \alpha_2(x)v)_x + \beta_2(x)v = g(x,t) \text{ in } Q_{2T}, \end{cases} \tag{S}$$

with which we associate initial conditions

$$u(x,0) = u_0(x),\ a \le x \le b,\quad v(x,0) = v_0(x),\ b \le x \le c, \tag{IC}$$

the following natural transmissions conditions at $b$:

$$\begin{cases} u(b,t) = v(b,t), \\ -\varepsilon u_x(b,t) + \alpha_1(b)u(b,t) = -\mu(b)v_x(b,t) + \alpha_2(b)v(b,t),\quad 0 \le t \le T, \end{cases} \tag{TC}$$

as well as one of the following boundary conditions:

$$u(a,t) = v(c,t) = 0,\quad 0 \le t \le T; \tag{BC.1}$$

$$u_x(a,t) = v(c,t) = 0,\quad 0 \le t \le T; \tag{BC.2}$$

$$u(a,t) = 0,\quad -v_x(c,t) = \gamma(v(c,t)),\quad 0 \le t \le T, \tag{BC.3}$$

where $Q_{1T} = (a,b) \times (0,T)$, $Q_{2T} = (b,c) \times (0,T)$, $a,\ b,\ c \in \mathbb{R}$, $a < b < c$, $T > 0$, and $\varepsilon$ is a small parameter, $0 < \varepsilon \ll 1$.

The following general assumptions will be required:

($i_1$) $\alpha_1 \in H^1(a,b)$, $\beta_1 \in L^2(a,b)$, $(1/2)\alpha_1' + \beta_1 \ge C_1$ a.e. on $(a,b)$, for some constant $C_1$;

($i_2$) $\alpha_2 \in H^1(b,c)$, $\beta_2 \in L^2(b,c)$, $(1/2)\alpha_2' + \beta_2 \ge C_2$ a.e. on $(b,c)$, for some constant $C_2$;

($i_3$) $\mu \in H^1(b,c)$, $\mu(x) \ge \mu_0 > 0$;

$(i_4)$ $f : \overline{Q}_{1T} \to \mathbb{R}$, $g : \overline{Q}_{2T} \to \mathbb{R}$, $u_0 : [a,b] \to \mathbb{R}$, $v_0 : [b,c] \to \mathbb{R}$;
$(i_5)$ $\alpha_1 > 0$ on $[a,b]$; or
$(i_5')$ $\alpha_1 < 0$ on $[a,b]$;
$(i_6)$ $\gamma : D(\gamma) = \mathbb{R} \to \mathbb{R}$ is a continuous nondecreasing function.

Denote by $(P.k)_\varepsilon$ the problem which consists of $(S)$, $(IC)$, $(TC)$, $(BC.k)$, for $k = 1,2,3$. Note that the problems studied in the previous chapter are stationary versions of these ones.

This chapter contains three sections. Each section addresses one of the three problems $(P.k)_\varepsilon$, $k = 1,2,3$.

In Section 1, problem $(P.1)_\varepsilon$ is investigated, under hypotheses $(i_1)$–$(i_5)$. As in the stationary case, this problem is singularly perturbed with respect to the uniform norm, with a boundary layer located on the left side of the line segment $\{(b,t);\ t \in [0,T]\}$. A first order asymptotic expansion of the solution will be constructed. To validate this expansion, we will prove existence and regularity results for the perturbed problem as well as for the problems satisfied by the two terms of the regular series which are present in our expansion. In addition, we will obtain estimates for the remainder components, with respect to the uniform convergence norm. Of course, smoothness and compatibility conditions should be imposed to the data.

In the second section, we investigate problem $(P.2)_\varepsilon$, under the same requirements, except for $(i_5)$ which is replaced by $(i_5')$. For the solution of this problem, we construct an asymptotic expansion of order zero with respect to the uniform norm. Note that $(P.2)_\varepsilon$ is regularly perturbed of order zero with respect to this norm, so there is no correction in the asymptotic expansion. But the problem is singularly perturbed of order one, with respect to the same norm. Indeed, in the first order asymptotic expansion we will construct, a first order boundary layer function will be present, corresponding to a boundary layer located near the right side of the line segment $\{(a,t);\ t \in [0,T]\}$. Again, we perform a detailed analysis to validate both these asymptotic expansions.

In the third and final section we deal with problem $(P.3)_\varepsilon$, which is nonlinear due to the nonlinear boundary condition at $x = c$. We restrict ourselves to the construction of a zeroth order asymptotic expansion, under assumptions $(i_1)$–$(i_6)$. As expected, $(P.3)_\varepsilon$ is singularly perturbed with respect to the uniform norm. We again perform a detailed asymptotic analysis, which is a bit more difficult due to the nonlinear character of the problem. As usual, we conclude our analysis with some estimates for the remainder components.

Throughout this chapter we will denote by $\|\cdot\|_1$ and $\|\cdot\|_2$ the norms of $L^2(a,b)$ and $L^2(b,c)$, respectively.

## 8.1 A first order asymptotic expansion for the solution of problem $(P.1)_\varepsilon$

In this section we examine problem $(P.1)_\varepsilon$ formulated above, for which assumptions $(i_1)$–$(i_5)$ are required. Using a similar justification as in the corresponding stationary case, one can show that $(P.1)_\varepsilon$ is singularly perturbed with respect to the uniform norm, with a boundary layer in the vicinity of the line segment $\Sigma = \{(b,t);\ t \in [0,T]\}$.

### 8.1.1 Formal expansion

Let us denote by $U_\varepsilon := \big(u_\varepsilon(x,t), v_\varepsilon(x,t)\big)$ the solution of problem $(P.1)\varepsilon$. In the following we derive a first order asymptotic expansion of this solution by using the classical perturbation theory presented in Chapter 1 and taking into account our comments on the stationary case (see Subsection 7.1.1). Thus, we seek the solution of our problem in the form

$$\begin{cases} u_\varepsilon(x,t) = X_0(x,t) + \varepsilon X_1(x,t) + i_0(\xi,t) + \varepsilon i_1(\xi,t) + R_{1\varepsilon}(x,t), \\ v_\varepsilon(x,t) = Y_0(x,t) + \varepsilon Y_1(x,t) + R_{2\varepsilon}(x,t), \end{cases} \tag{8.1}$$

where:

$\xi := \varepsilon^{-1}(b-x)$ is the fast variable associated with the left side of $\Sigma$;

$\big(X_k(x,t), Y_k(x,t)\big),\ k=0,1$, are the first two regular terms;

$i_k(\xi,t),\ k=0,1$, are the transition layer corrections;

$\big(R_{1\varepsilon}(x,t), R_{2\varepsilon}(x,t)\big)$ is the first order remainder.

As usual, we replace (8.1) into $(S)$ and get

$$\begin{cases} X_{kt} + (\alpha_1 X_k)_x + \beta_1 X_k = \widetilde{f}_k \ \text{ in } \ Q_{1T}, \\ Y_{kt} + (-\mu Y_{kx} + \alpha_2 Y_k)_x + \beta_2 Y_k = \widetilde{g}_k \ \text{ in } \ Q_{2T},\ k=0,1, \end{cases} \tag{8.2}$$

where

$$\widetilde{f}_k(x,t) = \begin{cases} f(x,t), & k=0, \\ X_{0xx}(x,t), & k=1, \end{cases}$$

$$\widetilde{g}_k(x,t) = \begin{cases} g(x,t), & k=0, \\ 0, & k=1. \end{cases}$$

The transitions layer functions $i_0, i_1$ satisfy the equations

$$\begin{cases} i_{0\xi\xi}(\xi,t) + \alpha_1(b) i_{0\xi}(\xi,t) = 0, \\ i_{1\xi\xi}(\xi,t) + \alpha_1(b) i_{1\xi}(\xi,t) = I_1(\xi,t), \quad t \in [0,T],\ \xi \geq 0, \end{cases}$$

where $I_1(\xi,t) = i_{0t}(\xi,t) + \xi\alpha_1'(b)i_{0\xi}(\xi,t) + (\alpha_1'(b) + \beta_1(b))i_0(\xi,t)$. By easy computations, we derive

$$\begin{cases} i_0(\xi,t) = \theta_0(t)e^{-\alpha_1(b)\xi}, \\ i_1(\xi,t) = \theta_1(t)e^{-\alpha_1(b)\xi} \\ +\frac{\xi e^{-\alpha_1(b)\xi}}{\alpha_1(b)}\left(-\theta_0'(t) - \beta_1(b)\theta_0(t) + \frac{\alpha_1(b)\alpha_1'(b)}{2}\theta_0(t)\xi\right), \end{cases} \tag{8.3}$$

where $\theta_0,\ \theta_1$ are as yet undetermined functions.

For the components of the remainder, $R_{1\varepsilon},\ R_{2\varepsilon}$, we derive the system

$$\begin{cases} R_{1\varepsilon t} + (-\varepsilon T_{1\varepsilon x} + \alpha_1 R_{1\varepsilon})_x + \beta_1 R_{1\varepsilon} = h_\varepsilon \ \text{ in } Q_{1T}, \\ R_{2\varepsilon t} + \big(-\mu R_{2\varepsilon x} + \alpha_2 R_{2\varepsilon}\big)_x + \beta_2 R_{2\varepsilon} = 0 \ \text{ in } Q_{2T}, \end{cases} \tag{8.4}$$

where

$$\begin{aligned} T_{1\varepsilon} =& u_\varepsilon - X_0 - i_0 - \varepsilon i_1 = \varepsilon X_1 + R_{1\varepsilon}, \\ h_\varepsilon(x,t) =& -\varepsilon i_{1t}(\xi,t) + [\varepsilon^{-1}(\alpha_1(x) - \alpha_1(b)) + \xi\alpha_1'(b)]i_{0\xi}(\xi,t) \\ &+ (\alpha_1(x) - \alpha_1(b))i_{1\xi}(\xi,t) - (\alpha_1'(x) - \alpha_1'(b))i_0(\xi,t) \\ &- (\beta_1(x) - \beta_1(b))i_0(\xi,t) - \varepsilon(\beta_1(x) + \alpha_1'(x))i_1(\xi,t) \ \text{ in } Q_{1T}. \end{aligned} \tag{8.5}$$

Next, from $(IC)$ and $(BC.1)$ one gets

$$X_k(x,0) = \begin{cases} u_0(x), \ k=0, \\ 0, \ k=1, \end{cases} \quad Y_k(x,0) = \begin{cases} v_0(x), \ k=0, \\ 0, \ k=1, \end{cases} \tag{8.6}$$

$$i_0(\xi,0) + \varepsilon i_1(\xi,0) = 0 \ \forall \varepsilon > 0 \Longleftrightarrow \theta_0(0) = \theta_0'(0) = \theta_1(0) = 0, \tag{8.7}$$

$$\begin{cases} R_{1\varepsilon}(x,0) = 0, \ \ a \le x \le b, \\ R_{2\varepsilon}(x,0) = 0, \ \ b \le x \le c, \end{cases} \tag{8.8}$$

$$X_k(a,t) = Y_k(c,t) = 0, \ \ 0 \le t \le T, \ k = 0,1, \tag{8.9}$$

$$\begin{cases} R_{1\varepsilon}(a,t) = P_\varepsilon(\xi(a),t), \\ R_{2\varepsilon}(c,t) = 0, \ \ 0 \le t \le T, \end{cases} \tag{8.10}$$

where $\xi(a) = \varepsilon^{-1}(b-a),\ P_\varepsilon(\zeta,t) = -i_0(\zeta,t) - \varepsilon i_1(\zeta,t)$.

Finally, on account of $(TC)$ we find

$$\theta_k(t) = Y_k(b,t) - X_k(b,t), \ \ 0 \le t \le T, \ k = 0,1, \tag{8.11}$$

$$-\mu(b)Y_{0x}(b,t) = \alpha_1(b)X_0(b,t) - \alpha_2(b)Y_0(b,t), \ \ 0 \le t \le T, \tag{8.12}$$

$$\begin{aligned} -\mu(b)Y_{1x}(b,t) + \alpha_2(b)Y_1(b,t) =& \alpha_1(b)X_1(b,t) \\ &- X_{0x}(b,t) - \alpha_1(b)^{-1}(\theta_0'(t) + \beta_1(b)\theta_0(t)), \ \ 0 \le t \le T, \end{aligned} \tag{8.13}$$

$$\begin{cases} R_{1\varepsilon}(b,t) = R_{2\varepsilon}(b,t), \\ -\varepsilon T_{1\varepsilon x}(b,t) + \alpha_1(b)R_{1\varepsilon}(b,t) \\ = -\mu(b)R_{2\varepsilon x}(b,t) + \alpha_2(b)R_{2\varepsilon}(b,t), \ 0 \le t \le T. \end{cases} \tag{8.14}$$

In conclusion, from what we have done so far we see that the components of the zeroth order regular term satisfy the reduced problem, say $(P.1)_0$, which consists of $(8.2)_{k=0}$, $(8.6)_{k=0}$, $(8.9)_{k=0}$ and (8.12), while the components of the first order regular term satisfy the problem $(P.1)_1$ which comprises $(8.2)_{k=1}$, $(8.6)_{k=1}$, $(8.9)_{k=1}$ and (8.13). The remainder components satisfy the problem (8.4), (8.8), (8.10) and (8.14).

Note that equations (8.7) will appear again in the next section as smoothness and compatibility conditions. It is worth pointing out that these conditions also guarantee that our corrections $i_0$, $i_1$ do not introduce discrepancies at point $(x,t) = (b,0)$.

We finally remark that $(P.1)_\varepsilon$ is a coupled parabolic-parabolic problem, while $(P.1)_k$, $k = 0,1$, are coupled hyperbolic-parabolic problems.

### 8.1.2 Existence, uniqueness and regularity of the solutions of problems $(P.1)_\varepsilon$, $(P.1)_0$ and $(P.1)_1$

In order to investigate problem $(P.1)_\varepsilon$, we choose as a basic setup the Hilbert space $H := L^2(a,b) \times L^2(b,c)$, endowed with the usual scalar product, denoted $\langle \cdot, \cdot \rangle$, and the corresponding induced norm, denoted $\|\cdot\|$. This problem can be expressed as the Cauchy problem in $H$:

$$\begin{cases} W_\varepsilon'(t) + J_\varepsilon W_\varepsilon(t) = F(t), \ \ 0 < t < T, \\ W_\varepsilon(0) = W_0, \end{cases} \tag{8.15}$$

where $J_\varepsilon : D(J_\varepsilon) \subset H \to H$,

$$\begin{aligned} D(J_\varepsilon) :=& \Big\{(h,k) \in H^2(a,b) \times H^2(b,c), \ h(b) = k(b), \\ h(a) =& k(c) = 0, \ \varepsilon h'(b) - \alpha_1(b)h(b) = \mu(b)k'(b) - \alpha_2(b)k(b)\Big\}, \\ J_\varepsilon(h,k) :=& \big((-\varepsilon h' + \alpha_1 h)' + \beta_1 h, \ (-\mu k' + \alpha_2 k)' + \beta_2 k\big), \\ W_\varepsilon(t) :=& \big(u_\varepsilon(\cdot,t), v_\varepsilon(\cdot,t)\big), \ W_0 := \big(u_0, v_0\big), \ F(t) := \big(f(\cdot,t), g(\cdot,t)\big). \end{aligned}$$

Regarding operator $J_\varepsilon$, one can prove that

**Lemma 8.1.1.** *Assume that $(i_1)$–$(i_3)$ and $(i_5)$ are satisfied. Then, there is a positive number $\omega$, independent of $\varepsilon$, such that $J_\varepsilon + \omega I$ is maximal monotone, where $I$ is the identity of $H$.*

*Proof.* Obviously, $J_\varepsilon$ is well defined and linear. To prove the monotonicity of $J_\varepsilon + \omega I$ for a suitable $\omega$, we can see that for $(h,k) \in D(J_\varepsilon)$

$$\begin{aligned}
&\langle J_\varepsilon\big((h,k)\big) + \omega(h,k), (h,k)\rangle \\
&\quad = -\varepsilon \int_a^b h'' h dx + \int_a^b (\alpha_1 h)' h dx + \int_a^b (\omega + \beta_1) h^2 dx \\
&\quad\quad - \int_b^c (\mu k')' k dx + \int_b^c (\alpha_2 k)' k dx + \int_b^c (\omega + \beta_2) k^2 dx \\
&\quad = \varepsilon \int_a^b (h')^2 dx + \int_b^c \mu (k')^2 dx + \frac{[\alpha(b)]}{2} k(b)^2 \\
&\quad\quad + \int_a^b \left(\omega + \beta_1 + \frac{\alpha_1'}{2}\right) h^2 dx + \int_b^c \left(\omega + \beta_2 + \frac{\alpha_2'}{2}\right) k^2 dx,
\end{aligned} \tag{8.16}$$

where $[\alpha(b)] := \alpha_2(b) - \alpha_1(b)$. If $[\alpha(b)] \geq 0$, there is an $\omega > 0$ big enough which makes $J_\varepsilon + \omega I$ monotone. If $[\alpha(b)] < 0$, then we can use Lemma 7.1.2 of Chapter 7 to prove the existence of such a number $\omega$. Indeed, we have

$$\frac{[\alpha(b)]}{2} k(b)^2 \geq \frac{[\alpha(b)]}{2} \left(\delta \int_b^c k^2 dx + \frac{1}{\delta} \int_b^c (k')^2 dx\right) \ \forall\, \delta > 0.$$

Taking $\delta = -[\alpha(b)]/\mu_0$ in this inequality and using (8.16) we obtain

$$\begin{aligned}
\langle J_\varepsilon\big((h,k)\big) + \omega(h,k), (h,k)\rangle &\geq \varepsilon \int_a^b (h')^2 dx + \frac{\mu_0}{2} \int_b^c (k')^2 dx \\
&+ \int_a^b \left(\omega + \beta_1 + \frac{\alpha_1'}{2}\right) h^2 dx + \int_b^c \left(\omega + \beta_2 + \frac{\alpha_2'}{2} - \frac{[\alpha(b)]^2}{2\mu_0}\right) k^2 dx \geq 0,
\end{aligned}$$

for $\omega$ big enough. Now, we are going to show that operator $J_\varepsilon + \omega I$ is maximal monotone or, equivalently (see Theorem 2.0.6), for all $(f_1, f_2) \in H$, there exists $(h,k) \in D(J_\varepsilon)$, such that $(h,k) + (J_\varepsilon + \omega I)(h,k) = (f_1, f_2)$, that is, the following problem

$$\begin{cases}
(-\varepsilon h' + \alpha_1 h)' + (\beta_1 + \omega + 1) h = f_1 \ \text{ in } L^2(a,b), \\
(-\mu k' + \alpha_2 k)' + (\beta_2 + \omega + 1) k = f_2 \ \text{ in } L^2(b,c), \\
h(a) = k(c) = 0, \ h(b) = k(b), \\
-\varepsilon h'(b) + \alpha_1(b) h(b) = -\mu(b) k'(b) + \alpha_2(b) k(b)
\end{cases}$$

has a solution $(h,k) \in H^2(a,b) \times H^2(b,c)$. This fact follows as in the proof of Theorem 7.1.1, if $\omega$ is sufficiently large. □

As far as problem $(P.1)_\varepsilon$ is concerned, we have the following result

**Theorem 8.1.2.** *Assume that* $(i_1)$–$(i_5)$ *are fulfilled and*

$$F \in W^{1,1}(0,T;H), \quad W_0 \in D(J_\varepsilon).$$

*Then problem* (8.15) *has a unique strong solution* $W_\varepsilon$ *which belongs to*

$$C^1([0,T];H)\bigcap W^{1,2}(0,T;H^1(a,b)\times H^1(b,c))\bigcap C([0,T];H^2(a,b)\times H^2(b,c)).$$

*If, moreover,*

$$F\in W^{2,1}(0,T;H); \tag{8.17}$$

$$W_0\in D(J_\varepsilon),\ \ F(0)-J_\varepsilon W_0\in D(J_\varepsilon), \tag{8.18}$$

*then* $W_\varepsilon$ *belongs to*

$$C^2([0,T];H)\bigcap W^{2,2}(0,T;H^1(a,b)\times H^1(b,c))\bigcap C^1([0,T];H^2(a,b)\times H^2(b,c)).$$

*Proof.* By Lemma 8.1.1, $J_\varepsilon+\omega I$ is maximal monotone for some $\omega>0$. Then, $-(J_\varepsilon+\omega I)$ generates a linear $C_0$-semigroup of contractions on $H$, say $\{S(t);\ t\geq 0\}$, i.e., $-J_\varepsilon$ generates the $C_0$-semigroup $\{S_\omega(t)=e^{\omega t}S(t);\ t\geq 0\}$. Therefore, according to Theorem 2.0.27 in Chapter 2, problem (8.15) has a unique strong solution $W_\varepsilon\in C^1([0,T];H)$. Since $J_\varepsilon W_\varepsilon\in C([0,T];H)$, one can easily see that $W_\varepsilon\in C([0,T];H^2(a,b)\times H^2(b,c))$.

Let us now prove that $W_\varepsilon\in W^{1,2}(0,T;H^1(a,b)\times H^1(b,c))$. By a standard computation, we find

$$\begin{aligned}\frac{1}{2}\frac{d}{dt}\parallel W_\varepsilon(t+h)-W_\varepsilon(t)\parallel^2 &+\varepsilon\int_a^b(u_{\varepsilon x}(x,t+h)-u_{\varepsilon x}(x,t))^2dx\\ &+\frac{\mu_0}{2}\int_b^c(v_{\varepsilon x}(x,t+h)-v_{\varepsilon x}(x,t))^2dx\\ &\leq\omega\parallel W_\varepsilon(t+h)-W_\varepsilon(t)\parallel^2\\ &\quad+\parallel W_\varepsilon(t+h)-W_\varepsilon(t)\parallel\cdot\parallel F(t+h)-F(t)\parallel,\end{aligned}$$

for $0\leq t<t+h\leq T$. By integration over $[0,T-h]$ this implies

$$\begin{aligned}\varepsilon\int_0^{T-h}&\parallel u_{\varepsilon x}(\cdot,t+h)-u_{\varepsilon x}(\cdot,t)\parallel_1^2dt\\ &+\frac{\mu_0}{2}\int_0^{T-h}\parallel v_{\varepsilon x}(\cdot,t+h)-v_x(\cdot,t)\parallel_2^2dt\leq Ch^2,\end{aligned}$$

for some $C>0$. We have used our condition $F\in W^{1,1}(0,T;H)$ as well as the fact that $W_\varepsilon\in C^1([0,T];H)$. By virtue of Theorem 2.0.3 this last inequality gives

$$u_{\varepsilon x}\in W^{1,2}(0,T;L^2(a,b)),\ \ v_{\varepsilon x}\in W^{1,2}(0,T;L^2(b,c)),$$

so the first part of theorem is proved.

In what follows we suppose that (8.17) and (8.18) hold. Obviously, $\overline{W}_\varepsilon=W_\varepsilon'$ is the strong solution of the following Cauchy problem in $H$

$$\begin{cases}\overline{W}_\varepsilon'(t)+J_\varepsilon\overline{W}_\varepsilon(t)=F'(t),\ \ 0<t<T,\\ \overline{W}_\varepsilon(0)=F(0)-W_0.\end{cases}$$

Therefore, according to the first part of the proof, $W_\varepsilon'$ belongs to

$$C^1([0,T];H)\bigcap W^{1,2}(0,T;H^1(a,b)\times H^1(b,c))\bigcap C([0,T];H^2(a,b)\times H^2(b,c)).$$

□

*Remark* 8.1.3. It is important to note that our asymptotic analysis works if Theorem 8.1.2 is valid for all $\varepsilon > 0$. Fortunately, assumptions (8.17) and (8.18) hold if the following sufficient assumptions (independent of $\varepsilon$) are fulfilled:

$$\begin{cases} f\in W^{2,1}(0,T;L^2(a,b)),\ g\in W^{2,1}(0,T;L^2(b,c)),\\ f(\cdot,0)\in H^2(a,b),\ g(\cdot,0)\in H^2(b,c),\ \alpha_1\in H^3(a,b),\ \alpha_2\in H^3(b,c)\\ \beta_1\in H^2(a,b),\ \beta_2\in H^2(b,c),\ \mu\in H^3(b,c),\ u_0\in H^4(a,b),\ v_0\in H^4(b,c),\end{cases}$$

$$\begin{cases} u_0(b)=v_0(b),\ u_0'(b)=0,\ -\mu(b)v_0'(b)=\alpha_1(b)u_0(b)-\alpha_2(b)v_0(b),\\ u_0(a)=v_0(c)=0,\ u_0''(a)=0,\ f(a,0)=\alpha_1(a)u_0'(a),\\ g(c,0)+(\mu v_0')'(c)=\alpha_2(c)v_0'(c),\ u_0''(b)=0,\\ f(b,0)-(\alpha_1u_0)'(b)-(\beta_1u_0)(b)\\ =g(b,0)+(\mu v_0')'(b)-(\alpha_2v_0)'(b)-(\beta_2v_0)(b),\\ u_0^{(3)}(b)=0,\ f_x(b,0)=(\alpha_1u_0)''(b)+(\beta_1u_0)'(b),\\ -\mu(b)[g_x(b,0)+(\mu v_0')''(b)-(\alpha_2v_0)''-(\beta_2v_0)'(b)]\\ =(\alpha_1(b)-\alpha_2(b))[f(b,0)-(\alpha_1u_0)'(b)-(\beta_1u_0)(b)].\end{cases} \tag{8.19}$$

We continue with problems $(P.1)_0$ and $(P.1)_1$. Our aim is to obtain existence, uniqueness and sufficient regularity for the solutions of these problems, which will allow us to validate our asymptotic expansion and, even more, to obtain estimates for the remainder components.

We start with $(P.1)_1$ for which we need a solution $(X_1,Y_1)$ satisfying

$$X_1\in W^{1,2}(0,T;H^1(a,b)),\ Y_1\in W^{1,2}(0,T;H^1(b,c)).$$

In order to homogenize the boundary conditions at $b$, we set

$$\overline{X}_1(x,t)=X_1(x,t)+B(t)x+B_1(t),\ (x,t)\in\overline{Q}_{1T},$$

where

$$\begin{cases} B(t)=\rho(t)/[\alpha_1(b)(a-b)],\ B_1(t)=-aB(t),\\ \rho(t)=X_{0x}(b,t)+\alpha_1(b)^{-1}(\theta_0'(t)+\beta_1(b)\theta_0(t)),\ 0\le t\le T.\end{cases}$$

A straightforward computation shows that $(\overline{X}_1,Y_1)$ satisfies the problem

$$\begin{cases} \overline{X}_{1t}+(\alpha_1\overline{X}_1)_x+\beta_1\overline{X}_1=\overline{f}_1 \text{ in } Q_{1T},\\ Y_{1t}-(\mu Y_{1x}-\alpha_2Y_1)_x+\beta_2Y_1=\widetilde{g}_1 \text{ in } Q_{2T},\\ \overline{X}_1(x,0)=\sigma(x),\ x\in[a,b],\ Y(x,0)=0,\ x\in[b,c],\\ \overline{X}_1(a,t)=0,\ Y_1(c,t)=0,\\ -\mu(b)Y_{1x}(b,t)+\alpha_2(b)Y_1(b,t)=\alpha_1(b)\overline{X}_1(b,t),\ 0<t<T,\end{cases} \tag{8.20}$$

where

$$\begin{aligned}\sigma(x) =& B(0)(x-a),\\ \overline{f}_1(x,t) =& \widetilde{f}_1(x,t) + B'(t)x + B_1'(t) + B(t)\alpha_1(x)\\ &+ (\alpha_1'(x) + \beta_1(x))(B(t)x + B_1(t)).\end{aligned}$$

Now, we associate with this problem the following Cauchy problem in $H$

$$\begin{cases} Z_1'(t) + A_1 Z_1(t) = F_1(t), \quad 0 < t < T, \\ Z_1(0) = z_1, \end{cases} \tag{8.21}$$

where $A_1 : D(A_1) \subset H \to H$,

$$\begin{aligned} D(A_1) :=& \{(p,q) \in H^1(a,b) \times H^2(b,c);\ p(a) = q(c) = 0,\\ &- \alpha_1(b)p(b) + \alpha_2(b)q(b) = \mu(b)q'(b)\},\\ A_1(p,q) :=& \left((\alpha_1 p)' + \beta_1 p, (-\mu q' + \alpha_2 q)' + \beta_2 q\right),\\ Z_1(t) :=& (\overline{X}_1(\cdot,t), Y_1(\cdot,t)),\ F_1(t) := (\overline{f}_1(\cdot,t), \widetilde{g}_1(\cdot,t)),\ z_1(x) := (\sigma(x), 0). \end{aligned}$$

Concerning operator $A_1$ we have the following result:

**Lemma 8.1.4.** *Assume that $(i_1)$–$(i_3)$ and $(i_5)$ are satisfied. Then, operator $A_1 + \omega I$ is maximal monotone in $H$ for $\omega > 0$ sufficiently large, where $I$ is the identity of $H$.*

*Proof.* For all $(p,q) \in D(A_1)$, we have

$$\begin{aligned}\langle A_1(p,q) + \omega(p,q), (p,q)\rangle \geq& \mu_0 \int_b^c (q')^2 dx + \int_a^b \left(\omega + \beta_1 + \frac{\alpha_1'}{2}\right) p^2 dx\\ &+ \int_b^c \left(\omega + \beta_2 + \frac{\alpha_2'}{2}\right) q^2 dx + \frac{[\alpha(b)]}{2} q(b)^2 + \frac{\alpha_1(b)}{2}(p(b) - q(b))^2.\end{aligned}$$

By Lemma 7.1.2, $A_1 + \omega I$ is monotone for $\omega > 0$ large enough.
Now, let us prove that $A_1 + \omega I$ is maximal monotone, i.e., for every $(f_1, f_2) \in H$, there exists a pair $(p,q) \in D(A_1)$, such that $(p,q) + (A_1 + \omega)((p,q)) = (f_1, f_2)$, that is:

$$\begin{cases} (\alpha_1 p)' + (\beta_1 + \omega + 1)p = f_1 \text{ in } L^2(a,b), \\ (-\mu q' + \alpha_2 q)' + (\beta_2 + \omega + 1)q = f_2 \text{ in } L^2(b,c), \\ p(a) = q(c) = 0,\ p(b) = q(b),\ \alpha_1(b)p(b) = -\mu(b)q'(b) + \alpha_2(b)q(b). \end{cases} \tag{8.22}$$

First, we note that there exists a unique $p \in H^1(a,b)$ which satisfies $(8.22)_1$ and condition $p(a) = 0$. On the other hand, on account of Theorem 7.1.4 and Remark 7.1.5 in Chapter 7, we derive the existence of $q \in H^2(b,c)$ which satisfies $(8.22)_2$ and the above boundary conditions at $b$ and $c$. □

The next result is related to problem $P_1$.

**Theorem 8.1.5.** *Assume that* $(i_1)$*–*$(i_5)$ *are fulfilled and, in addition,*

$$F_1 \in W^{1,1}(0,T;H), \ z_1 \in D(A_1). \tag{8.23}$$

*Then problem* (8.21) *has a unique strong solution* $Z_1 \in C^1([0,T];H)$*,*

$$\overline{X}_1 \in C([0,T];H^1(a,b)), \ Y_1 \in W^{1,2}(0,T;H^1(b,c)) \bigcap C([0,T];H^2(b,c)).$$

*Proof.* By Lemma 8.1.4, $A_1$ generates a $C_0$-semigroup on $H$. Then, taking into account (8.23), we derive from Theorem 2.0.27 that problem (8.21) admits a unique strong solution $Z_1 \in C^1([0,T];H)$. Therefore, using equations $(8.20)_{1,2}$, we get

$$\overline{X}_1 \in C([0,T];H^1(a,b)), \ Y_1 \in C([0,T];H^2(b,c)).$$

It remains to show that $Y_1 \in W^{1,2}(0,T;H^1(b,c))$. From the inequality

$$\begin{aligned}
\frac{\mu_0}{2} &\parallel Y_{1x}(\cdot,t+h) - Y_{1x}(\cdot,t) \parallel_2^2 \\
&\le \langle (A_1 + \omega I)(Z_1(t+h) - Z_1(t)), Z_1(t+h) - Z_1(t)\rangle \\
&= \frac{1}{2}\frac{d}{dt} \parallel Z_1(t+h) - Z_1(t) \parallel^2 \\
&\quad + \langle (F_1 + \omega Z_1)(t+h) - (F_1 + \omega Z_1)(t), Z_1(t+h) - Z_1(t)\rangle,
\end{aligned}$$

for $0 \le t < t+h \le T$, we find by integration over $[0, T-h]$

$$\int_0^{T-h} \parallel Y_{1x}(\cdot,t+h) - Y_{1x}(\cdot,t) \parallel_2^2 \le \text{const. } h^2$$

(here, we have used condition $F_1 \in W^{1,1}(0,T;H)$ as well as the fact that $Z_1 \in C^1([0,T];H)$). Therefore, according to Theorem 2.0.3 of Chapter 2,

$$Y_{1x} \in W^{1,2}(0,T;L^2(b,c)),$$

which concludes the proof. □

*Remark* 8.1.6. Assumptions (8.23) are implied by the following sufficient conditions

$$\begin{aligned}
X_{0xx} &\in W^{1,1}(0,T;L^2(a,b)), \ \rho \in W^{2,1}(0,T), \\
\alpha_1(b)u_0'(b) &= -\theta_0'(0) - \beta_1(b)\theta_0(0).
\end{aligned} \tag{8.24}$$

Thus, our next goal is to show that $\rho \in W^{2,1}(0,T)$, therefore it is enough to obtain that $\theta_0 \in W^{3,1}(0,T)$.

As we said before, we want to have $X_1 \in W^{1,2}(0,T;H^1(a,b))$. To achieve this, we consider the separate problem for $X_1$

$$\begin{cases} X_{1t} + (\alpha_1 X_1)_x + \beta_1 X_1 = \widetilde{f}_1 \text{ in } Q_{1T}, \\ X_1(x,0) = 0, \ a < x < b, \\ X_1(a,t) = 0, \ 0 < t < T. \end{cases} \tag{8.25}$$

In order to study this problem, we consider the Hilbert space $H_1 := L^2(a,b)$, which is equipped with the usual inner product denoted $\langle \cdot, \cdot \rangle_1$ and the induced norm, denoted $\| \cdot \|_1$. Define the operator $A_2 : D(A_2) \subset H_1 \to H_1$,

$$D(A_2) := \Big\{ l \in H^1(a,b);\ l(a) = 0 \Big\},\ A_2 l := (\alpha_1 l)' + \beta_1 l \ \ \forall\, l \in D(A_2).$$

It is quite easy to show that under hypothesis $(i_1)$ and $(i_5)$ operator $A_2 + \omega I$ is (linear) maximal monotone, for some $\omega > 0$, where $I$ stands for the identity operator of $H_1$. Clearly, problem (8.25) can be written as the following Cauchy problems in $H_1$

$$\begin{cases} \mathcal{X}_1'(t) + A_2 \mathcal{X}_1(t) = \mathcal{F}_1(t), \quad 0 < t < T, \\ \mathcal{X}_1(0) = 0, \end{cases} \tag{8.26}$$

where $\mathcal{X}_1(t) := X_1(\cdot, t),\ \mathcal{F}_1(t) := \widetilde{f}_1(\cdot, t),\ 0 < t < T$.
We have

**Theorem 8.1.7.** *Assume that* $(i_1), (i_5)$ *are fulfilled and, in addition,*

$$\mathcal{F}_1 \in W^{2,1}(0,T;H_1); \tag{8.27}$$

$$\mathcal{F}_1(0) \in D(A_2). \tag{8.28}$$

*Then, problem* (8.26) *has a unique strong solution*

$$\mathcal{X}_1 \in C^2([0,T];H_1) \bigcap C^1([0,T];H^1(a,b)).$$

*Proof.* Since $A_2 + \omega I$ is linear maximal monotone, it follows that $A_2$ generates a $C_0$-semigroup on $H_1$, say $\{S(t);\ t \geq 0\}$. By Theorem 2.0.27, problem (8.26) admits a unique strong solution $\mathcal{X}_1$,

$$\mathcal{X}_1(t) = \int_0^t S(t-s)\mathcal{F}_1(s)ds.$$

In fact, under assumptions (8.27), (8.28), $\mathcal{X}_1 \in C^2([0,T];H_1)$, and

$$A_2 \mathcal{X}_1,\ A_2 \mathcal{X}_1' \in C([0,T];H_1).$$

This concludes the proof. □

*Remark* 8.1.8. It may readily be checked that the following conditions imply (8.27) and (8.28):

$$X_0 \in W^{2,1}(0,T;H^2(a,b)),\ u_0 \in H^3(a,b),$$
$$u_0''(a) = 0. \tag{8.29}$$

Thus, as we want to have $\theta_0 \in W^{3,1}(0,T)$ (see Remark 8.1.6 above), it is sufficient to prove that the solution $(X_0, Y_0)$ of problem $(P.1)_0$ belongs to

$$\big(W^{2,1}(0,T;H^2(a,b)) \bigcap W^{3,1}(0,T;H^1(a,b))\big) \times W^{3,1}(0,T;H^1(b,c)).$$

First, we see that problem $(P.1)_0$ can be expressed as the following Cauchy problem in $H$

$$\begin{cases} Z_0'(t) + A_1 Z_0(t) = F(t), \quad 0 < t < T, \\ Z_0(0) = z_0, \end{cases} \tag{8.30}$$

where $A_1$ is the same as before,

$$Z_0(t) := (X_0(\cdot,t), Y_0(\cdot,t)),\ F(t) := (f(\cdot,t), g(\cdot,t)),\ z_0 := (u_0, v_0).$$

For this problem we are able to prove the following result:

**Theorem 8.1.9.** *Assume that* $(i_1)$–$(i_5)$ *are fulfilled and, in addition,*

$$\begin{aligned} &F \in W^{2,1}(0,T;H),\ z_0 \in D(A_1),\ \widetilde{z}_0 := F(0) - A_1 z_0 \in D(A_1) \\ &F'(0) - A_1\widetilde{z}_0 \in D(A_1). \end{aligned} \tag{8.31}$$

*Then, problem* (8.30) *has a unique strong solution*

$$Z_0 \in C^3([0,T];H),\ X_0 \in C^2([0,T];H^1(a,b)),\ Y_0 \in W^{3,2}(0,T;H^1(b,c)).$$

*Proof.* One can use again Theorem 2.0.27. □

*Remark* 8.1.10. Here are some specific conditions (independent of $\varepsilon$) which imply $(8.31)_1$:

$$\begin{cases} (f,g) \in W^{3,1}(0,T;L^2(a,b) \times L^2(b,c)),\ f(\cdot,0) \in H^1(a,b),\ g(\cdot,0) \in H^2(b,c), \\ u_0 \in H^2(a,b),\ v_0 \in H^4(b,c),\ \mu \in H^3(b,c), \\ \alpha_1 \in H^2(a,b),\ \beta_1 \in H^1(a,b),\ \alpha_2 \in H^3(b,c),\ \beta_2 \in H^2(b,c), \end{cases}$$

$$\begin{cases} u_0(a) = v_0(c) = 0,\ (\alpha_1 u_0)(b) = (\alpha_2 v_0)(b) - (\mu v_0')(b), \\ f(a,0) - (\alpha_1 u_0)'(a) - (\beta_1 u_0)(a) = 0, \\ g(c,0) + (\mu v_0')'(c) - (\alpha_2 v_0)'(c) - (\beta_2 v_0)(c) = 0, \\ \alpha_1(b)\big(f(b,0) - (\alpha_1 u_0)'(b) - (\beta_1 u_0)(b)\big) \\ -\alpha_2(b)\big(g(b,0) + (\mu v_0')'(b) - (\alpha_2 v_0)'(b) - (\beta_2 v_0)(b)\big) \\ = -\mu(b)\big(g_x(b,0) + (\mu v_0')''(b) - (\alpha_2 v_0)''(b) - (\beta_2 v_0)'(b)\big). \end{cases} \tag{8.32}$$

Moreover, one can formulate additional sufficient conditions (independent of $\varepsilon$), in terms of $f$, $g$, $u_0$, $v_0$, $\alpha_i$, $\beta_i$ $(i = 1,2)$, $\mu$, which imply $(8.31)_2$.

Finally, since we need higher regularity for $X_0$, we consider the following separate problem for $X_0$

$$\begin{cases} X_{0t} + (\alpha_1 X_0)_x + \beta_1 X_0 = f \text{ in } Q_{1T}, \\ X_0(x,0) = u_0(x),\ a < x < b, \\ X_0(a,t) = 0,\ 0 < t < T, \end{cases} \tag{8.33}$$

which admits the abstract formulation

$$\begin{cases} \mathcal{X}_0'(t) + A_2 \mathcal{X}_0(t) = \mathcal{F}_0(t),\ \ 0 < t < T, \\ \mathcal{X}_0(0) = u_0, \end{cases} \tag{8.34}$$

where $A_2$ is the operator defined above, and

$$\mathcal{X}_0(t) := X_0(\cdot,t),\ \mathcal{F}_0(t) := f(\cdot,t),\ 0 < t < T.$$

We have

**Theorem 8.1.11.** *If* $(i_1),(i_5)$ *hold and, in addition,*

$$\alpha_1 \in H^2(a,b),\ \beta_1 \in H^1(a,b);$$

$$f \in W^{4,1}(0,T;H_1) \bigcap W^{2,1}(0,T;H^1(a,b)); \tag{8.35}$$

$$u_0 \in D(A_2),\ u_k := \mathcal{F}_0^{(k-1)}(0) - A_2(u_{k-1}) \in D(A_2),\ k = \overline{1,3}, \tag{8.36}$$

*then problem* (8.34) *has a unique strong solution*

$$X_0 \in W^{2,1}(0,T;H^2(a,b)) \bigcap C^3([0,T];H^1(a,b)).$$

The proof is very similar to the proof of Theorem 8.1.7. We omit it.

*Remark* 8.1.12. The above conditions for $u_0$ and $u_1$ in (8.36) are fulfilled if

$$\begin{cases} u_0,\ f(\cdot,0) - (\alpha_1 u_0)' - \beta_1 u_0 \in H^1(a,b), \\ u_0(a) = 0,\ f(a,0) = \alpha_1(a) u_0'(a). \end{cases} \tag{8.37}$$

Conditions $(8.36)_{k=2,3}$ can also be explicitly expressed, but this leads to complicated equations. We also require the additional conditions

$$u_0 \in H^4(a,b),\ \alpha_1 \in H^4(a,b),\ \beta_1 \in H^3(a,b),$$

since they appear as natural sufficient conditions in order that (8.36) are fulfilled.

Let us point out that all the assumptions we have used in this section so far for different problems form a compatible system of assumptions. Note that the conditions $u_0(b) = v_0(b)$, $u_0'(b) = 0$ (see Remark 8.1.3) and $\alpha_1(b)u_0'(b) = -\theta_0'(0) - \beta_1(b)\theta_0(0)$ (see Remark 8.1.6) imply that $\theta_0(0) = \theta_0'(0) = 0$. Therefore, in particular, we reobtain (8.7).

By virtue of what we have proved so far, we can formulate the following concluding result:

**Corollary 8.1.13.** *Assume that* $(i_1)$–$(i_5)$ *are satisfied. If* $f, g, \alpha_1, \alpha_2, \beta_1, \beta_2, \mu, u_0$ *and* $v_0$ *are smooth enough and the compatibility conditions* (8.19), $(8.23)_2$, (8.28), (8.32) (*plus additional conditions which imply* $(8.31)_2$), (8.36) *are all satisfied, then, for every* $\varepsilon > 0$, *problem* $(P.1)_\varepsilon$ *has a unique strong solution*

$$(u_\varepsilon, v_\varepsilon) \in C^2([0,T]; L^2(a,b) \times L^2(b,c)) \bigcap W^{2,2}(0,T; H^1(a,b) \times H^1(b,c)) \bigcap C^1([0,T]; H^2(a,b) \times H^2(b,c)),$$

*and problems* $(P.1)_0$, $(P.1)_1$ *admit unique solutions*

$$(X_0,Y_0) \in \left(W^{2,1}(0,T; H^2(a,b)) \bigcap C^3([0,T]; H^1(a,b))\right) \times \left(C^3([0,T]; L^2(b,c)) \bigcap W^{3,2}(0,T; H^1(b,c))\right),$$

$$(X_1,Y_1) \in \left(C^1([0,T]; H^1(a,b)) \bigcap C^2([0,T]; L^2(a,b))\right) \times \left(C^1([0,T]; L^2(b,c)) \bigcap W^{1,2}(0,T; H^1(b,c)) \bigcap C([0,T]; H^2(b,c))\right).$$

### 8.1.3 Estimates for the remainder components

In this section we are going to establish estimates with respect to the uniform convergence norm for the two components of the first order remainder. Of course, these estimates should be of order $\mathcal{O}(\varepsilon^r)$, $r > 1$, since the final goal is to validate completely our first order expansion.

**Theorem 8.1.14.** *Assume that all the assumptions of Corollary* 8.1.13 *are fulfilled. Then, for every* $\varepsilon > 0$, *the solution of problem* $(P.1)_\varepsilon$ *admits an asymptotic expansion of the form* (8.1) *and the following estimates hold:*

$$\|R_{1\varepsilon}\|_{C(\overline{Q}_{1T})} = \mathcal{O}(\varepsilon^{\frac{17}{16}}), \quad \|R_{2\varepsilon}\|_{C(\overline{Q}_{2T})} = \mathcal{O}(\varepsilon^{\frac{21}{16}}).$$

*Proof.* Throughout this proof we denote by $M_k$ ($k = 1, 2, \dots$) different positive constants which depend on the data, but are independent of $\varepsilon$. By Corollary 8.1.13 and (8.1) we can see that $R_\varepsilon(t) := \big(R_{1\varepsilon}(\cdot,t), R_{2\varepsilon}(\cdot,t)\big)$ is a strong solution of problem (8.4), (8.8), (8.10) and (8.14). Taking the scalar product in $H = L^2(a,b) \times L^2(b,c)$ of (8.4) and $R_\varepsilon(t)$, we get (as in the proof of Lemma 8.1.1)

$$\begin{aligned}\frac{1}{2}\frac{d}{dt}\|R_\varepsilon(\cdot,t)\|^2 &+ \varepsilon\|R_{1\varepsilon x}(\cdot,t)\|_1^2 + \frac{\mu_0}{2}\|R_{2\varepsilon x}(\cdot,t)\|_2^2 - \omega\|R_\varepsilon(t)\|^2 \\ &\le \| h_\varepsilon(\cdot,t) \|_1 \cdot \| R_{1\varepsilon}(\cdot,t) \|_1 + \frac{\alpha_1(a)}{2} R_{1\varepsilon}(a,t)^2 \\ &\quad + \varepsilon^2 \| X_{1x}(\cdot,t) \|_1 \cdot \| R_{1\varepsilon x}(\cdot,t) \|_1 \\ &\quad + \varepsilon \, | R_{1\varepsilon}(a,t) T_{1\varepsilon x}(a,t) | \quad \text{for all } t \in [0,T].\end{aligned}$$

If we multiply the above inequality by $e^{-2\omega t}$, integrate on $[0,t]$, and then multiply the resulting inequality by $e^{2\omega t}$, we find

$$\begin{aligned}\frac{1}{2}\|R_\varepsilon(t)\|^2 &+ \frac{\varepsilon}{2}\int_0^t e^{2\omega(t-s)}\|R_{1\varepsilon x}(\cdot,s)\|_1^2 ds + \frac{\mu_0}{2}\int_0^t e^{2\omega(t-s)}\|R_{2\varepsilon x}(\cdot,s)\|_2^2 ds\\
\le&\frac{\varepsilon^3}{2}\int_0^t e^{2\omega(t-s)} \parallel X_{1x}(\cdot,s)\parallel_1^2 ds\\
&+\frac{\alpha_1(a)}{2}\int_0^t e^{2\omega(t-s)}R_{1\varepsilon}(a,s)^2 ds\\
&+\varepsilon\int_0^t e^{2\omega(t-s)} \mid T_{1\varepsilon x}(a,s)R_{1\varepsilon}(a,s)\mid ds\\
&+\int_0^t e^{2\omega(t-s)}\|h_\varepsilon(\cdot,s)\|_1\cdot\|R_{1\varepsilon}(\cdot,s)\|_1 ds \;\; \forall\, t\in[0,T]. \qquad (8.38)\end{aligned}$$

Next, if we look at the structure of $h_\varepsilon$, we can see that

$$\int_0^t \parallel h_\varepsilon(\cdot,s)\parallel_1 ds \le M_1\varepsilon^{\frac{3}{2}} \;\; \forall\, t\in[0,T]. \qquad (8.39)$$

Note also that $\forall k\ge 1$ there exists a positive constant $M_2$ such that

$$\mid R_{1\varepsilon}(a,t)\mid=\mid P_\varepsilon(\xi(a),t)\mid\le M_2\varepsilon^k \quad \forall\, t\in[0,T]. \qquad (8.40)$$

In the following we are going to derive some estimates for $u_{\varepsilon tt}$, $u_{\varepsilon x}(a,\cdot)$, $u_{\varepsilon tx}(a,\cdot)$ and $v_{\varepsilon tt}$ which will be used in the next part of the proof. Since the solution $W_\varepsilon=(u_\varepsilon,v_\varepsilon)$ of problem $(P.1)_\varepsilon$ belongs to $C^2([0,T];H)$, we can differentiate two times in the corresponding variation of constants formula (see Theorem 2.0.27) and then derive the estimate

$$\|W_\varepsilon''(t)\|\le e^{\omega t}\Big(\|F'(0)-J_\varepsilon(F(0)-J_\varepsilon W_0)\|+\int_0^t e^{-\omega s}\|F''(s)\|ds\Big)\le M_3,$$

for all $t\in[0,T]$. Consequently,

$$\begin{aligned}\|R_\varepsilon'\|_{C([0,T];H)}&=\mathcal{O}(1),\;\; \|W_\varepsilon''\|_{C([0,T];H)}=\mathcal{O}(1),\\
\|W_\varepsilon'\|_{C([0,T];H)}&=\mathcal{O}(1),\;\; \|W_\varepsilon\|_{C([0,T];H)}=\mathcal{O}(1).\end{aligned} \qquad (8.41)$$

To establish an estimate for $u_{\varepsilon x}(a,\cdot)$, we start with the obvious inequality (see the proof of Lemma 8.1.1)

$$\begin{aligned}\langle W_\varepsilon'(t),W_\varepsilon(t)\rangle+\varepsilon\parallel u_{\varepsilon x}(\cdot,t)\parallel_1^2+\frac{\mu_0}{2}\parallel v_{\varepsilon x}(\cdot,t)\parallel_2^2&\\
\le\langle F(t)+\omega W_\varepsilon(t),W_\varepsilon(t)\rangle \;\; \forall\, t\in[0,T],&\end{aligned}$$

from which, in view of estimates $(8.41)_{3,4}$, we derive

$$\| u_{\varepsilon x} \|_{C([0,T];L^2(a,b))} = \mathcal{O}(\varepsilon^{-1/2}). \tag{8.42}$$

From the first equation of system $(S)$ we also obtain

$$\| u_{\varepsilon xx} \|_{C([0,T];L^2(a,b))} = \mathcal{O}(\varepsilon^{-3/2}).$$

This together with (8.42) implies $\| u_{\varepsilon x} \|_{C(\overline{Q}_{1T})} = \mathcal{O}(\varepsilon^{-1})$. Indeed, by (8.42) and the mean value theorem, for each $t \in [0,T]$ and $\varepsilon > 0$, there exists an $x_{t,\varepsilon} \in [a,b]$ such that $u_{\varepsilon x}(x_{t,\varepsilon})^2 \leq M_4\varepsilon^{-1}$, so our assertion follows from

$$u_{\varepsilon x}(x,t)^2 = 2\int_{x_{t,\varepsilon}}^{x} u_{\varepsilon x}(\xi,t)u_{\varepsilon xx}(\xi,t)d\xi + u_{\varepsilon x}(x_{t,\varepsilon},t)^2.$$

Consequently,

$$\| T_{1\varepsilon x}(a,\cdot) \|_{C[0,T]} = \mathcal{O}(\varepsilon^{-1}). \tag{8.43}$$

Similarly, starting from the problem which comes out by formal differentiation with respect to $t$ of problem $(P.1)_\varepsilon$, we derive $\| u_{\varepsilon tx} \|_{C(\overline{Q}_{1T})} = \mathcal{O}(\varepsilon^{-1})$, from which it follows

$$\| T_{1\varepsilon tx}(a,\cdot) \|_{L^2(0,T)} = \mathcal{O}(\varepsilon^{-1}). \tag{8.44}$$

Now, making use of $(8.38) - -(8.40)$ and (8.43), we find

$$\big\|R_{1\varepsilon}(\cdot,t)\big\|_1 \leq M_5\varepsilon^{\frac{3}{2}}, \ \big\|R_{2\varepsilon}(\cdot,t)\big\|_2 \leq M_5\varepsilon^{\frac{3}{2}} \ \forall\, t \in [0,T]. \tag{8.45}$$

Let us prove some additional estimates for $R_{1\varepsilon t}$ and $R_{2\varepsilon t}$. Since we do not have sufficient regularity for $i_1$ and $R_{2\varepsilon}$ ($Y_1 \in C^1([0,T];L^2(b,c))$), we cannot use the standard method based on differentiation with respect to $t$ of the problem satisfied by the remainder. Denote $S_{1\varepsilon} = R_{1\varepsilon} + \varepsilon i_1$. Note that

$$S_{1\varepsilon} = u_\varepsilon - X_0 - i_0 - \varepsilon X_1 \in W^{2,2}(0,T;L^2(a,b))$$

and from (8.4) we obtain the system

$$\begin{cases} S_{1\varepsilon t} + (-\varepsilon T_{1\varepsilon x} + \alpha_1 R_{1\varepsilon})_x + \beta_1 R_{1\varepsilon} = \overline{h}_\varepsilon \ \text{ in } Q_{1T}, \\ R_{2\varepsilon t} + \big( - \mu R_{2\varepsilon x} + \alpha_2 R_{2\varepsilon}\big)_x + \beta_2 R_{2\varepsilon} = 0 \ \text{ in } Q_{2T}, \end{cases} \tag{8.46}$$

where

$$\begin{aligned} \overline{h}_\varepsilon(x,t) =& [\varepsilon^{-1}\big(\alpha_1(x) - \alpha_1(b)\big) + \xi\alpha_1'(b)]i_{0\xi}(\xi,t) \\ &+ (\alpha_1(x) - \alpha_1(b))i_{1\xi}(\xi,t) - (\alpha_1'(x) - \alpha_1'(b))i_0(\xi,t) \\ &- (\beta_1(x) - \beta_1(b))i_0(\xi,t) - \varepsilon(\beta_1(x) + \alpha_1'(x))i_1(\xi,t) \ \text{ in } Q_{1T}, \\ \overline{h}_\varepsilon \in& W^{1,2}(0,T;L^2(a,b)). \end{aligned}$$

Taking the scalar product in $H$ of system (8.46) with $(R_{1\varepsilon}, R_{2\varepsilon})$, we derive in a standard manner the inequality

$$\begin{aligned} &\frac{1}{2}\left[\frac{d}{dt}\left(\|S_{1\varepsilon}(\cdot,t)\|_1^2 + \|R_{2\varepsilon}(\cdot,t)\|_2^2\right)\right] - \omega\|R_\varepsilon(t)\|^2 \\ &\quad \leq \| \overline{h}_\varepsilon(\cdot,t) \|_1 \cdot \| R_{1\varepsilon}(\cdot,t) \|_1 + \frac{\alpha_1(a)}{2} R_{1\varepsilon}(a,t)^2 \\ &\quad + \frac{\varepsilon^3}{2} \| X_{1x}(\cdot,t) \|_1^2 + \varepsilon \| S_{1\varepsilon t}(\cdot,t) \|_1 \cdot \| i_1(\cdot,t) \|_1 \\ &\quad + \varepsilon \mid R_{1\varepsilon}(a,t) T_{1\varepsilon x}(a,t) \mid \text{ for all } t \in [0,T]. \end{aligned} \tag{8.47}$$

Now, from $(8.41)_1$ it follows

$$\| R_\varepsilon(t_1) - R_\varepsilon(t_2) \| \leq M_6 \mid t_1 - t_2 \mid \ \forall\, t_1,\ t_2 \in [0,T].$$

This Lipschitz property together with (8.8) and (8.45) implies

$$\| R_\varepsilon(t) \|^2 \leq M_7 t \varepsilon^{3/2} \ \forall\, t \in [0,T].$$

Obviously, for every function $\zeta \in W^{1,2}(0,T;L^2(a,b))$ we have the inequality

$$\| \zeta(\cdot,t_1) - \zeta(\cdot,t_2) \|_1^2 \leq \mid t_1 - t_2 \mid \cdot \| \zeta_t \|_{L^2(Q_{1T})}^2 \quad \forall\, t_1, t_2 \in [0,T],$$

which in the particular case $\zeta(x,t) = \zeta_1(x)\zeta_2(t)$ becomes

$$\| \zeta(\cdot,t_1) - \zeta(\cdot,t_2) \|_1^2 \leq \mid t_1 - t_2 \mid \cdot \| \zeta_1 \|_1^2 \cdot \| \zeta_2' \|_{L^2(0,T)}^2 \quad \forall\, t_1, t_2 \in [0,T].$$

Since $i_1$, $\overline{h}_\varepsilon$ are functions of this form, and $\forall k \geq 1$ there exists a positive constant $M_8$ such that

$$\mid R_{1\varepsilon}(a,t) \mid = \mid P_\varepsilon(\xi(a),t) \mid \leq M_8 \sqrt{t} \varepsilon^k \ \forall\, t \in [0,T],$$

we have

$$\begin{aligned} &\mid R_{1\varepsilon}(a,t) T_{1\varepsilon x}(a,t) \mid \leq M_9 t \varepsilon^{k-1} \text{ (cf. (8.44))}, \\ &\| \overline{h}_\varepsilon(\cdot,t) \|_1 \cdot \| R_{1\varepsilon}(\cdot,t) \|_1 \leq M_{10} t \varepsilon^{3/2}, \\ &\| S_{1\varepsilon t}(\cdot,t) \|_1 \cdot \| i_1(\cdot,t) \|_1 \leq M_{11} t \varepsilon^{1/2} \ \forall\, t \in [0,T] \text{ (cf.}(8.41)_2), \end{aligned}$$

where $M_9$ depends on $k$. For these estimates we have used that

$$\overline{h}_\varepsilon(\cdot,0) = 0,\ R_{1\varepsilon}(\cdot,0) = 0,\ i_1(\cdot,0) = 0,\ S_{1\varepsilon t}(\cdot,0) = 0$$

(cf. (8.7), $(8.2)_1$, $k = 0,1$, and $(S)_1$). Thus, it follows from (8.47)

$$\begin{aligned} &\frac{1}{2}\left[\frac{d}{dt}\left(\|S_{1\varepsilon}(\cdot,t)\|_1^2 + \|R_{2\varepsilon}(\cdot,t)\|_2^2\right)\right] \\ &\quad \leq \frac{\varepsilon^3}{2} t \| X_{1tx} \|_{L^2(Q_{1T})}^2 + M_{12} t \varepsilon^{3/2} \text{ for all } t \in [0,T]. \end{aligned} \tag{8.48}$$

If we integrate (8.48) over $[0,\delta]$, we get

$$\big\|S_{1\varepsilon}(\cdot,\delta)\big\|_1 + \big\|R_{2\varepsilon}(\cdot,\delta)\big\|_2 \le M_{13}\delta\varepsilon^{3/4} \text{ for all } \delta\in[0,T]. \tag{8.49}$$

Now, we write (8.46) for $t$ and $t+\delta$, $\delta>0$, such that $t,\ t+\delta\in[0,T]$, subtract the two equations, and then take the scalar product in $H$ of the resulting equation with $R_\varepsilon(t+\delta)-R_\varepsilon(t)$. Then, by our standard computations, we arrive at

$$\begin{aligned}\big\|S_{1\varepsilon}(\cdot,t+\delta)-S_{1\varepsilon}(\cdot,t)\big\|_1^2 + \big\|R_{2\varepsilon}(\cdot,t+\delta)-R_{2\varepsilon}(\cdot,t)\big\|_2^2 \\ \le \big\|S_{1\varepsilon}(\cdot,\delta)\big\|_1^2 + \big\|R_{2\varepsilon}(\cdot,\delta)\big\|_2^2 + M_{14}\delta^2\varepsilon^{3/2}\end{aligned}$$

for all $0\le t<t+\delta\le T$.
Thus, according to (8.49), we have

$$\big\|S_{1\varepsilon}(\cdot,t+\delta)-S_{1\varepsilon}(\cdot,t)\big\|_1 + \big\|R_{2\varepsilon}(\cdot,t+\delta)-R_{2\varepsilon}(\cdot,t)\big\|_2 \le M_{15}\delta\varepsilon^{3/4}$$

for all $0\le t<t+\delta\le T$.
After dividing this last inequality by $\delta$, since

$$S_{1\varepsilon}\in W^{2,2}(0,T;L^2(a,b)),\ \ R_{2\varepsilon}\in C^1([0,T];L^2(b,c)),$$

we can pass to the limit as $\delta\to 0$ to obtain

$$\| S_{1\varepsilon t}(\cdot,t)\|_1 + \| R_{2\varepsilon t}(\cdot,t)\|_2 \le M_{15}\varepsilon^{3/4} \text{ for all } t\in[0,T]. \tag{8.50}$$

Now, from the estimate

$$\begin{aligned}\langle (S_{1\varepsilon t}(\cdot,t), R_{2\varepsilon t}(\cdot,t)), R_\varepsilon(t)\rangle + \varepsilon \| R_{1\varepsilon x}(\cdot,t)\|_1^2 + \frac{\mu_0}{2}\| R_{2\varepsilon x}(\cdot,t)\|_2^2 \\ \le \omega \| R_\varepsilon(t)\|^2 + \frac{\alpha_1(a)}{2} R_{1\varepsilon}(a,t)^2 \\ + \varepsilon^2 \| X_{1x}(\cdot,t)\|_1 \cdot \| R_{1\varepsilon x}(\cdot,t)\|_1 \\ + \varepsilon\, | T_{1\varepsilon x}(a,t) P_\varepsilon(\xi(a),t) | \\ + \| \overline{h}_\varepsilon(\cdot,t)\|_1 \cdot \| R_{1\varepsilon}(\cdot,t)\|_1\end{aligned}$$

for all $t\in[0,T]$, on account of (8.45) and (8.50), we easily find

$$\varepsilon\| R_{1\varepsilon x}(\cdot,t)\|_1^2 \le M_{16}\varepsilon^{9/4},\ \ \| R_{2\varepsilon x}(\cdot,t)\|_2^2 \le M_{16}\varepsilon^{9/4}$$

(we have also used the inequality

$$\varepsilon^2 \| X_{1x}(\cdot,t)\|_1 \cdot \| R_{1\varepsilon x}(\cdot,t)\|_1 \le \frac{\varepsilon^3}{2}\| X_{1x}(\cdot,t)\|_1^2 + \frac{\varepsilon}{2}\| R_{1\varepsilon x}(\cdot,t)\|_1^2).$$

Therefore,

$$R_{2\varepsilon}(x,t)^2 \le 2\| R_{2\varepsilon}(\cdot,t)\|_2 \cdot \| R_{2\varepsilon x}(\cdot,t)\|_2 \le M_{17}\varepsilon^{\frac{21}{8}}\ \forall\,(x,t)\in Q_{2T}.$$

Similarly, we get $\| R_{1\varepsilon}\|_{C(\overline{Q}_{1T})} = \mathcal{O}(\varepsilon^{\frac{17}{16}})$. □

*Remark* 8.1.15. If $i_1$, $X_1$, $Y_1$ were more regular (this is possible under additional assumptions), we would be able to derive better estimates for the two components of the remainder. Indeed, in this case we can differentiate with respect to $t$ problem (8.4), (8.8), (8.10), (8.14), and then, by computations similar to those which led to estimates (8.38), we find

$$\| R_{1\varepsilon t} \|_{C([0,T];L^2(a,b))} = \mathcal{O}(\varepsilon^{3/2}), \quad \| R_{2\varepsilon t} \|_{C([0,T];L^2(b,c))} = \mathcal{O}(\varepsilon^{3/2}),$$

which give the estimates

$$\|R_{1\varepsilon}\|_{C(\overline{Q}_{1T})} = \mathcal{O}(\varepsilon^{\frac{5}{4}}), \ \|R_{2\varepsilon}\|_{C(\overline{Q}_{2T})} = \mathcal{O}(\varepsilon^{\frac{3}{2}}).$$

## 8.2 A first order asymptotic expansion for the solution of problem $(P.2)_\varepsilon$

In this section we investigate problem $(P.2)_\varepsilon$ (which has been introduced at the beginning of this chapter), under assumptions $(i_1)$–$(i_4)$ and $(i_5')$. Since the coefficient $\alpha_1$ is a negative function, the boundary layer phenomenon is now present near the line segment $\{(a,t);\ t \in [0,T]\}$. Moreover, since we have a Neumann boundary condition at $x = a$, we have no boundary layer of order zero, even with respect to the uniform topology, as in the stationary case considered in Section 7.2. In particular, the first order asymptotic expansion we are going to establish includes a first order correction only (i.e., problem $(P.2)_\varepsilon$ is singularly perturbed of order one with respect to the uniform topology).

### 8.2.1 Formal expansion

By examining examples of problem $(P.2)_\varepsilon$, we are able to guess the form of the general first order asymptotic expansion for the solution. In particular, one can see that there is no boundary layer inside the rectangle $Q_T$, except for the neighborhood of the line segment $\{(a,t);\ t \in [0,T]\}$. Therefore, we seek a first order asymptotic expansion for the solution $U_\varepsilon := \big(u_\varepsilon(x,t), v_\varepsilon(x,t)\big)$ of problem $(P.2)_\varepsilon$ in the form:

$$\begin{cases} u_\varepsilon(x,t) = X_0(x,t) + \varepsilon X_1(x,t) + i_0(\zeta,t) + \varepsilon i_1(\zeta,t) + R_{1\varepsilon}(x,t), \\ v_\varepsilon(x,t) = Y_0(x,t) + \varepsilon Y_1(x,t) + R_{2\varepsilon}(x,t), \end{cases} \tag{8.51}$$

where:

$\zeta := \varepsilon^{-1}(x-a)$ is the fast variable;

$\big(X_k(x,t), Y_k(x,t)\big)$, $k = 0,1$, are the first two regular terms;

$i_0$, $i_1$ are the corresponding boundary layer functions;

$\big(R_{1\varepsilon}(x,t), R_{2\varepsilon}(x,t)\big)$ represents the remainder of order one.

Like in Subsection 8.1.1, substituting (8.51) into system $(S)$, we find:

$$\begin{cases} X_{kt} + (\alpha_1 X_k)_x + \beta_1 X_k = \widetilde{f}_k \ \text{ in } \ Q_{1T}, \\ Y_{kt} + (-\mu Y_{kx} + \alpha_2 Y_k)_x + \beta_2 Y_k = \widetilde{g}_k \ \text{ in } \ Q_{2T}, \ k = 0, 1, \end{cases} \tag{8.52}$$

where

$$\widetilde{f}_k(x,t) = \begin{cases} f(x,t), \ \ k = 0, \\ X_{0xx}(x,t), \ \ k = 1, \end{cases}$$

$$\widetilde{g}_k(x,t) = \begin{cases} g(x,t), \ \ k = 0, \\ 0, \ \ k = 1, \end{cases}$$

$$\begin{cases} i_{0\zeta\zeta} - \alpha_1(a) i_{0\zeta} = 0, \\ i_{1\zeta\zeta} - \alpha_1(a) i_{1\zeta} = i_{0t} + \zeta \alpha_1'(a) i_{0\zeta} + (\alpha_1'(a) + \beta_1(a)) i_0. \end{cases} \tag{8.53}$$

From $(BC.2)$ we derive:

$$i_{0\zeta}(a,t) = 0, \ i_{1\zeta}(a,t) = -X_{0x}(a,t), \ 0 < t < T, \tag{8.54}$$

$$\begin{cases} Y_0(c,t) = 0, \\ Y_1(c,t) = 0, \ 0 < t < T, \end{cases} \tag{8.55}$$

$$\begin{cases} R_{1\varepsilon x}(a,t) = -\varepsilon X_{1x}(a,t), \\ R_{2\varepsilon}(c,t) = 0, \ 0 < t < T. \end{cases} \tag{8.56}$$

Now, (8.53) and (8.54) lead us to

$$i_0 \equiv 0, \ i_1(\zeta,t) = -\frac{X_{0x}(a,t)}{\alpha_1(a)} e^{\alpha_1(a)\zeta}. \tag{8.57}$$

Therefore, the remainder components satisfy the system

$$\begin{cases} R_{1\varepsilon t} + (-\varepsilon T_{1\varepsilon x} + \alpha_1 R_{1\varepsilon})_x + \beta_1 R_{1\varepsilon} = h_\varepsilon \ \text{ in } Q_{1T}, \\ R_{2\varepsilon t} + \big(-\mu R_{2\varepsilon x} + \alpha_2 R_{2\varepsilon}\big)_x + \beta_2 R_{2\varepsilon} = 0 \ \text{ in } Q_{2T}, \end{cases} \tag{8.58}$$

where

$$T_{1\varepsilon} = u_\varepsilon - X_0 - \varepsilon i_1 = R_{1\varepsilon} + \varepsilon X_1,$$
$$h_\varepsilon(x,t) = -\varepsilon i_{1t}(\zeta,t) - (\alpha_1(x) - \alpha_1(a)) i_{1\zeta}(\zeta,t) - \varepsilon(\beta_1(x) + \alpha_1'(x)) i_1(\zeta,t).$$

Imposing the transmissions conditions at $b$ we get:

$$\begin{cases} X_0(b,t) = Y_0(b,t), \\ -\mu(b) Y_{0x}(b,t) = \alpha_1(b) X_0(b,t) - \alpha_2(b) Y_0(b,t), \ 0 \le t \le T, \end{cases} \tag{8.59}$$

$$\begin{cases} X_1(b,t) = Y_1(b,t), \\ -\mu(b)Y_{1x}(b,t) + \alpha_2(b)Y_1(b,t) = \alpha_1(b)X_1(b,t) - X_{0x}(b,t),\ 0 \le t \le T, \end{cases} \tag{8.60}$$

$$\begin{cases} R_{1\varepsilon}(b,t) + \varepsilon i_1(\zeta(b),t) = R_{2\varepsilon}(b,t), \\ -\varepsilon T_{1\varepsilon x}(b,t) + \alpha_1(b)R_{1\varepsilon}(b,t) - \varepsilon\left(i_{1\zeta}(\zeta(b),t) - \alpha_1(b)i_1(\zeta(b),t)\right) \\ = -\mu(b)R_{2\varepsilon x}(b,t) + \alpha_2(b)R_{2\varepsilon}(b,t),\ 0 \le t \le T, \end{cases} \tag{8.61}$$

where $\zeta(b) := \varepsilon^{-1}(b-a)$.
Finally, from $(IC)$ we have

$$X_k(x,0) = \begin{cases} u_0(x),\ k=0, \\ 0,\ k=1, \end{cases} \quad Y_k(x,0) = \begin{cases} v_0(x),\ k=0, \\ 0,\ k=1, \end{cases} \tag{8.62}$$

$$\begin{cases} R_{1\varepsilon}(x,0) = 0,\quad a \le x \le b, \\ R_{2\varepsilon}(x,0) = 0,\quad b \le x \le c, \end{cases} \tag{8.63}$$

$$i_1(\zeta,0) = 0 \Longleftrightarrow X_{0x}(a,0) = 0. \tag{8.64}$$

The last condition says that $i_1$ does not introduce any discrepancy at the corner boundary point $(x,t) = (a,0)$ of our domain $Q_T$. This condition will appear again in the next section as one of the compatibility conditions the data have to satisfy in order that problem $(P.2)_\varepsilon$ admit a sufficiently smooth solution.

Summarizing, we can see that the components of the zeroth order regular term satisfy the reduced problem $(P.2)_0$ which consists of $(8.52)_{k=0}$, $(8.55)_1$, $(8.59)$ and $(8.62)_{k=0}$; the components of first order regular term satisfy the problem $(P.2)_1$ which consists of $(8.52)_{k=1}$, $(8.55)_2$, $(8.60)$, $(8.62)_{k=1}$; and the remainder components satisfy (8.56), (8.58), (8.61) and (8.63).

### 8.2.2 Existence, uniqueness and regularity of the solutions of problems $(P.2)_\varepsilon$, $(P.2)_1$ and $(P.2)_0$

To investigate problem $(P.2)_\varepsilon$ we can follow a device similar to that of Subsection 8.1.2. We denote $H := L^2(a,b) \times L^2(b,c)$. This is equipped with the usual scalar product, denoted $\langle \cdot, \cdot \rangle$, and the corresponding induced norm, denoted $\|\cdot\|$. We define the operator $L_\varepsilon : D(L_\varepsilon) \subset H \to H$ by

$$D(L_\varepsilon) := \Big\{(h,k) \in H^2(a,b) \times H^2(b,c);\ h'(a) = k(c) = 0,$$
$$h(b) = k(b),\ \varepsilon h'(b) - \alpha_1(b)h(b) = \mu(b)k'(b) - \alpha_2(b)k(b)\Big\},$$
$$L_\varepsilon(h,k) := \big((-\varepsilon h' + \alpha_1 h)' + \beta_1 h,\ (-\mu k' + \alpha_2 k)' + \beta_2 k\big).$$

As far as this operator is concerned we can prove the following result which is similar to Lemma 8.1.1:

**Lemma 8.2.1.** *Assume that* $(i_1)$*–*$(i_3)$ *and* $(i'_5)$ *are satisfied. Then, there is a positive number* $\omega$*, independent of* $\varepsilon$*, such that* $L_\varepsilon + \omega I$ *is maximal monotone, where* $I$ *is the identity of* $H$*.*

*Proof.* It is similar to the proof of Lemma 8.1.1, so we just outline it. Using Lemma 7.1.2 in Chapter 7, one can see that operator $L_\varepsilon + \omega I$ is monotone for $\omega > 0$ sufficiently large. To prove the maximality of this operator one can use Theorems 2.0.6 and 7.2.1. □

Now, we write problem $(P.2)_\varepsilon$ in the form of the following Cauchy problem in $H$ :

$$\begin{cases} W'_\varepsilon(t) + L_\varepsilon W_\varepsilon(t) = F(t), \quad 0 < t < T, \\ W_\varepsilon(0) = W_0, \end{cases} \tag{8.65}$$

where

$$W_\varepsilon(t) := \big(u_\varepsilon(\cdot,t), v_\varepsilon(\cdot,t)\big), \ W_0 := \big(u_0, v_0\big), \ F(t) := \big(f(\cdot,t), g(\cdot,t)\big), \quad 0 < t < T.$$

Making use of the above lemma we are able to prove that:

**Theorem 8.2.2.** *Assume that* $(i_1)$*–*$(i_4)$*,* $(i'_5)$ *are fulfilled and*

$$F \in W^{1,1}(0,T;H), \quad W_0 \in D(L_\varepsilon). \tag{8.66}$$

*Then problem* (8.65) *has a unique strong solution* $W_\varepsilon$ *which belongs to*

$$C^1([0,T];H) \bigcap W^{1,2}(0,T;H^1(a,b) \times H^1(b,c)) \bigcap C([0,T];H^2(a,b) \times H^2(b,c)).$$

*If, in addition,*

$$F \in W^{2,1}(0,T;H), \ F(0) - L_\varepsilon(W_0) \in D(L_\varepsilon), \tag{8.67}$$

*then* $W_\varepsilon$ *belongs to*

$$C^2([0,T];H) \bigcap W^{2,2}(0,T;H^1(a,b) \times H^1(b,c)) \bigcap C^1([0,T];H^2(a,b) \times H^2(b,c)).$$

The proof of this result is based on arguments similar to those we used in the proof of Theorem 8.1.2, so we omit it.

*Remark* 8.2.3. The following conditions are sufficient to insure that (8.66) are fulfilled for every $\varepsilon > 0$ :

$$f \in W^{1,1}(0,T;L^2(a,b)), \ g \in W^{1,1}(0,T;L^2(b,c)), \ u_0 \in H^2(a,b), \ v_0 \in H^2(b,c);$$

$$\begin{cases} u'_0(a) = v_0(c) = 0, \ u_0(b) = v_0(b), \ u'_0(b) = 0, \\ -\mu(b)v'_0(b) = \alpha_1(b)u_0(b) - \alpha_2(b)v_0(b). \end{cases} \tag{8.68}$$

Obviously, condition $u'_0(a) = 0$ implies (8.64).

If (8.68) hold and, in addition, we assume that

$$\begin{cases} f \in W^{2,1}(0,T;L^2(a,b)),\ g \in W^{2,1}(0,T;L^2(b,c)), \\ f(\cdot,0) \in H^2(a,b),\ g(\cdot,0) \in H^2(b,c),\ \alpha_1 \in H^3(a,b),\ \alpha_2 \in H^3(b,c), \\ \beta_1 \in H^2(a,b),\ \beta_2 \in H^2(b,c),\ \mu \in H^3(b,c),\ u_0 \in H^4(a,b),\ v_0 \in H^4(b,c), \end{cases}$$

$$\begin{cases} {u_0}^{(3)}(a) = 0,\ f_x(a,0) = (\alpha_1 u_0)''(a) + (\beta_1 u_0)'(a), \\ g(c,0) + (\mu v_0')'(c) = \alpha_2(c)v_0'(c),\ u_0''(b) = 0, \\ f(b,0) - (\alpha_1 u_0)'(b) - (\beta_1 u_0)(b) \\ = g(b,0) + (\mu v_0')'(b) - (\alpha_2 v_0)'(b) - (\beta_2 v_0)(b), \\ {u_0}^{(3)}(b) = 0,\ f_x(b,0) = (\alpha_1 u_0)''(b) + (\beta_1 u_0)'(b), \\ -\mu(b)[g_x(b,0) + (\mu v_0')''(b) - (\alpha_2 v_0)'' - (\beta_2 v_0)'(b)] \\ = (\alpha_1(b) - \alpha_2(b))[f(b,0) - (\alpha_1 u_0)'(b) - (\beta_1 u_0)(b)], \end{cases} \tag{8.69}$$

then (8.66) and (8.67) are all satisfied. Note that (8.68) and (8.69) are all independent of $\varepsilon$.

Next we deal with problems $(P.2)_1$ and $(P.2)_0$. Since we want to obtain estimates for the remainder with respect to the uniform convergence norm, it would be enough for the regular term $(X_1, Y_1)$ to belong to the space

$$W^{1,2}(0,T;H^1(a,b)) \times W^{1,2}(0,T;H^1(b,c)).$$

First of all, let us homogenize the boundary condition at $b$, $(8.60)_2$. Consider the substitution

$$Y_1(x,t) := \overline{Y}_1(x,t) + l(x,t),\ (x,t) \in Q_{2T},$$

where

$$l(x,t) = M(t)x^2 + N(t)x + P(t),\ M(t) = \frac{\sigma(t)}{(b-c)\mu(b)},\ N(t) = -(b+c)M(t),$$
$$P(t) = -M(t)b^2 - N(t)b,\ \sigma(t) = X_{0x}(b,t).$$

It is easily seen that $(X_1, \overline{Y}_1)$ satisfies the problem

$$\begin{cases} X_{1t} + (\alpha_1 X_1)_x + \beta_1 X_1 = \widetilde{f}_1 \ \text{in} \ \ Q_{1T}, \\ \overline{Y}_{1t} + (-\mu \overline{Y}_{1x} + \alpha_2 \overline{Y}_1)_x + \beta_2 \overline{Y}_1 = \overline{g}_1 \ \text{in} \ \ Q_{2T}, \\ X_1(x,0) = 0,\ x \in [a,b],\ \overline{Y}_1(x,0) = \overline{y}_1(x),\ x \in [b,c], \\ X_1(b,t) = \overline{Y}_1(b,t),\ \overline{Y}_1(c,t) = 0, \\ -\mu(b)\overline{Y}_{1x}(b,t) + \alpha_2(b)\overline{Y}_1(b,t) = \alpha_1(b)X_1(b,t),\ 0 \le t \le T, \end{cases} \tag{8.70}$$

where $\overline{y}_1(x) = -l(x,0),\ b \le x \le c$, and

$$\overline{g}_1 = -l_t + (\mu l_x - \alpha_2 l)_x - \beta_2 l \text{ in } Q_{2T}.$$

This problem can be written as the following Cauchy problem in $H$

$$\begin{cases} Z_1'(t) + B_1 Z_1(t) = F_1(t), & 0 < t < T, \\ Z_1(0) = z_1, \end{cases} \tag{8.71}$$

where $B_1 : D(B_1) \subset H \to H$,

$$\begin{aligned} D(B_1) &:= \big\{(p,q) \in H^1(a,b) \times H^2(b,c);\ p(b) = q(b),\ q(c) = 0, \\ &\quad - \alpha_1(b)p(b) + \alpha_2(b)q(b) = \mu(b)q'(b)\big\}, \\ B_1(p,q) &:= \big((\alpha_1 p)' + \beta_1 p, (-\mu q' + \alpha_2 q)' + \beta_2 q\big), \\ Z_1(t) &:= (X_1(\cdot,t), \overline{Y}_1(\cdot,t)),\ F_1(t) := (\widetilde{f}_1(\cdot,t), \overline{g}_1(\cdot,t)),\ z_1 := (0, \overline{y}_1). \end{aligned}$$

Regarding operator $B_1$ we have the following result:

**Lemma 8.2.4.** *Assume that $(i_1)$–$(i_3)$ and $(i_5')$ are satisfied. Then, operator $B_1 + \omega I$ is maximal monotone in $H$ for $\omega$ sufficiently large, where $I$ is the identity of $H$.*

The proof is similar to that of Lemma 8.1.4.

Using the above lemma, we can derive existence, uniqueness and smoothness of the solution of problem $(P.2)_1$. More precisely, we have:

**Theorem 8.2.5.** *Assume that $(i_1)$–$(i_4)$, $(i_5')$ are fulfilled and, in addition,*

$$F_1 \in W^{2,1}(0,T;H); \tag{8.72}$$

$$z_1,\ F_1(0) - B_1 z_1 \in D(B_1). \tag{8.73}$$

*Then problem (8.71) has a unique strong solution $Z_1 \in C^2([0,T];H)$,*

$$\begin{aligned} X_1 &\in C^2([0,T];L^2(a,b)) \bigcap C^1([0,T];H^1(a,b)), \\ Y_1 &\in W^{2,2}(0,T;H^1(b,c)) \bigcap C^1([0,T];H^2(b,c)). \end{aligned}$$

*Proof.* As in the proof of Theorem 8.1.5, we show that problem (8.71) has a unique strong solution $Z_1 \in C^1([0,T];H)$, and moreover $X_1 \in C([0,T];H^1(a,b))$, $Y_1 \in W^{1,2}(0,T;H^1(b,c)) \bigcap C([0,T];H^2(b,c))$.

Note that, under assumptions (8.72) and (8.73), $Z_1'$ is a strong solution of the Cauchy problem obtained by differentiating (8.71) with respect to $t$. Then, reasoning as before, we can see that $Z_1' \in C^1([0,T];H)$,

$$\begin{aligned} X_{1t} &\in C^1([0,T];L^2(a,b)) \bigcap C([0,T];H^1(a,b)), \\ Y_{1t} &\in W^{1,2}(0,T;H^1(b,c)) \bigcap C([0,T];H^2(b,c)). \end{aligned}$$

□

*Remark* 8.2.6. It is easy to formulate sufficient specific conditions for the data which insure that (8.72) and (8.73) are fulfilled. We see that it is enough for $X_0$ to satisfy the following conditions

$$X_{0xx} \in W^{2,1}(0,T;L^2(a,b)),\ X_{0x}(b,\cdot) \in W^{3,1}(0,T). \tag{8.74}$$

Thus, our next goal is to show that $X_0 \in W^{3,1}(0,T;H^2(a,b))$.

The reduced problem $(P.2)_0$ can be expressed as the following Cauchy problem in $H$

$$\begin{cases} Z_0'(t) + B_1 Z_0(t) = F(t), & 0 < t < T, \\ Z_0(0) = z_0, \end{cases} \tag{8.75}$$

where $Z_0(t) := (X_0(\cdot,t), Y_0(\cdot,t)),\ F(t) := (f(\cdot,t), g(\cdot,t)),\ z_0 := (u_0, v_0)$.

Obviously, for $X_0$ we need more regularity than for $Y_0$. However, in this case we cannot split the reduced problem in two distinct problems, since $X_0$ and $Y_0$ are connected by the transmission condition $X_0(b,t) = Y_0(b,t)$. Thus, in the next result we derive for $X_0$ as much regularity as we need, and incidentally we get for $Y_0$ more regularity than necessary.

**Theorem 8.2.7.** *Assume that $(i_1)$–$(i_4)$, $(i_5')$ are fulfilled and*

$$F \in W^{1,1}(0,T;H),\ z_0 \in D(B_1). \tag{8.76}$$

*Then problem (8.75) has a unique strong solution*

$$Z_0 \in C^1([0,T];H) \bigcap \left(C([0,T];H^1(a,b)) \times W^{1,2}(0,T;H^1(b,c))\right).$$

*If, in addition,*

$$\begin{aligned} F \in & W^{5,1}(0,T;H),\ z_k := F^{(k-1)}(0) - B_1 z_{k-1} \in D(B_1),\ k = 1,2,3,4, \\ f \in & W^{3,1}(0,T;H^1(a,b)),\ \alpha_1 \in H^2(a,b),\ \beta_1 \in H^1(a,b), \end{aligned} \tag{8.77}$$

*then $Z_0 \in C^5([0,T];H)$,*

$$\begin{aligned} X_0 &\in W^{3,1}(0,T;H^2(a,b)) \bigcap C^4([0,T];H^1(a,b)), \\ Y_0 &\in W^{5,2}(0,T;H^1(b,c)). \end{aligned}$$

*Proof.* According to Theorem 2.0.27 in Chapter 2, problem (8.75) has a unique strong solution $Z_0 \in C^1([0,T];H)$. Moreover, by a reasoning similar to that used in the proof of Theorem 8.1.5, one can see that

$$(X_0, Y_0) \in C([0,T];H^1(a,b)) \times \left(W^{1,2}(0,T;H^1(b,c)) \bigcap C([0,T];H^2(b,c))\right).$$

Next, by our usual argument based on differentiation of problem (8.75) with respect to $t$, we obtain higher regularity of the solution. Thus, $Z_0^{(k)}$ is a strong solution of the problem

$$\begin{cases} Z_0^{(k+1)}(t) + B_1 Z_0^{(k)}(t) = F^{(k)}(t), \ 0 < t < T, \\ Z_0^{(k)}(0) = z_k, \ k = 1, 2, 3, 4. \end{cases} \tag{8.78}$$

Consequently,

$$Z_0 \in C^5([0,T]; H), \ X_0 \in C^4([0,T]; H^1(a,b)), \ Y_0 \in W^{5,2}(0,T; H^1(b,c)).$$

Moreover, $(8.78)_1$ for $k = 3$ reads

$$(\alpha_1 X_{0ttt})_x = f_{ttt} - X_{0tttt} - \beta_1 X_{0ttt} \ \text{ in } Q_{1T},$$

from which, according to assumptions (8.77), we infer $X_0 \in W^{3,1}(0,T; H^2(a,b))$. □

*Remark* 8.2.8. We can formulate specific conditions implying (8.76) as follows:

$$\begin{cases} (f,g) \in W^{1,1}(0,T; H), \ u_0 \in H^1(a,b), \ v_0 \in H^2(b,c), \\ u_0(b) = v_0(b), \ v_0(c) = 0, \ (\alpha_1 u_0)(b) = (\alpha_2 v_0)(b) - (\mu v_0')(b). \end{cases}$$

Similarly, one can formulate sufficient conditions in terms of the data which imply (8.77). This is left to the reader.

It is worth pointing out that all the conditions we have required (see (8.68), (8.69), (8.73), $(8.76)_2$, (8.77)) are not contradictory, i.e., they form a compatible system of assumptions.

By virtue of what we have proved so far, we can formulate the following two concluding results:

**Corollary 8.2.9.** *Assume that* $(i_1)$–$(i_4)$, $(i_5')$ *are satisfied and, in addition,*

$$\begin{aligned} f \in W^{1,1}(0,T; L^2(a,b)), \ & g \in W^{1,1}(0,T; L^2(b,c)), \\ u_0 \in H^2(a,b), \ & v_0 \in H^2(b,c), \\ u_0'(a) = u_0'(b) = v_0(c) = 0, \ & u_0(b) = v_0(b), \\ \mu(b) v_0'(b) = -\alpha_1(b) u_0(b) + \alpha_2(b) v_0(b). & \end{aligned}$$

*Then problems* $(P.2)_\varepsilon$, $\varepsilon > 0$, *and* $(P.2)_0$ *have unique solutions*

$$\begin{aligned} (u_\varepsilon, v_\varepsilon) \in C^1([0,T]; H) & \bigcap W^{1,2}(0,T; H^1(a,b) \times H^1(b,c)) \\ & \bigcap C([0,T]; H^2(a,b) \times H^2(b,c)), \\ (X_0, Y_0) \in C^1([0,T]; H) & \bigcap (C([0,T]; H^1(a,b)) \times W^{1,2}(0,T; H^1(b,c)). \end{aligned}$$

**Corollary 8.2.10.** *Assume that:* $(i_1)$–$(i_4)$, $(i'_5)$ *are satisfied;* $f$, $g$, $\alpha_1$, $\alpha_2$, $\beta_1$, $\beta_2$, $\mu$, $u_0$, $v_0$ *are smooth enough; and conditions* (8.68), (8.69), (8.73), $(8.76)_2$, (8.77) *are fulfilled. Then, for every* $\varepsilon > 0$, *problem* $(P.2)_\varepsilon$ *has a unique strong solution*

$$(u_\varepsilon, v_\varepsilon) \in C^2([0,T]; L^2(a,b) \times L^2(b,c)) \bigcap W^{2,2}(0,T; H^1(a,b) \times H^1(b,c))$$
$$\bigcap C^1([0,T]; H^2(a,b) \times H^2(b,c)),$$

*and problems* $(P.2)_0$, $(P.2)_1$ *admit unique solutions*

$$(X_0, Y_0) \in (W^{3,1}(0,T; H^2(a,b)) \bigcap C^4([0,T]; H^1(a,b))) \times W^{5,2}(0,T; H^1(b,c)),$$
$$(X_1, Y_1) \in (C^2([0,T]; L^2(a,b)) \bigcap C^1([0,T]; H^1(a,b))) \times W^{2,2}(0,T; H^1(b,c)).$$

### 8.2.3 Estimates for the remainder components

Here we establish estimates for the remainder components with respect to the uniform convergence norm. These estimates validate completely our first order expansion. Under weaker assumptions, we prove that problem $(P.2)_\varepsilon$ is regularly perturbed of order zero with respect to the same norm (see the first part of Theorem 8.2.11 below).

**Theorem 8.2.11.** *Assume that all the assumptions of Corollary* 8.2.9 *are fulfilled. Then, for every* $\varepsilon > 0$, *the following estimates hold true*

$$\|u_\varepsilon - X_0\|_{C(\overline{Q}_{1T})} = \mathcal{O}(\varepsilon^{\frac{1}{8}}), \ \|v_\varepsilon - Y_0\|_{C(\overline{Q}_{2T})} = \mathcal{O}(\varepsilon^{\frac{3}{8}}).$$

*If, moreover, all the assumptions of Corollary* 8.2.10 *are fulfilled, then the solution of problem* $(P.2)_\varepsilon$ *admits an asymptotic expansion of the form* (8.51) *and the following estimates are valid:*

$$\|R_{1\varepsilon}\|_{C(\overline{Q}_{1T})} = \mathcal{O}(\varepsilon^{\frac{5}{4}}), \ \|R_{2\varepsilon}\|_{C(\overline{Q}_{2T})} = \mathcal{O}(\varepsilon^{\frac{3}{2}}).$$

*Proof.* To prove the first part of the theorem, we denote

$$S_{1\varepsilon} = u_\varepsilon - X_0, \ S_{2\varepsilon} = v_\varepsilon - Y_0, \ S_\varepsilon(t) = (S_{1\varepsilon}(\cdot, t), S_{2\varepsilon}(\cdot, t)).$$

By Corollary 8.2.9, we have

$$(S_{1\varepsilon}, S_{2\varepsilon}) \in C^1([0,T]; H) \bigcap (C([0,T]; H^1(a,b)) \times W^{1,2}(0,T; H^1(b,c)).$$

Taking into account problems $(P.2)_\varepsilon$ and $(P.2)_0$, we derive

$$\begin{cases} S_{1\varepsilon t} + (-\varepsilon u_{\varepsilon x} + \alpha_1 S_{1\varepsilon})_x + \beta_1 S_{1\varepsilon} = 0 \text{ in } Q_{1T}, \\ S_{2\varepsilon t} + \big( -\mu S_{2\varepsilon x} + \alpha_2 S_{2\varepsilon} \big)_x + \beta_2 S_{2\varepsilon} = 0 \text{ in } Q_{2T}, \end{cases} \tag{8.79}$$

$$S_{1\varepsilon}(x,0) = 0, \ a \le x \le b, \ S_{2\varepsilon}(x,0) = 0, \ b \le x \le c, \tag{8.80}$$

$$\begin{cases} u_{\varepsilon x}(a,t) = 0, \\ S_{2\varepsilon}(c,t) = 0, \ 0 \le t \le T, \end{cases} \tag{8.81}$$

$$\begin{cases} S_{1\varepsilon}(b,t) = S_{2\varepsilon}(b,t), \\ -\varepsilon u_{\varepsilon x}(b,t) + \alpha_1(b) S_{1\varepsilon}(b,t) \\ = -\mu(b) S_{2\varepsilon x}(b,t) + \alpha_2(b) S_{2\varepsilon}(b,t), \ 0 \le t \le T. \end{cases} \tag{8.82}$$

In the next part of the proof we denote by $M_k$ $(k = 1, 2, \dots)$ different positive constants which depend on the data, but are independent of $\varepsilon$.
By a standard computation (see Theorem 8.1.14), we derive

$$\begin{aligned} \frac{1}{2}\|S_\varepsilon(t)\|^2 + \frac{\varepsilon}{2}\int_0^t \|S_{1\varepsilon x}(\cdot,s)\|_1^2 ds + \frac{\mu_0}{2}\int_0^t \|S_{2\varepsilon x}(\cdot,s)\|_2^2 ds \\ \le \frac{\varepsilon}{2} e^{2\omega T}\int_0^t \|X_{0x}(\cdot,s)\|_1^2 ds \ \forall\, t \in [0,T], \end{aligned} \tag{8.83}$$

which implies $\| S_\varepsilon(t) \|^2 \le M_1\varepsilon$, and therefore

$$\| S_{1\varepsilon}(\cdot,t) \|_1 \le \sqrt{M_1}\varepsilon^{1/2}, \ \ \| S_{2\varepsilon}(\cdot,t) \|_2 \le \sqrt{M_1}\varepsilon^{1/2} \ \ \forall\, t \in [0,T]. \tag{8.84}$$

Since $W_\varepsilon$ is a strong solution of problem (8.65), it follows by Theorem 2.0.27 (see also (2.8) in Chapter 2)

$$\|W_\varepsilon'(t)\| \le e^{\omega t}\Big(\|F(0) - L_\varepsilon W_0\| + \int_0^t e^{-\omega s}\|F'(s)\| ds\Big) \le M_2 \tag{8.85}$$

for all $t \in [0,T]$, and hence

$$\| u_{\varepsilon t} \|_{C([0,T];L^2(a,b))} = \mathcal{O}(1), \ \ \| v_{\varepsilon t} \|_{C([0,T];L^2(b,c))} = \mathcal{O}(1). \tag{8.86}$$

Using again system (8.79), we get after scalar multiplication in $H$ by $S_\varepsilon(t)$

$$\begin{aligned} \varepsilon\|S_{1\varepsilon x}(\cdot,t)\|_1^2 + \frac{\mu_0}{2}\|S_{2\varepsilon x}(\cdot,t)\|_2^2 - \omega \| S_\varepsilon(t) \|^2 \\ \le \| S_\varepsilon'(t) \| \cdot \| S_\varepsilon(t) \| + \varepsilon \| X_{0x}(\cdot,t) \|_1 \cdot \| S_{1\varepsilon x}(\cdot,t) \|_1 \end{aligned}$$

for all $t \in [0,T]$. Combining this inequality with (8.84) and (8.86), we derive

$$\|S_{1\varepsilon x}(\cdot,t)\|_1 \le M_3\varepsilon^{-1/4}, \ \ \|S_{2\varepsilon x}(\cdot,t)\|_2 \le M_4\varepsilon^{1/4} \ \ \forall\, t \in [0,T]. \tag{8.87}$$

On the other hand, using (8.84) and the mean value theorem, we can associate with each $(t,\varepsilon)$ a number $x_{t\varepsilon} \in [a,b]$ such that $| S_{1\varepsilon}(x_{t\varepsilon},t) | \le M_5\varepsilon^{1/2}$. Now, taking into account the following obvious formulas

$$S_{1\varepsilon}(x,t)^2 = 2\int_{x_{t\varepsilon}}^x S_{1\varepsilon}(y,t)S_{1\varepsilon y}(y,t)dy + S_{1\varepsilon}(x_{t\varepsilon},t)^2 \ \ \forall\, (x,t) \in \overline{Q}_{1T},$$

$$S_{2\varepsilon}(x,t)^2 = -2\int_x^c S_{2\varepsilon}(y,t)S_{2\varepsilon y}(y,t)dy \ \ \forall\, (x,t) \in \overline{Q}_{2T}$$

as well as estimates (8.84) and (8.87), we easily derive the first two estimates of our theorem.

*Remark* 8.2.12. If $X_0$ were more regular, for instance $X_0 \in W^{1,2}(0,T;H^1(a,b))$, then it would follow by a reasoning similar to that used in the proof of Theorem 8.1.14 that $\| S_{\varepsilon t} \|_{C([0,T];H)} = \mathcal{O}(\sqrt{\varepsilon})$, and therefore the following stronger estimates would hold true

$$\|u_\varepsilon - X_0\|_{C(\overline{Q}_{1T})} = \mathcal{O}(\varepsilon^{\frac{1}{4}}), \ \|v_\varepsilon - Y_0\|_{C(\overline{Q}_{2T})} = \mathcal{O}(\varepsilon^{\frac{1}{2}}).$$

Let us now prove the last part of the theorem. Obviously, all terms of our expansion (8.51) are well defined. Thus, we only need to prove that the remainder components satisfy the corresponding estimates. Denote

$$\widetilde{R}_{1\varepsilon}(x,t) = R_{1\varepsilon}(x,t) + \varepsilon i_1(\zeta(b),t), \ (x,t) \in Q_{1T}.$$

It is easily seen that $\widetilde{R}_\varepsilon(t) := (\widetilde{R}_{1\varepsilon}(\cdot,t), \ R_{2\varepsilon}(\cdot,t))$ satisfies the problem

$$\begin{cases} \widetilde{R}_{1\varepsilon t} + \big(-\varepsilon T_{1\varepsilon x} + \alpha_1 \widetilde{R}_{1\varepsilon}\big)_x + \beta_1 \widetilde{R}_{1\varepsilon} = \widetilde{h}_\varepsilon \text{ in } Q_{1T}, \\ R_{2\varepsilon t} + \big(-\mu R_{2\varepsilon x} + \alpha_2 R_{2\varepsilon}\big)_x + \beta_2 R_{2\varepsilon} = 0 \text{ in } Q_{2T}, \\ \widetilde{R}_{1\varepsilon}(x,0) = 0, \ a \le x \le b, \ R_{2\varepsilon}(x,0) = 0, \ b \le x \le c, \\ T_{1\varepsilon x}(a,t) = 0, \ R_{2\varepsilon}(c,t) = 0, \\ \widetilde{R}_{1\varepsilon}(b,t) = R_{2\varepsilon}(b,t), \\ \varepsilon T_{1\varepsilon x}(b,t) - \alpha_1(b)\widetilde{R}_{1\varepsilon}(b,t) \\ = \mu(b)R_{2\varepsilon x}(b,t) - \alpha_2(b)R_{2\varepsilon}(b,t) - \varepsilon i_{1\zeta}(\zeta(b),t), \ \ 0 \le t \le T, \end{cases} \tag{8.88}$$

where

$$\widetilde{h}_\varepsilon(x,t) = h_\varepsilon(x,t) + \varepsilon i_{1t}(\zeta(b),t) + \varepsilon(\alpha_1' + \beta_1)(x) i_1(\zeta(b),t).$$

Next, by a computation similar to that used in the proof of Theorem 8.1.14, we get

$$\begin{aligned} \frac{1}{2}\|\widetilde{R}_\varepsilon(t)\|^2 &+ \varepsilon \int_0^t \|\widetilde{R}_{1\varepsilon x}(\cdot,s)\|_1^2 ds + \frac{\mu_0}{2}\int_0^t \|R_{2\varepsilon x}(\cdot,s)\|_2^2 ds \\ &\le e^{2\omega T}\Big(\varepsilon^2 \int_0^t \| X_{1x}(\cdot,s)\|_1 \cdot \| \widetilde{R}_{1\varepsilon x}(\cdot,s)\|_1 \, ds \\ &+ \varepsilon \int_0^t | i_{1\zeta}(\zeta(b),t) R_{2\varepsilon}(b,s) | \, ds \\ &+ \int_0^t \|\widetilde{h}_\varepsilon(\cdot,s)\|_1 \cdot \|\widetilde{R}_{1\varepsilon}(\cdot,s)\|_1 ds\Big) \ \forall\, t \in [0,T]. \end{aligned} \tag{8.89}$$

Arguing as in the proof of Theorem 8.1.14, one can show that $\| v_{\varepsilon t} \|_{C(\overline{Q}_{2T})} = \mathcal{O}(1)$, which implies that

$$\| R_{2\varepsilon t}(b,\cdot) \|_{C[0,T]} = \mathcal{O}(1), \ \ \| R_{2\varepsilon}(b,\cdot) \|_{C[0,T]} = \mathcal{O}(1). \tag{8.90}$$

Note that $\widetilde{h}_\varepsilon \in W^{1,2}(0,T;L^2(a,b))$. We have

$$\| \widetilde{h}_\varepsilon(\cdot,t) \|_1 \leq M_6 \varepsilon^{\frac{3}{2}} \quad \forall\, t \in [0,T].$$

On the other hand, for every $k \geq 1$ there exists a positive constant $M_7$ such that

$$| i_{1\zeta}(\zeta(b),t) | \leq M_7 \varepsilon^k \ \forall t \in [0,T].$$

Taking into account the last two estimates, we infer from (8.89) and $(8.90)_2$

$$\|\widetilde{R}_{1\varepsilon}(\cdot,t)\|_1 \leq M_8 \varepsilon^{\frac{3}{2}}, \ \|R_{2\varepsilon}(\cdot,t)\|_2 \leq M_9 \varepsilon^{\frac{3}{2}} \ \forall\, t \in [0,T]. \tag{8.91}$$

Now we are going to prove some estimates for $\widetilde{R}_{1\varepsilon t}$ and $R_{2\varepsilon t}$. Denote $\overline{R}_{1\varepsilon} = \widetilde{R}_{1\varepsilon t}$, $\overline{R}_{2\varepsilon} = R_{2\varepsilon t}$. Note that the problem satisfied by the remainder components can be differentiated with respect to $t$ and thus $\overline{R}_{1\varepsilon}$, $\overline{R}_{2\varepsilon}$ satisfy the following problem

$$\begin{cases} \overline{R}_{1\varepsilon t} + \big( -\varepsilon T_{1\varepsilon t x} + \alpha_1 \overline{R}_{1\varepsilon}\big)_x + \beta_1 \overline{R}_{1\varepsilon} = \widetilde{h}_{\varepsilon t} \text{ in } Q_{1T}, \\ \overline{R}_{2\varepsilon t} + \big( -\mu \overline{R}_{2\varepsilon x} + \alpha_2 \overline{R}_{2\varepsilon}\big)_x + \beta_2 \overline{R}_{2\varepsilon} = 0 \text{ in } Q_{2T}, \\ \overline{R}_{1\varepsilon}(x,0) = 0, \ a \leq x \leq b, \ \overline{R}_{2\varepsilon}(x,0) = 0, \ b \leq x \leq c, \\ T_{1\varepsilon t x}(a,t) = 0, \ \overline{R}_{2\varepsilon}(c,t) = 0, \\ \overline{R}_{1\varepsilon}(b,t) = \overline{R}_{2\varepsilon}(b,t), \\ \varepsilon T_{1\varepsilon t x}(b,t) - \alpha_1(b) \overline{R}_{1\varepsilon}(b,t) \\ = \mu(b) \overline{R}_{2\varepsilon x}(b,t) - \alpha_2(b) \overline{R}_{2\varepsilon}(b,t) - \varepsilon i_{1t\zeta}(\zeta(b),t)), \quad 0 \leq t \leq T. \end{cases} \tag{8.92}$$

A standard computation leads us to the estimate

$$\begin{aligned} \frac{1}{2}\|\overline{R}_\varepsilon(t)\|^2 \leq e^{2\omega T}\Big( & \frac{\varepsilon^3}{2} \int_0^t \|X_{1tx}(\cdot,s)\|_1^2 ds \\ & + \int_0^t \|\widetilde{h}_{\varepsilon t}(\cdot,s)\|_1 \cdot \|\overline{R}_{1\varepsilon}(\cdot,s)\|_1 ds \\ & + \varepsilon \int_0^t | \overline{R}_{2\varepsilon}(b,s) i_{1t\zeta}(\zeta(b),s) | \, ds \Big) \end{aligned} \tag{8.93}$$

for all $t \in [0,T]$. Since

$$\int_0^t \| \widetilde{h}_{\varepsilon t}(\cdot,s) \|_1 \, ds \leq M_{10} \varepsilon^{\frac{3}{2}} \ \forall\, t \in [0,T],$$

and for every $k \geq 1$ there exists a positive constant $M_{11}$ such that

$$| i_{1t\zeta}(\zeta(b),t) | \leq M_{11} \varepsilon^k \text{ for all } t \in [0,T],$$

it follows from $(8.90)_1$ and (8.93) that

$$\| \widetilde{R}_{1\varepsilon t}(\cdot,t) \|_1 \leq M_{12} \varepsilon^{\frac{3}{2}}, \ \| R_{2\varepsilon t}(\cdot,t) \|_2 \leq M_{13} \varepsilon^{\frac{3}{2}} \ \forall\, t \in [0,T]. \tag{8.94}$$

Next, combining the obvious estimate

$$
\begin{aligned}
\langle \widetilde{R}'_\varepsilon(t), \widetilde{R}_\varepsilon(t)\rangle + \frac{\varepsilon}{2} \parallel \widetilde{R}_{1\varepsilon x}(\cdot,t) \parallel_1^2 + \frac{\mu_0}{2} \parallel R_{2\varepsilon x}(\cdot,t) \parallel_2^2 \\
\leq \omega \parallel \widetilde{R}_\varepsilon(t) \parallel^2 + \varepsilon \mid R_{2\varepsilon}(b,t) i_{1\zeta}(\zeta(b),t) \mid \\
+ \frac{\varepsilon^3}{2} \parallel X_{1x}(\cdot,t) \parallel_1^2 + \parallel \widetilde{h}_\varepsilon(\cdot,t) \parallel_1 \cdot \parallel \widetilde{R}_{1\varepsilon}(\cdot,t) \parallel_1 \ \forall\, t \in [0,T]
\end{aligned}
$$

with (8.91) and (8.94), we infer that

$$
\parallel \widetilde{R}_{1\varepsilon x}(\cdot,t) \parallel_1^2 \leq M_{14}\varepsilon^2, \ \parallel R_{2\varepsilon x}(\cdot,t) \parallel_2^2 \leq M_{15}\varepsilon^3 \ \forall\, t \in [0,T].
$$

The rest of the proof is similar to that of Theorem 8.1.14. □

## 8.3 A zeroth order asymptotic expansion for the solution of problem $(P.3)_\varepsilon$

In this section we investigate the nonlinear coupled problem $(P.3)_\varepsilon$ (which comprises $(S)$, $(IC)$, $(TC)$, and $(BC.3)$), under assumptions $(i_1)$–$(i_6)$. Since $\alpha_1 > 0$, a boundary (transition) layer of order zero is expected to occur near the line segment $\{(b,t);\ t \in [0,T]\}$ with respect to the uniform norm, like in the linear case investigated in Section 8.1. This is indeed the case, as proved below.

### 8.3.1 Formal expansion

Since the problem under investigation is nonlinear, we confine ourselves to constructing a zeroth order asymptotic expansion for the solution. We could determine higher order asymptotic expansions, but much more difficult computations are needed and additional assumptions on the data should be required. Having in mind the linear case studied in Section 8.1, we suggest a zeroth order asymptotic expansion for the solution $U_\varepsilon := \big(u_\varepsilon(x,t), v_\varepsilon(x,t)\big)$ of problem $(P.3)\varepsilon$ in the form:

$$
\begin{cases}
u_\varepsilon(x,t) = X_0(x,t) + i_0(\xi,t) + R_{1\varepsilon}(x,t), \\
v_\varepsilon(x,t) = Y_0(x,t) + R_{2\varepsilon}(x,t),
\end{cases} \tag{8.95}
$$

where the terms of the expansion have the same meanings as in Subsection 8.1.1.

By the standard identification procedure, we can see that the components of the zeroth order regular term, $X_0$, $Y_0$, satisfy the reduced problem, denoted $(P.3)_0$

$$
\begin{cases}
X_{0t} + (\alpha_1 X_0)_x + \beta_1 X_0 = f \ \text{ in } \ Q_{1T}, \\
Y_{0t} + (-\mu Y_{0x} + \alpha_2 Y_0)_x + \beta_2 Y_0 = g \ \text{ in } \ Q_{2T},
\end{cases} \tag{8.96}
$$

$$\begin{cases} X_0(x,0) = u_0(x), \ x \in [a,b], \\ Y_0(x,0) = v_0(x), \ x \in [b,c], \end{cases} \tag{8.97}$$

$$\begin{cases} X_0(a,t) = 0, \\ -\mu(b)Y_{0x}(b,t) = \alpha_1(b)X_0(b,t) - \alpha_2(b)Y_0(b,t), \\ -Y_{0x}(c,t) = \gamma(Y_0(c,t)), \ \ 0 \le t \le T, \end{cases} \tag{8.98}$$

while the (zeroth order) transition layer function (correction) is given by

$$i_0(\xi,t) = (Y_0(b,t) - X_0(b,t))e^{-\alpha(b)\xi}.$$

Finally, for the remainder components we obtain the problem

$$\begin{cases} R_{1\varepsilon t} + (-\varepsilon\,(R_{1\varepsilon x} + X_{0x}) + \alpha_1 R_{1\varepsilon})_x + \beta_1 R_{1\varepsilon} = h_\varepsilon \ \text{ in } Q_{1T}, \\ R_{2\varepsilon t} + \big( - \mu R_{2\varepsilon x} + \alpha_2 R_{2\varepsilon}\big)_x + \beta_2 R_{2\varepsilon} = 0 \ \text{ in } Q_{2T}, \\ R_{1\varepsilon}(x,0) = 0, \ \ a \le x \le b, \ R_{2\varepsilon}(x,0) = 0, \ \ b \le x \le c, \\ R_{1\varepsilon}(a,t) = -i_0(\xi(a),t), \ R_{1\varepsilon}(b,t) = R_{2\varepsilon}(b,t), \\ -\varepsilon\,(R_{1\varepsilon}(b,t) + X_0(b,t))_x + \alpha_1(b)R_{1\varepsilon}(b,t) \\ = -\mu(b)R_{2\varepsilon x}(b,t) + \alpha_2(b)R_{2\varepsilon}(b,t), \\ -R_{2\varepsilon x}(c,t) = \gamma(v_\varepsilon(c,t)) - \gamma(Y_0(c,t)), \ \ 0 \le t \le T, \end{cases} \tag{8.99}$$

where $\xi(a) = (b-a)/\varepsilon$,

$$h_\varepsilon(x,t) = -i_{0t}(\xi,t) - (\beta_1(x) + \alpha_1'(x))i_0(\xi,t) + \varepsilon^{-1}(\alpha_1(x) - \alpha_1(b))i_{0\xi}(\xi,t).$$

We also obtain the condition

$$i_0(\xi,0) = 0 \ \Leftrightarrow \ u_0(b) = v_0(b),$$

which will appear again in the next subsection among the compatibility conditions that are required for the data to achieve higher regularity of the solutions.

### 8.3.2 Existence, uniqueness and regularity of the solutions of problems $(P.3)_\varepsilon$ and $(P.3)_0$

To investigate problem $(P.3)_\varepsilon$, we consider as a basic framework the Hilbert space $H := L^2(a,b) \times L^2(b,c)$ we have already introduced in Subsection 8.1.2, endowed with the usual scalar product, denoted $\langle \cdot, \cdot \rangle$, and the corresponding induced norm, denoted $\| \cdot \|$. We define the nonlinear operator $J_\varepsilon : D(J_\varepsilon) \subset H \to H$,

$$\begin{aligned} D(J_\varepsilon) :=& \Big\{ (h,k) \in H, \ (h,k) \in H^2(a,b) \times H^2(b,c), \ h(b) = k(b), \ h(a) = 0, \\ -k'(c) =& \gamma(k(c)), \ \varepsilon h'(b) - \alpha_1(b)h(b) = \mu(b)k'(b) - \alpha_2(b)k(b) \Big\}, \\ J_\varepsilon(h,k) :=& \big( (-\varepsilon h' + \alpha_1 h)' + \beta_1 h, \ (-\mu k' + \alpha_2 k)' + \beta_2 k \big). \end{aligned}$$

Obviously, problem $(P.3)_\varepsilon$ can be written as the following Cauchy problem in $H$

$$\begin{cases} W_\varepsilon'(t) + J_\varepsilon W_\varepsilon(t) = F(t), \quad 0 < t < T, \\ W_\varepsilon(0) = W_0, \end{cases} \tag{8.100}$$

where $W_\varepsilon(t) := \big(u_\varepsilon(\cdot,t), v_\varepsilon(\cdot,t)\big)$, $W_0 := \big(u_0, v_0\big)$, $F(t) := \big(f(\cdot,t), g(\cdot,t)\big)$. Regarding $J_\varepsilon$, we have the following result

**Lemma 8.3.1.** *Assume that* $(i_1)$–$(i_3)$ *and* $(i_6)$ *are satisfied. Then, there is a positive number* $\omega$*, independent of* $\varepsilon$*, such that* $J_\varepsilon + \omega I$ *is maximal monotone, where* $I$ *is the identity of* $H$*.*

*Proof.* It is similar to the proof of Lemma 8.1.1, so we will just sketch it.
By a standard computation involving Lemma 7.1.2 in Chapter 7, it follows that operator $J_\varepsilon + \omega I$ is monotone for some $\omega$ large enough.
To prove the maximality of this operator one can use Theorem 2.0.6 and Theorem 7.3.1 (see also Remark 7.3.2 in Chapter 7). □

Using the above lemma, we are able to derive existence and uniqueness for problem $(P.3)_\varepsilon$, as follows:

**Theorem 8.3.2.** *Assume that* $(i_1)$–$(i_6)$ *are fulfilled and*

$$F \in W^{1,1}(0,T;H), \quad W_0 \in D(J_\varepsilon). \tag{8.101}$$

*Then problem* (8.100) *has a unique strong solution* $W_\varepsilon$ *which belongs to*

$$W^{1,\infty}(0,T;H) \bigcap W^{1,2}(0,T;H^1(a,b) \times H^1(b,c)) \bigcap L^\infty(0,T;H^2(a,b) \times H^2(b,c)).$$

The proof is similar to that corresponding to the linear case. Here, we can use Theorem 2.0.20 instead of Theorem 2.0.27. See the first part of the proof of Theorem 8.1.2.

*Remark* 8.3.3. For our asymptotic analysis, it would be convenient to have assumptions which are independent of $\varepsilon$ and insure that Theorem 8.3.2 is valid for all $\varepsilon > 0$. This is indeed possible: conditions (8.101) hold if the following sufficient assumptions (independent of $\varepsilon$) are fulfilled:

$$\begin{cases} f \in W^{1,1}(0,T;L^2(a,b)), \ g \in W^{1,1}(0,T;L^2(b,c)), \\ u_0 \in H^2(a,b), \ v_0 \in H^2(b,c), \end{cases}$$

$$\begin{cases} u_0(a) = 0, \ u_0(b) = v_0(b), \ u_0'(b) = 0, \\ -\mu(b)v_0'(b) = \alpha_1(b)u_0(b) - \alpha_2(b)v_0(b), \\ -v_0'(c) = \gamma(v_0(c)). \end{cases} \tag{8.102}$$

Next we examine problem $(P.3)_0$. For our asymptotic analysis, we need a solution of this problem which satisfies at least the following smoothness conditions

$$X_0 \in W^{1,2}(0,T;H^1(a,b)),\ Y_0 \in W^{1,2}(0,T;H^1(b,c)).$$

We define the operator $A : D(A) \subset H \to H$,

$$\begin{aligned} D(A) &:= \big\{(p,q) \in H^1(a,b) \times H^2(b,c);\ p(a) = 0,\ \ -q'(c) = \gamma(q(c)), \\ &\quad -\alpha_1(b)p(b) + \alpha_2(b)q(b) = \mu(b)q'(b)\big\}, \\ A(p,q) &:= \big((\alpha_1 p)' + \beta_1 p, (-\mu q' + \alpha_2 q)' + \beta_2 q\big). \end{aligned}$$

As usual, let us express our problem $(P.3)_0$ as a Cauchy problem in $H$:

$$\begin{cases} Z'(t) + AZ(t) = F(t), & 0 < t < T, \\ Z(0) = z_0, \end{cases} \tag{8.103}$$

where $Z(t) := (X_0(\cdot,t), Y_0(\cdot,t)),\ F(t) := (f(\cdot,t), g(\cdot,t)),\ z_0 := (u_0, v_0)$. Concerning operator $A$ we have the following result:

**Lemma 8.3.4.** *Assume that* $(i_1)$–$(i_3)$ *and* $(i_6)$ *are satisfied. Then, operator* $A+\omega I$ *is maximal monotone in* $H$ *for* $\omega > 0$ *sufficiently large, where* $I$ *is the identity of* $H$.

The proof is similar to that of Lemma 8.1.4 and relies on Theorem 7.3.3 and Remark 7.3.4 in Chapter 7.

Using this lemma we derive the following

**Theorem 8.3.5.** *Assume that* $(i_1)$–$(i_6)$ *are fulfilled and, in addition,*

$$F \in W^{1,1}(0,T;H), z_0 \in D(A). \tag{8.104}$$

*Then problem* (8.103) *has a unique strong solution*

$$\begin{aligned} &Z \in W^{1,\infty}(0,T;H),\ X_0 \in L^\infty(0,T;H^1(b,c)), \\ &Y_0 \in W^{1,2}(0,T;H^1(b,c)) \bigcap L^\infty(0,T;H^2(b,c)). \end{aligned}$$

The proof is omitted since it is similar to that of Theorem 8.1.5.

As shown before, we need $X_0 \in W^{1,2}(0,T;H^1(a,b))$.

In order to achieve this regularity for $X_0$, we consider the following separate problem for $X_0$

$$\begin{cases} X_{0t} + (\alpha_1 X_0)_x + \beta_1 X_0 = f \ \text{ in } \ Q_{1T}, \\ X_0(x,0) = u_0(x),\ a < x < b, \\ X_0(a,t) = 0,\ 0 < t < T, \end{cases}$$

which is similar to problem (8.25) in Subsection 8.1.2. Thus, if we assume that

$$\begin{aligned} &f \in W^{2,1}(0,T;L^2(a,b)),\ f(\cdot,0), \beta_1 \in H^1(a,b),\ u_0, \alpha_1 \in H^2(a,b), \\ &u_0(a) = f(a,0) - (\alpha_1 u_0)'(a) = 0, \end{aligned}$$

then (see Theorem 8.1.7) $X_0 \in C^2([0,T];L^2(a,b)) \bigcap C^1([0,T];H^1(a,b))$.

Summarizing, we have

**Corollary 8.3.6.** *Assume that assumptions* $(i_1)$–$(i_6)$ *are satisfied. If* $f$, $g$, $\alpha_1$, $\alpha_2$, $\beta_1$, $\beta_2$, $\mu, u_0$ *and* $v_0$ *are smooth enough and the compatibility conditions*

$$\begin{cases} u_0(a) = 0, \;\; u_0(b) = v_0(b), \;\; u_0'(b) = 0, \\ -\mu(b)v_0'(b) = \alpha_1(b)u_0(b) - \alpha_2(b)v_0(b), \\ f(a,0) - (\alpha_1 u_0)'(a) = 0, \;\; -v_0'(c) = \gamma(v_0(c)) \end{cases}$$

*are all satisfied, then problem* $(P.3)_\varepsilon$ *has for every* $\varepsilon > 0$ *a unique strong solution*

$$\begin{aligned}(u_\varepsilon, v_\varepsilon) \in & W^{1,\infty}(0,T; L^2(a,b) \times L^2(b,c)) \bigcap W^{1,2}(0,T; H^1(a,b) \times H^1(b,c)) \\ & \bigcap L^\infty(0,T; H^2(a,b) \times H^2(b,c)),\end{aligned}$$

*and problem* $(P.3)_0$ *admits a unique solution*

$$\begin{aligned} X_0 &\in C^1([0,T]; H^1(a,b)) \bigcap C^2([0,T]; L^2(a,b)), \\ Y_0 &\in W^{1,\infty}(0,T; L^2(b,c)) \bigcap W^{1,2}(0,T; H^1(b,c)). \end{aligned}$$

### 8.3.3 Estimates for the remainder components

According to the above qualitative analysis, our expansion (8.95) is well defined. Here we state some estimates for the remainder components, which are more than enough to validate this expansion completely. More precisely, we have

**Theorem 8.3.7.** *Assume that all the assumptions of Corollary* 8.3.6 *are fulfilled. Then, for every* $\varepsilon > 0$, *the solution of problem* $(P.3)_\varepsilon$ *admits an asymptotic expansion of the form* (8.95) *and the following estimates hold true*

$$\|R_{1\varepsilon}\|_{C(\overline{Q}_{1T})} = \mathcal{O}(\varepsilon^{\frac{1}{8}}), \;\; \|R_{2\varepsilon}\|_{C(\overline{Q}_{2T})} = \mathcal{O}(\varepsilon^{\frac{3}{8}}).$$

The proof follows ideas similar to those we have already used in the proof of Theorem 8.1.14. It is left to the reader, as an exercise.

*Remark* 8.3.8. If $X_0$, $Y_0$ were more regular (this is possible under additional assumptions), we would be able to get better estimates for the remainder components, as follows

$$\|R_{1\varepsilon}\|_{C(\overline{Q}_{1T})} = \mathcal{O}(\varepsilon^{\frac{1}{4}}), \;\; \|R_{2\varepsilon}\|_{C(\overline{Q}_{2T})} = \mathcal{O}(\varepsilon^{\frac{1}{2}}).$$

# Part IV

# Elliptic and Hyperbolic Regularizations of Parabolic Problems

# Chapter 9

# Presentation of the Problems

While in Parts II and III of the book we discussed the possibility to replace singular perturbation problems with the corresponding reduced models, in what follows we aim at reversing the process in the sense that we replace given problems with singularly perturbed, higher order (with respect to $t$) problems, admitting solutions which are more regular and approximate the solutions of the original problems. More precisely, let us consider the classical heat equation

$$u_t - \Delta u = f(x,t), \ x \in \Omega, \ 0 < t < T, \tag{E}$$

with which we associate the Dirichlet boundary condition

$$u(x,t) = 0 \text{ for } x \in \partial\Omega, \ 0 < t < T, \tag{BC}$$

and the initial condition

$$u(x,0) = u_0(x), \ x \in \Omega, \tag{IC}$$

where $\Omega$ is a nonempty open bounded subset of $\mathbb{R}^n$ with smooth boundary $\partial\Omega$, and $\Delta$ is the Laplace operator with respect to $x = (x_1, \ldots, x_n)$, i.e., $\Delta u = \sum_{i=1}^{n} u_{x_i x_i}$. Denote by $P_0$ this initial-boundary value problem. If we add $\pm\varepsilon u_{tt}$, $0 < \varepsilon \ll 1$, to equation $(E)$, we obtain a hyperbolic or elliptic equation, respectively,

$$\varepsilon u_{tt} + u_t - \Delta u = f(x,t), \ x \in \Omega, \ 0 < t < T, \tag{HE}$$

$$\varepsilon u_{tt} + \Delta u = u_t - f(x,t), \ x \in \Omega, \ 0 < t < T. \tag{EE}$$

Now, if we associate with each of the resulting equations the original conditions $(BC)$ and $(IC)$, then we obtain new problems. Obviously, these problems are incomplete, since both $(EE)$ and $(HE)$ are of a higher order with respect to $t$ than the original heat equation. For each problem we need to add one additional condition to get a complete problem. We prefer to add a condition at $t = T$ for equation $(EE)$, either for $u$ or for $u_t$, and an initial condition at $t = 0$ for $u_t$

in the case of equation $(HE)$. So, depending on the case, we obtain an elliptic or hyperbolic regularization of the original problem, which is expected to have a more regular solution that approximates in some sense the solution of problem $P_0$.

In fact, what we have said so far can be expressed in an abstract form. Let $H$ be a real Hilbert space. Consider the following linear evolution problem in $H$, denoted again $P_0$ :

$$\begin{cases} u' + Lu = f(t), \ t \in (0, T), \\ u(0) = u_0, \end{cases}$$

where $L : D(L) \subset H \rightarrow H$ is assumed to be linear, self-adjoint and positive (or, more generally, $L + \omega I$ is positive for some $\omega > 0$), while $f$ is a given function, say $f \in L^2(0, T; H)$ (see the next chapter for more precise assumptions).

Now, let us consider the following second order equation, again denoted $(EE)$,

$$\varepsilon u'' - u' - Lu = -f(t), \ t \in (0, T), \tag{$EE$}$$

with which we associate either

$$u(0) = u_0, \quad u(T) = u_T, \tag{$BC.1$}$$

or

$$u(0) = u_0, \quad u'(T) = u_T. \tag{$BC.2$}$$

We denote by $(P.k)_\varepsilon$ the problem $(EE)$, $(BC.k)$, $k = 1, 2$. Both these problems are "elliptic regularizations" of the original problem $P_0$.

If instead of equation $(EE)$ we consider the equation, again denoted $(HE)$,

$$\varepsilon u'' + u' + Lu = f(t), \ t \in (0, T), \tag{$HE$}$$

together with the initial conditions

$$u(0) = u_0, \quad u'(0) = u_1, \tag{$IC.3$}$$

we obtain a "hyperbolic regularization" of problem $P_0$, denoted $(P.3)_\varepsilon$. Problems $(P.k)_\varepsilon$, $k = 1, 2, 3$, will be investigated in the next chapter. Note that we will require different assumptions in the case of the hyperbolic regularization. All the three problems are singularly perturbed of the boundary layer type (the last two are singularly perturbed of order one), with respect to the norm of $C([0, T]; H)$. Obviously, from our abstract setting we can derive results for the heat equation we started with. In fact, we will illustrate our results on a more general parabolic equation.

In Chapter 11 we shall investigate regularizations of the Dirichlet problem associated with the nonlinear heat equation

$$u_t - \Delta u + \beta(u) = f(x, t), \ x \in \Omega, \ 0 < t < T, \tag{$NE$}$$

where $\beta : \mathbb{R} \longrightarrow \mathbb{R}$ is a nonlinear function. More precisely, the starting problem $P_0$ will be the following

$$\begin{cases} u_t - \Delta u + \beta(u) = f(x,t), \ x \in \Omega, \ 0 < t < T, \\ u(x,t) = 0 \ \text{ for } \ x \in \partial\Omega, \ 0 < t < T, \\ u(x,0) = u_0(x), \ x \in \Omega. \end{cases}$$

Like in the linear case, we will investigate elliptic regularizations given by the nonlinear elliptic equation

$$\varepsilon u_{tt} + \Delta u = \beta(u) + u_t - f(x,t), \ x \in \Omega, \ 0 < t < T, \qquad (NEE)$$

with the Dirichlet boundary condition on $\partial\Omega$ as well as two-point boundary conditions in the form $(BC.1)$ or $(BC.2)$. Also, we will be interested in the hyperbolic regularization given by the nonlinear wave equation

$$\varepsilon u_{tt} + u_t - \Delta u + \beta(u) = f(x,t), \ x \in \Omega, \ 0 < t < T, \qquad (NHE)$$

together with the Dirichlet boundary condition on $\partial\Omega$ and initial conditions in the form $(IC.3)$. Since there is no danger of confusion, we will denote again the three nonlinear regularizations by $(P.k)_\varepsilon$, $k = 1, 2, 3$. We will show that, under suitable hypotheses, problem $(P.1)_\varepsilon$ is singularly perturbed of order zero, while problems $(P.k)_\varepsilon$, $k = 2, 3$, are singularly perturbed of order one, all of them with respect to the norm of $C([0,T]; H)$.

# Chapter 10

# The Linear Case

As promised in Chapter 9, we are going to study in this chapter elliptic and hyperbolic regularizations of the following linear evolution problem in $H$, denoted $P_0$ :

$$\begin{cases} u' + Lu = f(t), \ t \in (0,T), \\ u(0) = u_0, \end{cases}$$

where $H$ is a real Hilbert space. The scalar product of $H$ and the norm induced by it will be denoted $\langle \cdot, \cdot \rangle$ and $\| \cdot \|$, respectively.

The following assumptions will be used in what follows:

$(h_1)$ $L : D(L) \subset H \to H$ is linear, densely defined, self-adjoint, and positive, i.e., $\langle Lu, u \rangle \geq 0 \ \forall \ u \in D(L)$;

$(h_2)$ $u_0 \in D(L)$;

$(h_3)$ $f \in L^2(0,T;H)$.

According to our discussion in the previous chapter, we examine in what follows two elliptic regularizations of problem $P_0$, denoted $(P.1)_\varepsilon$ and $(P.2)_\varepsilon$, which are defined by the linear perturbed equation

$$\varepsilon u'' - u' - Lu = -f(t), \ t \in (0,T), \tag{EE}$$

with boundary conditions in the form $(BC.1)$ or $(BC.2)$, respectively.

If instead of equation $(EE)$ we consider the equation

$$\varepsilon u'' + u' + Lu = f(t), \ t \in (0,T), \tag{HE}$$

with initial conditions of the form $(IC.3)$, then we obtain a hyperbolic regularization of problem $P_0$, denoted $(P.3)_\varepsilon$.

In this last case, we shall require different assumptions:

$(i_1)$ $V$ and $H$ are two real Hilbert spaces, $V \subset H \subset V^*$, with dense and continuous inclusions, where $V^*$ stands for the dual of $V$, and $H$ is identified with its own dual.

In order to formulate further assumptions, let us denote by $(\cdot,\cdot)$ and $\langle\cdot,\cdot\rangle$ the scalar products of $V$ and $H$, respectively. The norms induced by these scalar products will be denoted $|\cdot|$ and $\|\cdot\|$, respectively. The duality between $V$ and $V^*$ will also be denoted by $\langle\cdot,\cdot\rangle$.

$(i_2)$ $L : V \to V^*$ is linear, continuous, symmetric, and

$$\exists \alpha \geq 0,\ \omega > 0 \text{ such that } \langle Lv, v\rangle + \alpha \| v \|^2 \geq \omega \mid v \mid^2 \quad \forall\, v \in V;$$

$(i_3)$ $u_0 \in V,\ u_1 \in H$;

$(i_4)$ $f \in L^2(0,T;H)$.

The present chapter comprises four sections. The first three sections are concerned with the asymptotic analysis of problems $(P.k)_\varepsilon$, $k = 1, 2, 3$, just formulated before. Each of the first three sections starts with the formal derivation of an $N$th order asymptotic expansion, $N \geq 0$, by employing the method discussed in Chapter 1. We shall see that in the case of the first two problems some boundary layers (of order zero and one, respectively) occur in a neighborhood of point $t = T$, with respect to the norm of $C([0,T];H)$. On the other hand, in the case of problem $(P.3)_\varepsilon$, a boundary layer (of order one) occurs in a right vicinity of point $t = 0$. We continue our treatment by validating these asymptotic expansions. We establish some estimates for the corresponding $N$th order remainders, which are more than enough to validate our asymptotic expansions completely. In the fourth and last section we exploit our theoretical results to discuss possible regularizations for a parabolic type equation with the homogeneous Dirichlet boundary condition.

Let us point out that elliptic regularizations of the type $(P.1)_\varepsilon$ and $(P.2)_\varepsilon$ have previously been studied in J.L. Lions [32], pp. 407–420, where asymptotic expansions (of order zero and one, respectively) have been established, including some boundary layer functions (correctors). However the methods used in that book are different from ours.

Note that J.L. Lions [32] (see pp. 491–495) also discussed a regularization of the form $(P.3)_\varepsilon$, determining a zeroth order expansion for the corresponding solution.

Moreover, an $N$th order asymptotic expansion for the solution of problem $(P.1)_\varepsilon$ was constructed in [4] in the case in which $L$ is self-adjoint and strongly monotone.

In addition, J.L. Lions investigated in [31] some abstract hyperbolic variational inequalities with singular perturbations.

# 10.1 Asymptotic analysis of problem $(P.1)_\varepsilon$

In this section we investigate problem $(P.1)_\varepsilon$ under assumptions $(h_1)$–$(h_3)$. In order to get some information about the nature of the problem, we start with a particular example. More precisely, we consider the following two-point boundary value problem associated with a scalar ordinary differential equation:

$$\begin{cases} \varepsilon u_\varepsilon'' - u_\varepsilon' - u_\varepsilon = 0, \ t \in [0,T], \\ u_\varepsilon(0) = a, \ u_\varepsilon(T) = b, \end{cases} \tag{10.1}$$

where $a, \ b \in \mathbb{R}$.

The solution of this problem is given by

$$u_\varepsilon(t) = \frac{a(e^{r_{2\varepsilon}T + r_{1\varepsilon}t} - e^{r_{1\varepsilon}T + r_{2\varepsilon}t})}{e^{r_{2\varepsilon}T} - e^{r_{1\varepsilon}T}} + \frac{b(e^{r_{2\varepsilon}t} - e^{r_{1\varepsilon}t})}{e^{r_{2\varepsilon}T} - e^{r_{1\varepsilon}T}}, \ t \in [0,T],$$

where

$$r_{1\varepsilon} = \frac{2}{\sqrt{1+4\varepsilon} - 1}, \ r_{2\varepsilon} = \frac{-2}{\sqrt{1+4\varepsilon}+1}.$$

It is worth pointing out that $u_\varepsilon(t)$ converges uniformly to $X_0(t) = ae^{-t}$, as $\varepsilon \to 0$, on every subinterval of the form $[0, T-\delta]$, $\delta \in (0,T)$, but not on the whole interval $[0,T]$. More precisely, $u_\varepsilon(t)$ can be written in the form

$$u_\varepsilon(t) = X_0(t) + i_0(\tau) + R_\varepsilon(t), \ t \in [0,T],$$

where

$$\tau = \frac{T-t}{\varepsilon}, \ i_0(\tau) = \left(b - ae^{-T}\right) e^{-\tau}.$$

One can easily check that $R_\varepsilon(t)$ converges uniformly to zero on $[0,T]$ as $\varepsilon \to 0$. Consequently, problem $(P.1)_\varepsilon$ is singularly perturbed with respect to the uniform convergence norm, and there is a boundary layer near the endpoint $t = T$. We also note that the corresponding rapid variable is $\tau = (T-t)/\varepsilon$.

As expected, function $X_0$ is the solution of the reduced problem

$$\begin{cases} X_0' + X_0 = 0, \ t \in [0,T], \\ X_0(0) = a. \end{cases}$$

Additional information related to a problem which is more general than (10.1) can be found in W. Eckhaus [18], pp. 51–63.

## 10.1.1 $N$th order asymptotic expansion

Taking into account the above example as well as the general theory developed in Chapter 1, we propose an asymptotic expansion of order $N$ for the solution $u_\varepsilon$ of problem $(P.1)_\varepsilon$ in the form

$$u_\varepsilon(t) = \sum_{k=0}^{N} X_k(t)\varepsilon^k + \sum_{k=0}^{N} i_k(\tau)\varepsilon^k + R_\varepsilon(t), \ t \in [0,T], \tag{10.2}$$

where:

$\tau := (T-t)/\varepsilon$ is the stretched (fast) variable;
$X_k,\ k=0,1,\ldots,N$, are the first $(N+1)$ regular terms;
$i_k,\ k=0,1,\ldots,N$, are the corresponding boundary layer functions;
$R_\varepsilon$ denotes the $N$th order remainder.

Following the usual procedure, we substitute (10.2) into $(P.1)_\varepsilon$ and then equate the coefficients of $\varepsilon^k$, $k=-1,0,\ldots,(N-1)$, separately those depending on $t$ from those depending on $\tau$, thus determining formally the terms of expansion (10.2). First of all, we see that the regular terms satisfy the following problems, denoted $(P.1)_k$:

$$\begin{cases} X_k'(t) + LX_k(t) = f_k(t),\ 0<t<T, \\ X_k(0) = \begin{cases} u_0,\ k=0, \\ 0,\ k=1,\ldots,N, \end{cases} \end{cases} \tag{10.3}$$

where

$$f_k(t) = \begin{cases} f(t),\ k=0, \\ X_{k-1}''(t),\ k=1,\ldots,N. \end{cases}$$

For the boundary layer functions we derive the equations

$$\begin{aligned} &i_0''(\tau) + i_0'(\tau) = 0,\ \tau \ge 0, \\ &i_0(0) = u_T - X_0(T), \end{aligned}$$

$$\begin{aligned} &i_k''(\tau) + i_k'(\tau) = Li_{k-1}(\tau),\ \tau \ge 0, \\ &i_k(0) = -X_k(T),\ k=1,\ldots,N. \end{aligned}$$

Taking into account the usual requirement that $\| i_k(\tau) \| \to 0$ as $\tau \to \infty$, we derive

$$\begin{cases} i_0(\tau) = (u_T - X_0(T))e^{-\tau}, \\ i_k(\tau) = P_k(\tau)e^{-\tau},\ k=1,\ldots,N, \end{cases} \tag{10.4}$$

where $P_k,\ k=1,\ldots,N$, are polynomials of degree at most $k$ whose coefficients belong to $H$. Finally, one can see that the remainder of order $N$ satisfies the boundary value problem

$$\begin{cases} \varepsilon(R_\varepsilon + \varepsilon^N X_N)'' - R_\varepsilon' - LR_\varepsilon = \varepsilon^N Li_N \text{ in } (0,T), \\ R_\varepsilon(0) = l_N(T/\varepsilon),\ R_\varepsilon(T) = 0, \end{cases} \tag{10.5}$$

where $l_N = -i_0 - \varepsilon i_1 - \cdots - \varepsilon^N i_N$.

Thus, all the terms of the suggested expansion (10.2) have been determined formally.

### 10.1.2 Existence, uniqueness and regularity of the solutions of problems $(P.1)_\varepsilon$ and $(P.1)_k$

We start with problem $(P.1)_\varepsilon$. Since operator $L$ satisfies assumption $(h_1)$, it is maximal cyclically monotone. Indeed, $L$ is the subdifferential of $\varphi : H \to (-\infty, +\infty]$,

$$\varphi(x) = \begin{cases} \frac{1}{2}\|L^{1/2}x\|^2 \text{ for } x \in D(L^{1/2}), \\ +\infty \text{ otherwise}, \end{cases}$$

which is convex and lower semicontinuous (see, e.g., [34], p. 43).
Now, in order to obtain higher order regularity of the solution of $(P.1)_\varepsilon$, we are going to examine the problem

$$\begin{cases} u'' = Lu + h, \ t \in (0,T), \\ u(0) = u_0, \ u(T) = u_T. \end{cases} \tag{10.6}$$

We have

**Lemma 10.1.1.** *Assume that $(h_1), (h_2)$ are satisfied and, in addition,*

$$h \in W^{1,2}(0,T;H), \ u_T \in D(L).$$

*Then, problem* (10.6) *has a unique solution $u \in W^{2,2}(0,T;H) \bigcap W^{3,2}_{\text{loc}}(0,T;H)$, and*

$$t^{3/2}(T-t)^{3/2}u''' \in L^2(0,T;H).$$

*Proof.* According to Theorem 2.0.35, problem (10.6) admits a unique solution $u$ which belongs to $W^{2,2}(0,T;H)$. In order to prove that this solution is as regular as stated, we consider the following problem which one derives by formal differentiation with respect to $t$ of problem (10.6):

$$\begin{cases} z'' = Lz + h'(t), \ t \in (0,T), \\ z(0) = u'(0), \ z(T) = u'(T). \end{cases}$$

According to Theorem 2.0.36, this problem has a unique solution $z \in W^{2,2}_{\text{loc}}(0,T;H)$, with $t^{3/2}(T-t)^{3/2}z'' \in L^2(0,T;H)$.
We shall prove that $z = u'$. We consider the following approximate problems

$$\begin{cases} z''_\lambda = L_\lambda z_\lambda + h'(t), \ t \in (0,T), \\ z_\lambda(0) = u'(0), \ z_\lambda(T) = u'(T), \end{cases} \tag{10.7}$$

and

$$\begin{cases} u''_\lambda = L_\lambda u_\lambda + h(t), \ t \in (0,T) \\ u_\lambda(0) = u_0, \ u_\lambda(T) = u_T, \end{cases} \tag{10.8}$$

where $L_\lambda$ denotes the Yosida approximation of operator $L$, $\lambda > 0$ (for details on Yosida's approximation, see Chapter 2, Theorem 2.0.10).
Since $L_\lambda$ is everywhere defined, positive and continuous, problems (10.7) and (10.8) have unique solutions

$$z_\lambda \in W^{2,2}(0,T;H),\ u_\lambda \in C^2([0,T];H)\ \forall\ \lambda > 0,$$

and (see [7], p. 306)

$$u_\lambda \to u,\ u'_\lambda \to u' \text{ in } C([0,T];H),\ as\ \lambda \longrightarrow 0^+, \tag{10.9}$$

as well as (see [12])

$$z_\lambda \to z \text{ in } C([\delta, T-\delta];H),\ as\ \lambda \longrightarrow 0^+, \tag{10.10}$$

for any $0 < \delta < T$. Therefore, if we denote $v_\lambda = u'_\lambda - z_\lambda$, then $v_\lambda \to u' - z$ in $C([\delta, T-\delta];H)$.

Moreover, since $L_\lambda$ is linear, continuous, and $h \in W^{1,2}(0,T;H)$, one can differentiate equation $(10.8)_1$, then subtract it from equation $(10.7)_1$ to derive

$$\begin{cases} v''_\lambda = L_\lambda v_\lambda \text{ for a.a. } t \in (0,T), \\ v_\lambda(0) = u'_\lambda(0) - u'(0),\ v_\lambda(T) = u'_\lambda(T) - u'(T). \end{cases} \tag{10.11}$$

Obviously, $v_\lambda \in W^{2,2}(0,T;H)$ and

$$\frac{d^2}{dt^2} \parallel v_\lambda \parallel^2 = 2\langle v_\lambda, v''_\lambda\rangle + 2 \parallel v'_\lambda \parallel^2 \geq 0 \text{ for a.a. } t \in (0,T).$$

Consequently, the function $t \to \parallel v_\lambda(t) \parallel^2$ is convex, which implies

$$\parallel v_\lambda(t) \parallel \leq \max\{\parallel u'_\lambda(0) - u'(0) \parallel,\ \parallel u'_\lambda(T) - u'(T) \parallel\}\ \forall\ t \in [0,T].$$

Therefore, in view of (10.9), one has $v_\lambda \to 0$ in $C([0,T];H)$, as $\lambda \longrightarrow 0^+$. Thus, $u' = z$, which concludes the proof of the lemma. □

Now, using Lemma 10.1.1, one can state

**Theorem 10.1.2.** *Assume that* $(h_1)$, $(h_2)$ *are satisfied, and*

$$f \in W^{1,2}(0,T;H),\ u_T \in D(L). \tag{10.12}$$

*Then, problem* $(P.1)_\varepsilon$ *has a unique solution* $u_\varepsilon \in W^{2,2}(0,T;H) \bigcap W^{3,2}_{\text{loc}}(0,T;H)$, *with*

$$t^{3/2}(T-t)^{3/2}u''' \in L^2(0,T;H).$$

*Proof.* First of all, note that $(P.1)_\varepsilon$ is a particular case of problem (2.19) in Chapter 2. Therefore, according to Theorem 2.0.37, it has a unique solution $u_\varepsilon \in W^{2,2}(0,T;H)$. Now, we apply Lemma 10.1.1 with $L := \varepsilon^{-1}L$ and $h := \varepsilon^{-1}(u'_\varepsilon - f) \in W^{1,2}(0,T;H)$. □

In what follows we deal with problems $(P.1)_k$, $k = 0, \dots, N$. Our aim is to show that, under appropriate assumptions, $X_N \in W^{1,2}(0,T;H)$, which will be needed for deriving estimates for the $N$th order remainder of expansion (10.2).

Denote $\alpha_{00} = u_0$, $\alpha_{0k} = 0 \ \forall \ k = 1, \dots, N$. For a fixed $k \in \{0, \dots, N-1\}$, we have

**Theorem 10.1.3.** *Assume that $N \geq 1$, $(h_1)$, $(h_2)$ are satisfied, and*

$$f_k \in W^{N-k,2}(0,T;H); \tag{10.13}$$

$$\begin{aligned} \alpha_{jk} &= f_k^{(j-1)}(0) - L\alpha_{j-1,k} \in D(L) \ \forall \ j = 1, \dots, N-k-1, \\ \alpha_{N-k,k} &= f_k^{(N-k-1)}(0) - L\alpha_{N-k-1,k} \in D(L^{1/2}). \end{aligned} \tag{10.14}$$

*Then, problem $(P.1)_k$ has a unique solution $X_k \in W^{N-k+1,2}(0,T;H)$.*

*Proof.* As noted before, by virtue of $(h_1)$, $L$ is a subdifferential: $L = \partial\varphi$. Therefore, by Theorem 2.0.24, $(P.1)_k$ has a unique strong solution $X_k \in W^{1,2}(0,T;H)$. Note that $\forall \ j = 1, \dots, N-k$, $X_k^{(j)}$ is the unique strong solution of the Cauchy problem

$$\begin{cases} Z_j' + LZ_j = f_k^{(j)}, \ t \in (0,T), \\ Z_j(0) = \alpha_{jk}. \end{cases}$$

Thus, $Z_j = X_k^{(j)} \in W^{1,2}(0,T;H) \ \forall \ j = 1, \dots, N-k$.

Therefore, $X_k \in W^{N-k+1,2}(0,T;H)$. □

*Remark* 10.1.4. We suppose that $N \geq 1$. As said before, we want to obtain $X_N \in W^{1,2}(0,T;H)$. Thus, taking into account the form of problems $(P.1)_k$, one should have

$$X_k \in W^{N-k+1,2}(0,T;H) \ \forall k = 1, \dots, N-1.$$

Hence, according to Theorem 10.1.3, if

$$\begin{aligned} & f \in W^{N,2}(0,T;H), \\ & \alpha_{jk} \in D(L), \quad f_k^{(N-k-1)}(0) - L\alpha_{N-k-1,k} \in D(\varphi) = D(L^{1/2}), \\ & \qquad \forall \ k = 0, \dots, N-1, \ \forall \ j = 0, \dots, N-k-1, \end{aligned} \tag{10.15}$$

then, $X_{N-1} \in W^{2,2}(0,T;H)$. Taking into account problem $(P.1)_N$, we obtain that $X_N \in W^{1,2}(0,T;H)$.

Note also that the following conditions

$$f^{(N-1)}(0) \in D(L^{1/2}), \ f^{(N-2)}(0) \in D(L), \dots, f(0) \in D(L^N), \ u_0 \in D(L^{N+1})$$

are sufficient to imply that

$$\begin{aligned} \alpha_{jk} \in & D(L), \quad f_k^{(N-k-1)}(0) - L\alpha_{N-k-1,k} \in D(L^{1/2}), \\ & \forall \ k = 0, \dots, N-1 \ \forall \ j = 0, \dots, N-k-1. \end{aligned}$$

If $N = 0$ and assumptions $(h_2)$, $(h_3)$ hold, then one can easily see that $X_0 \in W^{1,2}(0,T;H)$.

By employing what we have proved so far, we can state the following concluding result:

**Corollary 10.1.5.** *Let $N \geq 1$. Assume that $(h_1), (h_2)$, (10.15) are satisfied and $u_T \in D(L)$. Then problems $(P.1)_\varepsilon$, $\varepsilon > 0$, and $(P.1)_k$, $k = 0, \ldots, N$, have unique solutions*

$$u_\varepsilon \in W^{2,2}(0,T;H) \bigcap W^{3,2}_{\mathrm{loc}}(0,T;H), \; t^{3/2}(T-t)^{3/2}u''' \in L^2(0,T;H),$$
$$X_k \in W^{N-k+1,2}(0,T;H).$$

*If $N = 0$, $(h_1)$–$(h_3)$ hold, and $u_T \in D(L)$, then problems $(P.1)_\varepsilon$, $\varepsilon > 0$, and $(P.1)_0$ have unique solutions*

$$u_\varepsilon \in W^{2,2}(0,T;H), \; X_0 \in W^{1,2}(0,T;H).$$

*Remark* 10.1.6. The above results remain valid to some extent, if $(h_1)$ is replaced with the following weaker assumption:

$(h_1')$ $L : D(L) \subset H \to H$ is a linear operator, with the property that there is a constant $\omega > 0$ such that $L + \omega I$ is maximal monotone ($I$ denotes the identity operator).

Indeed, in our above investigation of problem $(P.1)_\varepsilon$ we have used $(h_1)$ to obtain maximal monotonicity for $L$. But, if the weaker assumption $(h_1')$ holds, then one can use the transformation $v(t) = e^{-\gamma t}u(t)$ to rewrite equation $(EE)$ in the form

$$-\varepsilon v'' + (1 - 2\varepsilon\gamma)v' + \gamma(1 - \varepsilon\gamma)v + Lv = e^{-\gamma t}f(t), \; t \in (0,T),$$

and operator $\overline{L} = L + \gamma(1-\varepsilon\gamma)I$ is maximal monotone for $\gamma = 2\omega$ and $\varepsilon \leq 1/(4\omega)$.

On the other hand, in the study of problems $(P.1)_k$ we have essentially used the fact that $L$ is a subdifferential, $L = \partial\varphi$. If the weaker assumption $(h_1')$ holds, one can apply Theorem 2.0.27 in Chapter 2 to derive a result similar to Theorem 10.1.3. However, our previous requirements (10.13) and $(10.14)_2$ should be replaced with the stronger assumptions $f_k \in W^{N-k+1,1}(0,T;H)$ and $f_k^{(N-k-1)}(0) - L\alpha_{N-k-1,k} \in D(L)$, respectively, and our requirements (10.15) should also be modified accordingly.

### 10.1.3 Estimates for the remainder

In what follows we establish an estimate in the norm of $C([0,T];H)$ for the $N$th order remainder of expansion (10.2). In particular, this estimate validates expansion (10.2) completely.

More precisely, we have:

**Theorem 10.1.7.** *Suppose that all the assumptions of Corollary* 10.1.5 *are satisfied. Then, for every* $\varepsilon > 0$, *the solution of problem* $(P.1)_\varepsilon$ *admits an asymptotic expansion of the form* (10.2) *and the following estimates hold*

$$\|R_\varepsilon\|_{C([0,T];H)} = O(\varepsilon^{N+1/4}), \quad \| R_\varepsilon' \|_{L^2(0,T;H)} = \mathcal{O}(\varepsilon^N),$$
$$\langle LR_\varepsilon, R_\varepsilon\rangle_{L^2(0,T;H)} = \mathcal{O}(\varepsilon^{2N+1}).$$

*For* $N=0$, $u_\varepsilon' \to X_0'$ *weakly in* $L^2(0,T;H)$.

*Proof.* From Corollary 10.1.5 and (10.2) we derive $R_\varepsilon + \varepsilon^N X_N \in W^{2,2}(0,T;H)$, $R_\varepsilon \in W^{1,2}(0,T;H)$.
Now, let us homogenize the first condition in $(10.5)_2$. Denote

$$\overline{R}_\varepsilon(t) = R_\varepsilon(t) - \frac{(T-t)}{T} l_N\Big(\frac{T}{\varepsilon}\Big), \ t \in [0,T].$$

Obviously, $\overline{R}_\varepsilon$ satisfies the boundary value problem

$$\begin{cases} \varepsilon\Big(\overline{R}_\varepsilon + \frac{(T-t)}{T} l_N\big(\frac{T}{\varepsilon}\big) + \varepsilon^N X_N\Big)'' - \overline{R}_\varepsilon' - L\overline{R}_\varepsilon = \overline{h}_\varepsilon \text{ in } (0,T), \\ \overline{R}_\varepsilon(0) = \overline{R}_\varepsilon(T) = 0, \end{cases} \tag{10.16}$$

where $\overline{h}_\varepsilon(t) = \varepsilon^N L i_N - \frac{1}{T} l_N\big(\frac{T}{\varepsilon}\big) + \frac{T-t}{T} L l_N\big(\frac{T}{\varepsilon}\big)$.

If we take the scalar product in $H$ of equation $(10.16)_1$ with $\overline{R}_\varepsilon(t)$, and then integrate the resulting equation over $[0,T]$, we get

$$\begin{aligned} &\varepsilon \int_0^T \Big[ \Big\langle \Big(\overline{R}_\varepsilon + \frac{(T-t)}{T} l_N\Big(\frac{T}{\varepsilon}\Big) + \varepsilon^N X_N\Big)', \overline{R}_\varepsilon \Big\rangle' - \| \overline{R}_\varepsilon' \|^2 \Big] dt \\ &\quad - \frac{1}{2} \int_0^T \frac{d}{dt} \| \overline{R}_\varepsilon \|^2 \, dt = \int_0^T \langle L\overline{R}_\varepsilon, \overline{R}_\varepsilon \rangle dt \\ &\quad + \int_0^T \langle \overline{h}_\varepsilon, \overline{R}_\varepsilon\rangle dt + \varepsilon^{N+1} \int_0^T \langle X_N', \overline{R}_\varepsilon' \rangle dt. \end{aligned}$$

Since $\overline{R}_\varepsilon \in H_0^1(0,T;H)$, this equation leads us to

$$\begin{aligned} &\langle L\overline{R}_\varepsilon, \overline{R}_\varepsilon\rangle_{L^2(0,T;H)} + \varepsilon \| \overline{R}_\varepsilon' \|^2_{L^2(0,T;H)} \\ &\quad \le \| \overline{R}_\varepsilon \|_{L^2(0,T;H)} \cdot \| \overline{h}_\varepsilon \|_{L^2(0,T;H)} \\ &\qquad + \varepsilon^{N+1} \| \overline{R}_\varepsilon' \|_{L^2(0,T;H)} \cdot \| X_N' \|_{L^2(0,T;H)} \\ &\quad \le K \| \overline{R}_\varepsilon' \|_{L^2(0,T;H)} \cdot \| \overline{h}_\varepsilon \|_{L^2(0,T;H)} \\ &\qquad + \varepsilon^{N+1} \| \overline{R}_\varepsilon' \|_{L^2(0,T;H)} \cdot \| X_N' \|_{L^2(0,T;H)}, \end{aligned} \tag{10.17}$$

where $K$ is a positive constant, independent of $\varepsilon$.

On the other hand, for every $k = 0, \dots, N$ and $j \in \mathbb{N}$, we have

$$\begin{aligned} \| i_k \|_{L^2(0,T;H)} &= \mathcal{O}(\varepsilon^{1/2}), \quad \| Li_N \|_{L^2(0,T;H)} = \mathcal{O}(\varepsilon^{1/2}), \\ \left\| l_N\left(\frac{T}{\varepsilon}\right)\right\| &= \mathcal{O}(\varepsilon^j), \end{aligned} \tag{10.18}$$

from which it follows that

$$\| \overline{h}_\varepsilon \|_{L^2(0,T;H)} = \mathcal{O}(\varepsilon^{N+1/2}).$$

From (10.17) and $(h_1)$ we easily derive the estimate

$$\| \overline{R}_\varepsilon' \|_{L^2(0,T;H)} = \mathcal{O}(\varepsilon^{N-1/2}). \tag{10.19}$$

In what follows we are going to obtain an estimate for $\| \overline{R}_\varepsilon \|_{L^2(0,T;H)}$. Denote $\widetilde{R}_\varepsilon = e^{-t}\overline{R}_\varepsilon$. Thus, equation $(10.16)_1$ reads

$$\varepsilon e^{-t}(R_\varepsilon + \varepsilon^N X_N)'' - \widetilde{R}_\varepsilon' - \widetilde{R}_\varepsilon - L\widetilde{R}_\varepsilon = \widetilde{h}_\varepsilon \text{ in } (0,T),$$

where $\widetilde{h}_\varepsilon = e^{-t}\overline{h}_\varepsilon$. If we multiply this equation by $\widetilde{R}_\varepsilon$ and then integrate the resulting equation over $[0,T]$, we get:

$$\begin{aligned} \| \widetilde{R}_\varepsilon \|^2_{L^2(0,T;H)} &+ \int_0^T \langle L\widetilde{R}_\varepsilon, \widetilde{R}_\varepsilon \rangle dt \\ &= \varepsilon \int_0^T e^{-t}\langle (R_\varepsilon + \varepsilon^N X_N)'', \widetilde{R}_\varepsilon\rangle dt - \int_0^T \langle \widetilde{h}_\varepsilon, \widetilde{R}_\varepsilon \rangle dt, \end{aligned}$$

from which we derive

$$\begin{aligned} &\| \widetilde{R}_\varepsilon \|^2{}_{L^2(0,T;H)} + \varepsilon \int_0^T e^{-2t} \| \overline{R}_\varepsilon' \|^2 dt + \langle L\widetilde{R}_\varepsilon, \widetilde{R}_\varepsilon \rangle_{L^2(0,T;H)} \\ &\quad \le \int_0^T | \langle \widetilde{h}_\varepsilon, \widetilde{R}_\varepsilon \rangle | dt + 2\varepsilon \int_0^T | \langle S_\varepsilon', \widetilde{R}_\varepsilon \rangle | dt \\ &\quad + \varepsilon \int_0^T e^{-2t} \left| \left\langle \varepsilon^N X_N' - \frac{1}{T} l_N\left(\frac{T}{\varepsilon}\right), \overline{R}_\varepsilon' \right\rangle \right| dt \\ &\quad \le \frac{1}{2} \| \widetilde{h}_\varepsilon \|^2_{L^2(0,T;H)} + \frac{1}{2} \| \widetilde{R}_\varepsilon \|^2_{L^2(0,T;H)} + 2\varepsilon \| \widetilde{R}_\varepsilon \|_{L^2(0,T;H)} \cdot \| S_\varepsilon' \|_{L^2(0,T;H)} \\ &\quad + \frac{\varepsilon}{2} \int_0^T e^{-2t} \| \overline{R}_\varepsilon' \|^2 dt + M_1 \varepsilon \left\| \varepsilon^N X_N' - \frac{1}{T} l_N\left(\frac{T}{\varepsilon}\right) \right\|^2_{L^2(0,T;H)} \end{aligned}$$

($M_1$ is a positive constant independent of $\varepsilon$).
Denoting $E_\varepsilon = \| \widetilde{R}_\varepsilon \|_{L^2(0,T;H)}$ and taking into account estimates (10.18) and (10.19), we get

$$\frac{1}{2} E_\varepsilon^2 \le M_2 \varepsilon^{N+1/2} E_\varepsilon + M_3 \varepsilon^{2N+1},$$

where $M_2$, $M_3$ are some positive constants, independent of $\varepsilon$. It follows that

$$\| \widetilde{R}_\varepsilon \|_{L^2(0,T;H)} = \mathcal{O}(\varepsilon^{N+1/2}) \Rightarrow \| \overline{R}_\varepsilon \|_{L^2(0,T;H)} = \mathcal{O}(\varepsilon^{N+1/2}).$$

This together with (10.17) implies

$$\langle LR_\varepsilon, R_\varepsilon \rangle_{L^2(0,T;H)} = \mathcal{O}(\varepsilon^{2N+1}), \quad \| \overline{R}'_\varepsilon \|_{L^2(0,T;H)} = \mathcal{O}(\varepsilon^N). \tag{10.20}$$

Thus, from the obvious formula

$$\| \overline{R}_\varepsilon(t) \|^2 = 2 \int_0^t (\overline{R}_\varepsilon(s), \overline{R}'_\varepsilon(s)) ds \ \forall\ t \in [0,T],$$

we derive

$$\|\overline{R}_\varepsilon\|_{C([0,T];H)} = O(\varepsilon^{N+1/4}).$$

Taking into account the definition of $\overline{R}_\varepsilon$, we obtain the desired estimates for $R_\varepsilon$. In the particular case $N = 0$, it follows from $(10.20)_2$ that $\{R'_\varepsilon\}_{\varepsilon>0}$ converges to zero weakly in $L^2(0,T;H)$. □

*Remark* 10.1.8. If we consider the particular case $N = 0$ and $u_T = X_0(T)$ (i.e., $i_0 \equiv 0$) then, the problem is regularly perturbed of order zero, and the following estimates hold true

$$\|u_\varepsilon - X_0\|_{C([0,T];H)} = O(\varepsilon^{1/4}), \ \|u_\varepsilon - X_0\|_{L^2(0,T;H)} = O(\varepsilon^{1/2}),$$
$$\langle L(u_\varepsilon - X_0), u_\varepsilon - X_0 \rangle_{L^2(0,T;H)} = \mathcal{O}(\varepsilon).$$

*Remark* 10.1.9. Theorem 10.1.7 is still valid if we require the weaker assumption $(h'_1)$ (instead of $(h_1)$). In this case, one can use similar ideas and computation, with $\widetilde{R}_\varepsilon = e^{-(1+\omega)t}\overline{R}_\varepsilon$.

## 10.2 Asymptotic analysis of problem $(P.2)_\varepsilon$

In this section we examine problem $(P.2)_\varepsilon$ (which has been formulated in Chapter 9), under assumptions $(h_1)$–$(h_3)$. This problem comprises equation $(E)$ and the boundary conditions $(BC.2)$.

If we considered again the example discussed in Section 10.1, but with the new condition $u'_\varepsilon(T) = b$ instead of $u_\varepsilon(T) = b$, we would easily conclude that the solution $u_\varepsilon$ of problem $(P.2)_\varepsilon$ converges uniformly on the whole interval $[0,T]$ to the solution $X_0$ of the corresponding reduced problem (which is exactly the same as in the previous case: see below). Thus, such an elliptic regularization is a regularly perturbed problem of order zero. However, we shall see that in general $(P.2)_\varepsilon$ is singularly perturbed of any order $\geq 1$, with respect to the norm of $C([0,T];H)$.

### 10.2.1 $N$th order asymptotic expansion

According to the above remark on that particular example, we shall try to determine a high order asymptotic expansion for the solution $u_\varepsilon$ of problem $(P.2)_\varepsilon$ in the form

$$u_\varepsilon(t) = \sum_{k=0}^{N} X_k(t)\varepsilon^k + \sum_{k=0}^{N} i_k(\tau)\varepsilon^k + R_\varepsilon(t),\ t \in [0,T]. \tag{10.21}$$

The terms involved in this expansion have the same meaning as in Subsection 10.1.1, and the rapid variable $\tau$ is the same. Moreover, one can see that $X_k(t)$ satisfy the same problems, which will be re-denoted here $(P.2)_k,\ k = 0,\dots,N$, for convenience. The boundary layer functions satisfy the problems

$$\begin{aligned} &i_0''(\tau) + i_0'(\tau) = 0,\ \tau \geq 0,\\ &i_0'(0) = 0,\ i_0(\infty) = 0, \end{aligned}$$

$$\begin{aligned} &i_k''(\tau) + i_k'(\tau) = Li_{k-1}(\tau),\ \tau \geq 0,\ k = 1,\dots,N,\\ &i_k'(0) = \begin{cases} X_0'(T) - u_T,\ k = 1,\\ X_{k-1}'(T),\ k = 2,\dots,N, \end{cases}\\ &i_k(\infty) = 0,\ k = 1,\dots,N, \end{aligned}$$

and therefore

$$\begin{aligned} &i_0(\tau) \equiv 0,\\ &i_k(\tau) = P_k(\tau)e^{-\tau},\ k = 1,\dots,N, \end{aligned} \tag{10.22}$$

where $P_k,\ k = 1,\dots,N$, are polynomials with coefficients in $H$, whose degrees do not exceed $k-1$.

Finally, the $N$th order remainder satisfies formally the following boundary value problem

$$\begin{cases} \varepsilon(R_\varepsilon + \varepsilon^N X_N)'' - R_\varepsilon' - LR_\varepsilon = \varepsilon^N Li_N \text{ in } (0,T),\\ R_\varepsilon(0) = l_N(T/\varepsilon),\\ R_\varepsilon'(T) = \begin{cases} u_T - X_0'(T),\ N = 0,\\ -\varepsilon^N X_N'(T),\ N \geq 1, \end{cases} \end{cases} \tag{10.23}$$

where $l_N = -\varepsilon i_1 - \cdots - \varepsilon^N i_N$.

It is worth mentioning that there is no boundary layer of order zero, since $i_0$ is the null function (see (10.22)). This assertion will be validated later.

### 10.2.2 Existence, uniqueness and regularity of the solutions of problems $(P.2)_\varepsilon$ and $(P.2)_k$

One can use ideas similar to those in Subsection 10.1.2 to establish similar results. We shall not go into details, but just state these results. For problem $(P.2)_\varepsilon$ we have

**Theorem 10.2.1.** *Assume that $(h_1)$, $(h_2)$ are satisfied and, in addition,*

$$f \in W^{1,2}(0,T;H), \ u_T \in D(L).$$

*Then, problem $(P.2)_\varepsilon$ has a unique solution $u_\varepsilon \in W^{2,2}(0,T;H)\bigcap W^{3,2}_{\text{loc}}(0,T;H)$, and*

$$t^{3/2}(T-t)^{3/2}u''' \in L^2(0,T;H).$$

The problems satisfied by the regular terms are exactly the same as in Subsection 10.1.1, thus Theorem 10.1.3 in Subsection 10.1.2 can be used here as such.

We conclude by stating the following

**Corollary 10.2.2.** *Assume that $N \geq 1$, $(h_1)$, $(h_2)$,* (10.15) *are satisfied, and $u_T \in D(L)$. Then problems $(P.2)_\varepsilon$, $\varepsilon > 0$, and $(P.2)_k$, $k = 0, \ldots, N$, have unique solutions*

$$u_\varepsilon \in W^{2,2}(0,T;H)\bigcap W^{3,2}_{\text{loc}}(0,T;H), \ t^{3/2}(T-t)^{3/2}u''' \in L^2(0,T;H),$$
$$X_k \in W^{N-k+1,2}(0,T;H).$$

*If $N = 0$, $(h_1)$–$(h_3)$, $u_T \in D(L)$ hold, then problems $(P.2)_\varepsilon$, $\varepsilon > 0$, and $(P.2)_0$ have unique solutions*

$$u_\varepsilon \in W^{2,2}(0,T;H), \ X_0 \in W^{1,2}(0,T;H).$$

### 10.2.3 Estimates for the remainder

As usual, we continue our asymptotic analysis with estimates for the remainder. In particular, our estimate below for $R_\varepsilon$, with respect to the norm of $C([0,T];H)$, is more than enough to validate our $N$th order asymptotic expansion (10.21).

**Theorem 10.2.3.** *Suppose that all the assumptions of Corollary* 10.2.2 *are satisfied. Then, for every $\varepsilon > 0$, the solution of problem $(P.2)_\varepsilon$ admits an asymptotic expansion of the form* (10.21) *and the following estimates hold*

$$\|R_\varepsilon\|_{C([0,T];H)} = O(\varepsilon^{N+1/4}), \ \| R'_\varepsilon \|_{L^2(0,T;H)} = \mathcal{O}(\varepsilon^N),$$

$$\langle LR_\varepsilon, R_\varepsilon\rangle_{L^2(0,T:H)} = \mathcal{O}(\varepsilon^{2N+1}) \ \forall \ N \geq 1.$$

*Moreover, in case $N = 0$ we have the estimates*

$$\|u_\varepsilon - X_0\|_{C([0,T];H)} = O(\varepsilon^{1/4}), \ \|u_\varepsilon - X_0\|_{L^2(0,T;H)} = O(\varepsilon^{1/2}),$$
$$\langle L(u_\varepsilon - X_0), u_\varepsilon - X_0\rangle_{L^2(0,T;H)} = \mathcal{O}(\varepsilon),$$

*and $u_\varepsilon' \to X_0'$ weakly in $L^2(0,T;H)$.*

*Proof.* It is similar to the proof of Theorem 10.1.7. We shall just provide the reader with the main steps of it. First, by the substitution

$$\overline{R}_\varepsilon(t) = R_\varepsilon(t) - \frac{(T-t)}{T} l_N\Big(\frac{T}{\varepsilon}\Big), \ t \in [0,T],$$

we homogenize the boundary condition $(10.23)_2$. Indeed, the new unknown $\overline{R}_\varepsilon$ satisfies the problem

$$\begin{cases} \varepsilon\Big(\overline{R}_\varepsilon + \frac{(T-t)}{T} l_N\big(\frac{T}{\varepsilon}\big) + \varepsilon^N X_N\Big)'' - \overline{R}_\varepsilon' - L\overline{R}_\varepsilon = \overline{h}_\varepsilon \text{ in } (0,T), \\ \overline{R}_\varepsilon(0) = 0, \ \overline{R}_\varepsilon'(T) = R_\varepsilon'(T) + \frac{1}{T} l_N\big(\frac{T}{\varepsilon}\big), \end{cases} \tag{10.24}$$

where

$$\overline{h}_\varepsilon(t) = \varepsilon^N L i_N - \frac{1}{T} l_N\Big(\frac{T}{\varepsilon}\Big) + \frac{T-t}{T} L l_N\Big(\frac{T}{\varepsilon}\Big).$$

If we repeat the computation which led to (10.17), we derive the inequality

$$\begin{aligned} \varepsilon \| \overline{R}_\varepsilon' \|^2_{L^2(0,T;H)} &+ \frac{1}{2} \| \overline{R}_\varepsilon(T) \|^2 + \langle L\overline{R}_\varepsilon, \overline{R}_\varepsilon\rangle_{L^2(0,T;H)} \\ \le & \| \overline{R}_\varepsilon \|_{L^2(0,T;H)} \cdot \| \overline{h}_\varepsilon \|_{L^2(0,T;H)} \\ & + \varepsilon \| \overline{R}_\varepsilon(T) \| \Big( \| \overline{R}_\varepsilon'(T) \| + T^{-1} \| l_N(\varepsilon^{-1}T) \| + \varepsilon^N \| X_N'(T) \| \Big) \\ & + \varepsilon^{N+1} \| \overline{R}_\varepsilon' \|_{L^2(0,T;H)} \cdot \| X_N' \|_{L^2(0,T;H)} \\ \le & M\varepsilon^{2N} + \frac{\varepsilon}{2} \| \overline{R}_\varepsilon' \|^2_{L^2(0,T;H)} + \frac{1}{4} \| \overline{R}_\varepsilon(T) \|^2, \end{aligned} \tag{10.25}$$

where $M$ is a positive constant which does not depend on $\varepsilon$. We have used estimates (10.18), $\| \overline{R}_\varepsilon'(T) \| = \mathcal{O}(\varepsilon^N)$, as well as the homogeneous condition $\overline{R}_\varepsilon(0) = 0$. From (10.25) it follows

$$\| \overline{R}_\varepsilon' \|_{L^2(0,T;H)} = \mathcal{O}(\varepsilon^{N-1/2}), \ \ \| \overline{R}_\varepsilon(T) \| = \mathcal{O}(\varepsilon^N).$$

In case $N = 0$, we have $\overline{h}_\varepsilon \equiv 0$, and so we find

$$\| \overline{R}_\varepsilon' \|_{L^2(0,T;H)} = \mathcal{O}(1), \ \ \| \overline{R}_\varepsilon(T) \| = \mathcal{O}(\varepsilon^{1/2}).$$

To derive convenient estimates for $\| \overline{R}_\varepsilon \|_{L^2(0,T;H)}$, we denote $\widetilde{R}_\varepsilon = e^{-t}\overline{R}_\varepsilon$, and follow our previous reasoning from the proof of Theorem 10.1.7 to reach the desired conclusions. □

*Remark* 10.2.4. It is worth pointing out that all the results of this section remain valid if we replace $(h_1)$ with $(h_1')$.

## 10.3 Asymptotic analysis of problem $(P.3)_\varepsilon$

In this section we deal with problem $(P.3)_\varepsilon$, which is a hyperbolic regularization of a parabolic problem. In our investigation we shall require assumptions $(i_1)$–$(i_4)$.

As for the previous cases, to get information on the existence of possible boundary layers and the form of the corresponding rapid variables, we examine the Cauchy problem associated with a simple ordinary differential equation:

$$\begin{cases} \varepsilon u_\varepsilon'' + u_\varepsilon' + u_\varepsilon = 0, \ t \in [0,T], \\ u_\varepsilon(0) = a, \ u_\varepsilon'(0) = b, \end{cases} \tag{10.26}$$

where $a, \ b \in \mathbb{R}$.
A straightforward computation leads us to

$$u_\varepsilon(t) = \frac{ar_{2\varepsilon} - b}{r_{2\varepsilon} - r_{1\varepsilon}} e^{r_{1\varepsilon}t} + \frac{b - ar_{1\varepsilon}}{r_{2\varepsilon} - r_{1\varepsilon}} e^{r_{2\varepsilon}t}, \ t \in [0,T],$$

where

$$r_{1\varepsilon} = -\left(\frac{1}{\varepsilon}\right) \cdot \frac{1+\sqrt{1-4\varepsilon}}{2}, \ r_{2\varepsilon} = \frac{-1+\sqrt{1-4\varepsilon}}{2\varepsilon} = -\frac{2}{\sqrt{1-4\varepsilon}+1}.$$

It is easily seen that $u_\varepsilon(t)$ can be written as

$$u_\varepsilon(t) = ae^{-t} + (a+b-at)e^{-t}\varepsilon - (a+b)e^{-t/\varepsilon}\varepsilon + R_\varepsilon(t), \ t \in [0,T],$$

where $\varepsilon^{-1}R_\varepsilon(t)$ approaches zero uniformly on $[0,T]$, as $\varepsilon \to 0$. This first order expansion shows that the problem is regularly perturbed of order zero with respect to the uniform norm, but singularly perturbed of order one (with respect to the same norm). We have a boundary layer of order one near the endpoint $t=0$, and the corresponding rapid variable is $\xi = t/\varepsilon$.

Taking into account this example, we guess that the general problem $(P.3)_\varepsilon$ is regularly perturbed of order zero with respect to the norm of $C([0,T];V)$, but singularly perturbed of order one, and of any order $N \geq 1$. We can also guess the form of the fast variable: $\xi = t/\varepsilon$.

In fact, one can also derive (heuristically) these things by transforming $(P.3)_\varepsilon$ into a first order problem by the usual substitution $v = u'$ and by comparing the latter with the hyperbolic problems examined in Part II.

### 10.3.1 $N$th order asymptotic expansion

Taking into account the above discussion, we shall attempt to determine an $N$th order asymptotic expansion for the solution $u_\varepsilon$ of problem $(P.3)_\varepsilon$ in the form

$$u_\varepsilon(t) = \sum_{k=0}^{N} X_k(t)\varepsilon^k + \sum_{k=0}^{N} i_k(\xi)\varepsilon^k + R_\varepsilon(t), \ t \in [0,T], \tag{10.27}$$

where $\xi = t/\varepsilon$ is the fast variable, while the terms of the expansion have the same meaning as in Subsection 10.1.1. In order to determine these terms we follow the standard procedure. We start with the following problem satisfied by $i_0$:

$$\begin{cases} i_0''(\xi) + i_0'(\xi) = 0, \ \xi \geq 0, \\ i_0'(0) = 0, \ i_0(+\infty) = 0. \end{cases} \tag{10.28}$$

Obviously, $i_0 \equiv 0$, as expected. Then, we see that $X_0(t)$ should satisfy the reduced problem $(P.3)_0$:

$$\begin{cases} X_0'(t) + LX_0(t) = f(t), \ 0 < t < T, \\ X_0(0) = u_0. \end{cases} \tag{10.29}$$

If $X_0'(0)$ makes sense, one can determine $i_1(\xi)$ as the solution of the problem

$$\begin{cases} i_1''(\xi) + i_1'(\xi) = 0, \ \xi \geq 0, \\ i_1'(0) = -X_0'(0) + u_1, \ i_1(+\infty) = 0, \end{cases} \tag{10.30}$$

i.e.,

$$i_1(\xi) = [X_0'(0) - u_1]e^{-\xi}.$$

For the other regular terms we derive the following problems, denoted $(P.3)_k$:

$$\begin{cases} X_k'(t) + LX_k(t) = -X_{k-1}''(t), \ 0 < t < T, \\ X_k(0) = -i_k(0), \ k = 1, \ldots, N. \end{cases} \tag{10.31}$$

The other boundary layer functions should satisfy the problems

$$\begin{cases} i_k''(\xi) + i_k'(\xi) = -Li_{k-1}(\xi), \ \xi \geq 0, \\ i_k'(0) = -X_{k-1}'(0), \ i_k(+\infty) = 0, \ k = 2, \ldots, N. \end{cases} \tag{10.32}$$

Since $i_1$ is known, problem $(P.3)_1$ is completely defined. Using (10.31) and (10.32) alternatively, we can determine all the corrections (see the next subsection). Moreover, all $(P.3)_k$ can be determined formally.

Finally, for the $N$th order reminder we can derive the Cauchy problem

$$\begin{cases} \varepsilon R_\varepsilon'' + R_\varepsilon' + LR_\varepsilon = -\varepsilon^N(\varepsilon X_N'' + Li_N) \text{ in } (0,T), \\ R_\varepsilon(0) = 0, \ R_\varepsilon'(0) = -\varepsilon^N X_N'(0). \end{cases} \tag{10.33}$$

### 10.3.2 Existence, uniqueness and regularity of the solutions of problems $(P.3)_\varepsilon$ and $(P.3)_k$

We start with problem $(P.3)_\varepsilon$. Without any loss of generality, we assume that $\varepsilon = 1$ and denote the unknown function by $u$ (instead of $u_\varepsilon$). Consider the product Hilbert space $H_1 = V \times H$ with the scalar product

$$\langle h_1, h_2 \rangle_1 = \langle Lu_1, u_2 \rangle + \alpha \langle u_1, u_2 \rangle + \langle v_1, v_2 \rangle \ \ \forall \ h_i = (u_i, v_i) \in H_1, \ i = 1, 2,$$

and operator $A : D(A) \subset H_1 \to H_1$,

$$D(A) = \{(u,v) \in H_1;\ Lu \in H\},\ A(u,v) = (-v, Lu + v).$$

Obviously, problem $(P.3)_\varepsilon$ can be written in the form of the following Cauchy problem in $H_1$:

$$\begin{cases} U'(t) + AU(t) = F(t),\ t \in (0,T), \\ U(0) = U_0, \end{cases} \tag{10.34}$$

where $U(t) = (u(t), u'(t))$, $F(t) = (0, f(t))$, $U_0 = (u_0, u_1)$. We have the following existence and regularity result (where $u_\varepsilon$ appears again for later use):

**Theorem 10.3.1.** *Assume that $(i_1)$–$(i_3)$ are satisfied and*

$$f \in W^{1,1}(0,T;H),\ Lu_0 \in H. \tag{10.35}$$

*Then, problem $(P.3)_\varepsilon$ has a unique solution $u_\varepsilon \in C^1([0,T];V) \bigcap C^2([0,T];H)$. If, in addition,*

$$f \in W^{2,1}(0,T;H),\ u_1 \in V,\ Lu_1 \in H, \tag{10.36}$$

*then $u_\varepsilon$ belongs to the space $C^2([0,T];V) \bigcap C^3([0,T];H)$.*

*Proof.* It is well known that under assumptions $(i_1)$, $(i_2)$ (see [7], p. 268) operator $A + \omega_1 I$ is maximal monotone, where $\omega_1$ is the positive constant defined by

$$\omega_1 = \sup\left\{ \frac{\alpha\langle u,v\rangle}{\langle Lu,u\rangle + \alpha \parallel u \parallel^2 + \parallel v \parallel^2};\ u \in V,\ v \in H,\ \mid u \mid + \parallel v \parallel \neq 0 \right\},$$

while $I$ is the identity operator on $H_1$. Thus, $-A$ generates a linear $C_0$-semigroups on $H_1$. On the other hand, assumptions $(i_3)$ and $(10.35)_2$ imply that $U_0 \in D(A)$. Therefore, according to Theorem 2.0.27 in Chapter 2, problem (10.34) has a unique strong solution $U \in C^1([0,T];H_1)$, $AU \in C([0,T];H_1)$, i.e.,

$$u_\varepsilon \in C^1([0,T];V) \bigcap C^2([0,T];H),\ Lu_\varepsilon \in C([0,T];H).$$

If, in addition, assumptions (10.36) hold, then $F(0) - AU_0 \in D(A)$ and $F' \in W^{1,1}(0,T;H_1)$, and hence $U'$ is the strong solution of the equation which is derived by formal differentiation of equation $(10.34)_1$, with $U'(0) = F(0) - AU_0$. Therefore, the last conclusion of the theorem follows immediately. $\square$

Let us continue with problems $(P.3)_k$, $k = 0, \ldots, N$. First of all, we may assume that $\alpha = 0$, since otherwise we can see that $\overline{X}_k = e^{-\alpha t} X_k$ satisfy the following problems

$$\begin{cases} \overline{X}_k'(t) + \overline{L}\,\overline{X}_k(t) = e^{-\alpha t} f_k(t),\ 0 < t < T, \\ \overline{X}_k(0) = X_k(0),\ k = 0, \ldots, N, \end{cases}$$

where

$$f_k(t) = \begin{cases} f(t), \ k = 0, \\ -X''_{k-1}(t), \ k = 1, \dots, N, \end{cases}$$

and $\overline{L} = L + \alpha I$ ($I$ is the identity operator on $V$) satisfying the inequality $\langle \overline{L}v, v\rangle \geq \omega \mid v \mid^2 \ \ \forall\, v \in V$.
Now, we define $L_H : D(L_H) \to H$ by

$$D(L_H) = \{u \in V;\ Lu \in H\},\ L_H u = Lu \ \ \forall\, u \in D(L_H),$$

which is a monotone operator in $H$. It is well known that $L_H$ is maximal monotone. In fact, $L_H$ is the subdifferential of $\varphi : H \to (-\infty, +\infty]$,

$$\varphi(x) = \begin{cases} \frac{1}{2}\langle Lx, x\rangle \text{ for } x \in V, \\ +\infty \quad \text{otherwise.} \end{cases}$$

Consider the Cauchy problems

$$\begin{cases} X'_k(t) + L_H X_k(t) = f_k(t), \ 0 < t < T, \\ X_k(0) = \begin{cases} u_0, \ k = 0, \\ -i_k(0), \ k = 1, \dots, N. \end{cases} \end{cases}$$

For $k = 0$, if we denote

$$\alpha_{00} = u_0,\ \alpha_{j0} = f^{(j-1)}(0) - L\alpha_{j-1,0} \ \forall\, j = 1, \dots, N+1,$$

then, under the following conditions

$$f \in W^{N+1,2}(0, T; H); \tag{10.37}$$

$$\alpha_{j0} \in V,\ L\alpha_{j0} \in H \ \forall\, j = 0, \dots, N,\ \alpha_{N+1,0} \in V, \tag{10.38}$$

we can show (see Theorem 10.1.3) that problem $(P.3)_0$ has a unique solution $X_0 \in W^{N+2,2}(0, T; H)$, and

$$X_0^{(l)}(t) \in D(L_H) \ \ \forall\, t \in [0, T] \ \forall\, l = 0, \dots, N.$$

On the other hand, from (10.30) we know that

$$i_1 = (X'_0(0) - u_1)e^{-\xi} = (f(0) - Lu_0 - u_1)e^{-\xi} \ \ \forall\, \xi \geq 0.$$

Let us assume that the above conditions (10.37) and (10.38) hold. We continue with the case $k = 1$, in which $f_1 = -X''_0 \in W^{N,2}(0, T : H)$. Denote

$$\alpha_{01} = -i_1(0) = -f(0) + Lu_0 + u_1,\ \alpha_{j1} = f_1^{(j-1)}(0) - L\alpha_{j-1,1} \ \forall\, j = 1, \dots, N.$$

Assume in addition that the following conditions are satisfied

$$\alpha_{j1} \in V,\ L\alpha_{j1} \in H\ \forall\, j = 0,\dots,N-1,\ \alpha_{N1} \in V. \tag{10.39}$$

Like in case $k = 0$, we can prove that problem $(P.3)_1$ has a unique solution

$$X_1 \in W^{N+1,2}(0,T;H),\ X_1^{(l)}(t) \in D(L_H)\ \ \forall\, t \in [0,T]\ \forall\, l = 0,\dots,N-1.$$

From (10.32) we infer that

$$i_2(\xi) = (X_1'(0) - L\alpha_{01} - \xi L\alpha_{01})e^{-\xi}\ \ \forall\, \xi \geq 0.$$

By a recurrent procedure, one can formulate conditions for $f$, $u_0$, $u_1$ (which are in fact regularity and compatibility conditions for the data) which ensure that all problems $(P.3)_k$ admit unique solutions $X_k \in W^{N-k+2,2}(0,T;H)$, and

$$i_k(\xi) = P_k(\xi)e^{-\xi}\ \forall\, \xi \geq 0\ \forall\, k = 1,\dots,N,$$

where $P_k$, $k = 1,\dots,N$, are polynomials of degree at most $k-1$ with coefficients in $V$.

Taking into account the above results, we can state the following

**Corollary 10.3.2.** *Assume that* $(i_1)$–$(i_2)$ *hold and that* $f$, $u_0$, $u_1$ *satisfy appropriate conditions as described above. Then, problems* $(P.3)_\varepsilon$, $\varepsilon > 0$, *and* $(P.3)_k$, $k = 0,\dots,N$, *have unique solutions*

$$u_\varepsilon \in C^2([0,T];V)\bigcap C^3([0,T];H),\ \ X_k \in W^{N-k+2,2}(0,T;H).$$

### 10.3.3 Estimates for the remainder

We conclude this chapter with some estimates for the $N$th order remainder of expansion (10.27):

**Theorem 10.3.3.** *Suppose that all the assumptions of Corollary* 10.3.2 *are satisfied. Then, for every* $\varepsilon > 0$, *the solution* $u_\varepsilon$ *of problem* $(P.3)_\varepsilon$ *admits an asymptotic expansion of the form* (10.27) *and the following estimates hold*

$$\|R_\varepsilon\|_{C([0,T];V)} = \mathcal{O}(\varepsilon^{N+1/2}),\ \ \| R_\varepsilon' \|_{L^2(0,T;H)} = \mathcal{O}(\varepsilon^{N+1/2}).$$

*Proof.* According to Corollary 10.3.2 we get from (10.27) that $R_\varepsilon \in W^{2,2}(0,T;H)$. In addition, $R_\varepsilon(t),\ R_\varepsilon'(t) \in V$, $LR_\varepsilon(t) \in H\ \forall\, t \in [0,T]$.

Now, we choose a positive constant $\gamma_0 \geq \alpha$ and denote

$$\overline{R}_\varepsilon(t) = e^{-\gamma_0 t}R_\varepsilon(t),\ t \in [0,T].$$

A straightforward computation shows that $\overline{R}_\varepsilon$ satisfies the following boundary value problem

$$\begin{cases} \varepsilon\overline{R}_\varepsilon'' + (1+2\varepsilon\gamma_0)\overline{R}_\varepsilon' + (\gamma_0+\varepsilon\gamma_0^2)\overline{R}_\varepsilon + L\overline{R}_\varepsilon = \overline{h}_\varepsilon \text{ in } (0,T), \\ \overline{R}_\varepsilon(0) = 0,\ \overline{R}_\varepsilon'(0) = -\varepsilon^N X_N'(0), \end{cases} \tag{10.40}$$

where $\overline{h}_\varepsilon = -\varepsilon^N e^{-\gamma_0 t}(\varepsilon X_N'' + Li_N)$, $\overline{h}_\varepsilon \in L^2(0,T;H)$.

We multiply equation $(10.40)_1$ by $\overline{R}'_\varepsilon$. Thus, we can derive

$$\frac{\varepsilon}{2}\frac{d}{dt}\Big(\| \overline{R}'_\varepsilon(t) \|^2 + \gamma_0^2 \| \overline{R}_\varepsilon(t) \|^2\Big) + \| \overline{R}'_\varepsilon(t) \|^2 + \frac{1}{2}\frac{d}{dt}\langle L\overline{R}_\varepsilon(t) + \gamma_0 \overline{R}_\varepsilon(t), \overline{R}_\varepsilon(t)\rangle$$
$$\leq \langle \overline{h}_\varepsilon(t), \overline{R}'_\varepsilon(t)\rangle \ \forall\, t \in [0,T].$$

If we integrate this inequality over $[0,t]$ and take into account that $\langle Lu+\gamma_0 u, u\rangle \geq \omega \mid u \mid^2 \ \forall u \in V$, we get

$$\begin{aligned}\frac{\varepsilon}{2} \| \overline{R}'_\varepsilon(t) \|^2 + \int_0^t \| \overline{R}'_\varepsilon(s) \|^2\, ds + \frac{\omega}{2} \mid \overline{R}_\varepsilon(t) \mid^2 \leq \frac{\varepsilon}{2} \| \overline{R}'_\varepsilon(0) \|^2 \\ + \frac{1}{2}\int_0^t \| \overline{h}_\varepsilon(s) \|^2\, ds + \frac{1}{2}\int_0^t \| \overline{R}'_\varepsilon(s) \|^2\, ds\end{aligned} \tag{10.41}$$

$\forall\, t \in [0,T]$. Since $\| \overline{h}_\varepsilon \|_{L^2(0,T;H)} = \mathcal{O}(\varepsilon^{N+1/2})$, $\| \overline{R}'_\varepsilon(0) \| = \mathcal{O}(\varepsilon^N)$, we can derive from (10.41) the following estimates

$$\begin{aligned}\| \overline{R}'_\varepsilon(t) \| &\leq M_1\varepsilon^N, \\ \mid \overline{R}_\varepsilon(t) \mid &\leq M_2\varepsilon^{N+1/2} \ \forall\, t \in [0,T], \\ \| \overline{R}'_\varepsilon \|_{L^2(0,T;H)} &= \mathcal{O}(\varepsilon^{N+1/2}),\end{aligned} \tag{10.42}$$

where $M_1$, $M_2$ are some constants, independent of $\varepsilon$. Finally, taking into account the definition of $\overline{R}_\varepsilon$, one can easily derive the desired estimates for $R_\varepsilon$. □

*Remark* 10.3.4. If we choose $u_1 = f(0) - Lu_0$, then $R'_\varepsilon(0) = 0$, and so inequality (10.41) leads us to

$$\mid \overline{R}_\varepsilon(t) \mid \leq C_1\varepsilon, \ \ \| \overline{R}'_\varepsilon(t) \| \leq C_2\varepsilon^{1/2} \ \forall\, t \in [0,T], \ \ \| \overline{R}'_\varepsilon \|_{L^2(0,T;H)} = \mathcal{O}(\varepsilon),$$

since $\| \overline{h}_\varepsilon \|_{L^2(0,T;H)} = \mathcal{O}(\varepsilon)$. Consequently, we can readily derive the following stronger estimates

$$\begin{aligned}&\|u_\varepsilon - X_0\|_{C([0,T];V)} = \mathcal{O}(\varepsilon), \ \| u'_\varepsilon - X'_0 \|_{C([0,T];H)} = \mathcal{O}(\varepsilon^{1/2}), \\ &\| u'_\varepsilon - X'_0 \|_{L^2(0,T;H)} = \mathcal{O}(\varepsilon).\end{aligned}$$

## 10.4 An Example

In this section we illustrate our previous abstract results on a specific parabolic problem.

Let $\Omega \subset \mathbb{R}^n$ be a bounded open set, with boundary $\partial\Omega$ sufficiently smooth, and let $T > 0$ be given. Denote $\Omega_T = \Omega \times (0,T]$, $\Sigma_T = \partial\Omega \times [0,T]$.

Consider the following linear differential operator

$$L : D(L) \subset H \to H,\ H = L^2(\Omega),\ D(L) = H^2(\Omega) \bigcap H_0^1(\Omega),$$

$$Lu = -\sum_{i,j=1}^{n} \frac{\partial}{\partial x_j}\Big(a_{ij}(x)\frac{\partial u}{\partial x_i}\Big) + \sum_{i=1}^{n} b_i(x)\frac{\partial u}{\partial x_i} + c(x)u \quad \forall\, u \in D(L), \tag{10.43}$$

where $a_{ij} : \overline{\Omega} \to \mathbb{R},\ a_{ij} \in C^1(\overline{\Omega}),\ a_{ij} = a_{ji};\ b_i,\ c \in L^\infty(\Omega),\ \ 1 \le i,\ j \le n.$
Assume in addition that operator $L$ is uniformly elliptic, i.e., there exists a constant $\theta > 0$ such that

$$\sum_{i,j=1}^{n} a_{ij}(x)\xi_i\xi_j \ge \theta \mid \xi \mid^2 \ \ \forall\, x \in \overline{\Omega}\ \ \forall\, \xi = (\xi_1, \dots, \xi_n) \in \mathbb{R}^n, \tag{10.44}$$

where $| \cdot |$ stands for the Euclidean norm of $\mathbb{R}^n$.
Consider the following problem, denoted $P_0$,

$$\begin{cases} u_t(x,t) + Lu(x,t) = f(x,t),\ (x,t) \in \Omega_T, \\ u(x,t) = 0 \ \text{ for } \ (x,t) \in \Sigma_T, \\ u(x,0) = u_0(x),\ x \in \Omega, \end{cases} \tag{10.45}$$

where $f : \Omega_T \to \mathbb{R},\ u_0 : \Omega \to \mathbb{R}$ are given functions.

Obviously, if $a_{ij} = \delta_{ij},\ b_i \equiv 0,\ c \equiv 0$, then $L = -\Delta$ and problem $P_0$ coincides with problem $(E)$, $(BC)$, $(IC)$ with which we started Chapter 9.

Suppose that $H = L^2(\Omega)$ is equipped with the usual scalar product $\langle \cdot, \cdot \rangle$ and the induced norm $\| \cdot \|$. We can write problem (10.45) as a Cauchy problem in $H$, as follows

$$\begin{cases} u' + Lu = f(t),\ t \in (0,T), \\ u(0) = u_0. \end{cases}$$

It is well known that, under the above hypotheses, there exists an $\omega > 0$ such that $L + \omega I$ is maximal monotone in $H$, where $I$ is the identity operator on $H$ (see L.C. Evans [19], pp. 421–422).

Let us consider elliptic regularizations of the above problem $P_0$ of the form presented in Chapter 10 and again denoted $(P.i)_\varepsilon$, $i = 1, 2$. A zeroth order asymptotic expansion of the solution $u_\varepsilon$ of problem $(P.1)_\varepsilon$ is

$$u_\varepsilon(x,t) = X_0(x,t) + i_0(x,\tau) + R_\varepsilon(x,t),\ (x,t) \in \Omega_T, \tag{10.46}$$

where $X_0$ is the solution of problem $P_0$, $i_0(x,\tau) = (u_T(x) - X_0(x,T))e^{-\tau}$, $\tau = (T-t)/\varepsilon$, $R_\varepsilon$ is the remainder of order zero.
For problem $(P.2)_\varepsilon$, we have the following first order asymptotic expansion

$$u_\varepsilon(x,t) = X_0(x,t) + \varepsilon X_1(x,t) + \varepsilon i_1(x,\tau) + R_\varepsilon(x,t),\ (x,t) \in \Omega_T, \tag{10.47}$$

where $X_0$ has the same meaning as before, $X_1$ is the solution of problem $(P.2)_1$ :

$$\begin{cases} X_{1t}(x,t) + LX_1(x,t) = X_0''(x,t), \ (x,t) \in \Omega_T, \\ X_1(x,t) = 0 \ \text{ for } \ (x,t) \in \Sigma_T, \\ X_1(x,0) = 0, \ x \in \Omega, \end{cases}$$

$i_1(x,\tau) = (u_T(x) - f(x,T) + LX_0(x,T))e^{-\tau}$, while $R_\varepsilon$ is the corresponding first order remainder.

From Corollary 10.1.5, Corollary 10.2.2 and Remark 10.1.6 we derive the following result:

**Proposition 10.4.1.** *Assume that*

$$f \in L^2(\Omega_T), \ u_0, \ u_T \in H_0^1(\Omega) \bigcap H^2(\Omega). \tag{10.48}$$

*Then, problems $(P.i)_\varepsilon$, $i = 1, 2$, $\varepsilon > 0$, and $P_0$ have unique solutions*

$$u_\varepsilon \in W^{2,2}(0,T;L^2(\Omega)),$$
$$X_0 \in W^{1,2}(0,T;L^2(\Omega)) \bigcap L^\infty(0,T;H_0^1(\Omega)) \bigcap L^2(0,T;H^2(\Omega)).$$

*If, in addition,*

$$f \in W^{1,2}(0,T;L^2(\Omega)), \ f(\cdot,0) - Lu_0 \in H_0^1(\Omega), \tag{10.49}$$

*then the solutions of problems $(P.2)_\varepsilon$, $P_0$ satisfy*

$$u_\varepsilon \in W^{2,2}(0,T;L^2(\Omega)) \bigcap W^{3,2}_{\text{loc}}(0,T;L^2(\Omega)),$$
$$X_0 \in W^{2,2}(0,T;L^2(\Omega)) \bigcap L^\infty(0,T;H^2(\Omega)) \bigcap W^{1,2}(0,T;H_0^1(\Omega)),$$

*while the solution $X_1$ of problem $(P.2)_1$ belongs to*

$$W^{1,2}(0,T;L^2(\Omega)) \bigcap L^\infty(0,T;H_0^1(\Omega)) \bigcap L^2(0,T;H^2(\Omega)).$$

*Remark* 10.4.2. More information on the regularity of the solutions of the above problems is available in the literature (see, e.g., L.C. Evans [19], pp. 360, 365).

According to Theorems 10.1.7, 10.2.3, we can state the following two results concerning the remainders (of order zero and one, respectively) of the asymptotic expansions (10.46), (10.47):

**Proposition 10.4.3.** *Suppose that all the assumptions* (10.48) *are satisfied. Then, for every $\varepsilon > 0$, the solution $u_\varepsilon$ of problem $(P.1)_\varepsilon$ admits an asymptotic expansion of the form* (10.46) *and the following estimates hold true*

$$\|R_\varepsilon\|_{C([0,T];L^2(\Omega))} = O(\varepsilon^{1/4}), \ \| u_\varepsilon - X_0 \|_{L^2(0,T;H_0^1(\Omega))} = \mathcal{O}(\varepsilon^{1/2}),$$

*and $u_{\varepsilon t} \to X_{0t}$ weakly in $L^2(\Omega_T)$.*

**Proposition 10.4.4.** *Suppose that all the assumptions* (10.48) *are satisfied. Then, for every* $\varepsilon > 0$,

$$\|u_\varepsilon - X_0\|_{C([0,T];L^2(\Omega))} = O(\varepsilon^{1/4}), \; \|u_\varepsilon - X_0\|_{L^2(0,T;H_0^1(\Omega))} = O(\varepsilon^{1/2}),$$

*and* $u_{\varepsilon t} \to X_{0t}$ *weakly in* $L^2(\Omega_T)$, *where* $u_\varepsilon$ *and* $X_0$ *are the solutions of problems* $(P.2)_\varepsilon$ *and* $P_0$, *respectively.*

*If, in addition, hypotheses* (10.49) *are satisfied, then* $u_\varepsilon$ *admits an asymptotic expansion of the form* (10.47), *and* $\|R_{\varepsilon t}\|_{L^2(\Omega_T)} = O(\varepsilon)$,

$$\|R_\varepsilon\|_{C([0,T];L^2(\Omega))} = O(\varepsilon^{5/4}), \; \| u_\varepsilon - X_0 - \varepsilon X_1 \|_{L^2(0,T;H_0^1(\Omega))} = \mathcal{O}(\varepsilon^{3/2}).$$

Finally, let us examine the hyperbolic regularization (in the sense described before) of problem (10.45), which will again be denoted $P_0$. Consider the Hilbert spaces $V = H_0^1(\Omega)$ and $H = L^2(\Omega)$. It is well known that $V \subset H \subset V'$ densely and continuously. Consider $L : V \to V'$ defined by (10.43). We admit the same assumptions on $\Omega$ and coefficients $a_{ij}$, $c$, but we suppose that $b_i \equiv 0$. This last assumption is required just for simplicity.

Denote as before by $(P.3)_\varepsilon$ the corresponding hyperbolic regularization:

$$\begin{cases} \varepsilon u'' + u' + Lu = f(t), \; t \in (0,T), \\ u(0) = u_0, \quad u'(0) = u_1. \end{cases}$$

It is well known that operator $L$ satisfies assumptions $(i_2)$ in this chapter (see L.C. Evans [19], pp. 300–301).

According to our previous treatment, we have the following first order asymptotic expansion for the solution of problem $(P.3)_\varepsilon$:

$$u_\varepsilon(x,t) = X_0(x,t) + \varepsilon X_1(x,t) + \varepsilon i_1(x,\xi) + R_\varepsilon(x,t), \; (x,t) \in \Omega_T, \tag{10.50}$$

where $X_0$ is the solution of problem $P_0$, $R_\varepsilon$ denotes the first order remainder, $X_1$ satisfies problem $(P.3)_1$:

$$\begin{cases} X_{1t}(x,t) + LX_1(x,t) = -X_0''(x,t), \; (x,t) \in \Omega_T, \\ X_1(x,t) = 0 \;\text{ for }\; (x,t) \in \Sigma_T, \\ X_1(x,0) = u_1(x) - f(x,0) + Lu_0(x), \; x \in \Omega, \end{cases}$$

while $i_1(x,\xi) = (f(x,0) - Lu_0(x) - u_1(x))e^{-\xi}$, $\xi = t/\varepsilon$. Our results from Subsection 10.3.2 read as follows:

**Proposition 10.4.5.** *Assume that*

$$f \in W^{1,2}(0,T;L^2(\Omega)), \; u_0, \; f(\cdot,0) - Lu_0 \in H_0^1(\Omega), \; u_1, \; Lu_0 \in L^2(\Omega). \tag{10.51}$$

*Then, problems* $(P.3)_\varepsilon$, $\varepsilon > 0$, *and* $P_0$ *have unique solutions*

$$u_\varepsilon \in C^1([0,T]; H_0^1(\Omega)) \bigcap C^2([0,T]; L^2(\Omega)),$$
$$X_0 \in W^{2,2}(0,T; L^2(\Omega)) \bigcap L^\infty(0,T; H^2(\Omega)) \bigcap W^{1,2}(0,T; H_0^1(\Omega)).$$

*If, in addition,*

$$f \in W^{2,2}(0,T; L^2(\Omega)), \tag{10.52}$$

$$u_1 \in H_0^1(\Omega) \bigcap H^2(\Omega),\ f(\cdot,0) - Lu_0 \in H^2(\Omega),$$
$$L(f(\cdot,0) - Lu_0),\ f_t(\cdot,0) \in H_0^1(\Omega), \tag{10.53}$$

*then* $u_\varepsilon \in C^2([0,T]; H_0^1(\Omega)) \cap C^3([0,T]; L^2(\Omega))$,

$$X_0 \in W^{3,2}(0,T; L^2(\Omega)),\ X_1 \in W^{2,2}(0,T; L^2(\Omega)).$$

*Remark* 10.4.6. In fact, the solutions of the above problems satisfy higher order regularity properties, since $L$ is a divergence operator (see, e.g., L.C. Evans [19], pp. 360, 365 and pp. 389, 391).

Let us also point out that assumptions (10.51)–(10.53) imply (10.36)–(10.39).

Finally, Theorem 10.3.3 and Remark 10.3.4 lead us to the following results:

**Proposition 10.4.7.** *Suppose that all the assumptions* (10.51) *are satisfied. Then, for every* $\varepsilon > 0$, *the following estimates hold*

$$\|u_\varepsilon - X_0\|_{C([0,T];H_0^1(\Omega))} = O(\varepsilon^{1/2}),\quad \| u_{\varepsilon t} - X_{0t} \|_{L^2(\Omega_T)} = \mathcal{O}(\varepsilon^{1/2}),$$

*where* $u_\varepsilon$ *and* $X_0$ *are the solutions of problems* $(P.3)_\varepsilon$ *and* $P_0$, *respectively. If, in addition,* $u_1(x) = f(x,0) - Lu_0(x)$ *for* a.a. $x \in \Omega$, *then*

$$\|u_\varepsilon - X_0\|_{C([0,T];H_0^1(\Omega))} = O(\varepsilon),\quad \| u_{\varepsilon t} - X_{0t} \|_{C([0,T];L^2(\Omega))} = \mathcal{O}(\varepsilon^{1/2}),$$
$$\|u_{\varepsilon t} - X_{0t}\|_{L^2(\Omega_T)} = O(\varepsilon).$$

*Moreover, if assumptions* (10.52), (10.53) *are also satisfied, then we have* (10.50), *as well as*

$$\|R_\varepsilon\|_{C([0,T];H_0^1(\Omega))} = O(\varepsilon^{3/2}),\quad \| R_{\varepsilon t} \|_{L^2(\Omega_T)} = \mathcal{O}(\varepsilon^{3/2}).$$

# Chapter 11

# The Nonlinear Case

In this chapter we address elliptic and hyperbolic regularizations of the nonlinear parabolic problem introduced in Chapter 9 and denoted here $P_0$ :

$$\begin{cases} u_t(x,t) - \Delta u(x,t) + \beta(u(x,t)) = f(x,t), \ (x,t) \in \Omega_T, \\ u(x,t) = 0 \ \text{ for } \ (x,t) \in \Sigma_T, \\ u(x,0) = u_0(x), \ x \in \Omega, \end{cases}$$

where $\Omega \subset \mathbb{R}^n$ is a bounded open set with boundary $\partial\Omega$ sufficiently smooth; $T > 0$ is a given time; $\Omega_T = \Omega \times (0,T]$; $\Sigma_T = \partial\Omega \times [0,T]$; $f : \Omega_T \to \mathbb{R}$, $u_0 : \Omega \to \mathbb{R}$ are given functions, $f \in L^2(\Omega_T)$, $u_0 \in L^2(\Omega)$.

Concerning the nonlinear function $\beta$ we assume that

$(h)$ $\beta : D(\beta) = \mathbb{R} \to \mathbb{R}$ is continuous, nondecreasing, and $\beta(0) = 0$.

As promised before (see Chapter 9), we consider here two kinds of elliptic regularization of the above problem $P_0$, which are defined by the nonlinear equation

$$\varepsilon u_{tt}(x,t) - u_t(x,t) + \Delta u(x,t) - \beta(u(x,t)) = -f(x,t), \ (x,t) \in \Omega_T, \qquad (NEE)$$

with the homogeneous Dirichlet boundary condition

$$u(x,t) = 0 \ \text{ for } \ (x,t) \in \Sigma_T,$$

and with conditions in the form $(BC.1)$ or $(BC.2)$, respectively, i.e.,

$$u(x,0) = u_0(x), \ u(x,T) = u_T(x), \ x \in \Omega, \qquad (BC.1)$$

or

$$u(x,0) = u_0(x), \ u_t(x,T) = u_T(x), \ x \in \Omega, \qquad (BC.2)$$

where $u_T \in L^2(\Omega)$. These problems will be denoted again, as in the linear case, by $(P.1)_\varepsilon$ and $(P.2)_\varepsilon$, respectively. We shall also examine the hyperbolic regularization, denoted by $(P.3)_\varepsilon$, and defined by the nonlinear hyperbolic equation

$$\varepsilon u_{tt}(x,t) + u_t(x,t) - \Delta u(x,t) + \beta(u(x,t)) = f(x,t),\ (x,t) \in \Omega_T, \qquad (NHE)$$

with the homogeneous Dirichlet boundary condition

$$u(x,t) = 0 \ \text{ for } \ (x,t) \in \Sigma_T,$$

and initial conditions of the form $(IC.3)$, i.e.,

$$u(x,0) = u_0(x),\ u_t(x,0) = u_1(x),\ x \in \Omega, \qquad (IC.3)$$

where $u_1 \in L^2(\Omega)$.

In this case the nonlinear function $\beta$ is supposed to satisfy the Lipschitz condition, i.e.,
$(i)$ $\beta : D(\beta) = \mathbb{R} \to \mathbb{R}$ and $\exists l > 0$ such that $\mid \beta(x) - \beta(y) \mid \le l \mid x - y \mid \ \forall\, x, y \in \mathbb{R}$.
As in the linear case, we shall investigate the three types of regularization in the following three sections of this chapter. Obviously, the nonlinear case is more delicate and requires much more effort.

In the case of problem $(P.1)_\varepsilon$, we shall determine only a zeroth order asymptotic expansion. Like in the linear case, this is a singularly perturbed problem of order zero with respect to the norm of $C([0,T]; L^2(\Omega))$, with a boundary layer located near the boundary set $\Omega \times \{T\}$. Under suitable assumptions, we shall prove results on the existence, uniqueness and regularity of the solutions of problems $(P.1)_\varepsilon$ and $P_0$, and establish estimates for the remainder term. The boundary layer function (of order zero) is determined explicitly. Thus the zeroth order asymptotic expansion is completely determined and validated.

We continue with the elliptic regularization $(P.2)_\varepsilon$. This problem is regularly perturbed of order zero, but singularly perturbed of any order $N \ge 1$, with respect to the norm of $C([0,T]; L^2(\Omega))$. In fact, we shall restrict our investigation to the case of an asymptotic expansion of order one, which is enough to describe the situation.

In the third and last section, we shall pay attention to the hyperbolic regularization $(P.3)_\varepsilon$. For its solution we shall try to determine a first order asymptotic expansion, since $(P.3)_\varepsilon$ is regularly perturbed of order zero, but singularly perturbed of order one, with respect to the norm of $C([0,T]; H_0^1(\Omega))$, with a boundary layer located near $\Omega \times \{0\}$. As usual, we determine formally the corresponding asymptotic expansion, then, in the case $n \le 3$, we prove existence and smoothness results for the problems involved, and conclude with estimates for the remainder of order one with respect to the norm of $C([0,T]; H_0^1(\Omega))$.

Note that an elliptic regularizations of the form $(P.1)_\varepsilon$ can be found in J.L. Lions [32], pp. 424–427, where the nonlinear function $\beta$ has the form $\beta(u) = u^3 \ \forall\, u \in \mathbb{R}$.

Moreover, a zeroth order asymptotic expansion for the solution of problem $(P.1)_\varepsilon$ has been constructed in [4].

# 11.1 Asymptotic analysis of problem $(P.1)_\varepsilon$

In this section we study the elliptic regularization $(P.1)_\varepsilon$ formulated above. Having in mind the information we have acquired about the linear case (see Section 10.1), we guess that the present $(P.1)_\varepsilon$ is singularly perturbed of order zero with respect to the norm of $C([0,T];L^2(\Omega))$. We shall see that this is indeed the case.

## 11.1.1 A zeroth order asymptotic expansion for the solution of problem $(P.1)_\varepsilon$

As in the linear case (see Subsection 10.1.1), one can determine a zeroth order asymptotic expansion for the solution of problem $(P.1)_\varepsilon$ in the form

$$u_\varepsilon(x,t) = X_0(x,t) + i_0(x,\tau) + R_\varepsilon(x,t),\ (x,t) \in \Omega_T, \tag{11.1}$$

where:

$\tau := (T-t)/\varepsilon$ is the stretched (fast) variable, $X_0$ is the regular term, $i_0$ is the corresponding correction, while $R_\varepsilon$ denotes the remainder of order zero.

By the standard matching procedure, one can see that $X_0$ satisfies the reduced problem $P_0$, while $i_0$ satisfies the following problem

$$\begin{cases} i_{0\tau\tau}(x,\tau) + i_{0\tau}(x,\tau) = 0,\ (x,\tau) \in \Omega \times (0,\infty), \\ i_0(x,0) = u_T(x) - X_0(x,T),\ x \in \Omega,\ \ \| i_0(\cdot,\tau) \| \to 0 \text{ as } \tau \to \infty, \end{cases}$$

where $\| \cdot \|$ denotes the usual norm of $L^2(\Omega)$. Therefore,

$$i_0(x,\tau) = (u_T(x) - X_0(x,T))e^{-\tau}. \tag{11.2}$$

Finally, one can see that the remainder term satisfies the problem

$$\begin{cases} \varepsilon(R_\varepsilon + X_0)_{tt} - R_{\varepsilon t} + \Delta R_\varepsilon - \beta(u_\varepsilon) + \beta(X_0) = -\Delta i_0 \text{ in } \Omega_T, \\ R_\varepsilon = -i_0 \text{ on } \Sigma_T, \\ R_\varepsilon(x,0) = -i_0(x,T/\varepsilon),\ R_\varepsilon(x,T) = 0,\ x \in \Omega. \end{cases} \tag{11.3}$$

## 11.1.2 Existence, uniqueness and regularity of the solutions of problems $(P.1)_\varepsilon$ and $P_0$

In order to investigate the existence of solutions to problem $(P.1)_\varepsilon$, we consider as our framework the Hilbert space $H = L^2(\Omega)$ with the usual scalar product and norm, denoted $\langle \cdot,\cdot \rangle$ and $\| \cdot \|$, respectively. Define $\overline{\beta} : D(\overline{\beta}) \subset H \to H$ by

$$D(\overline{\beta}) = \{u \in H;\ x \to \beta(u(x)) \text{ belongs to } H\},$$
$$\left(\overline{\beta}(u)\right)(x) = \beta(u(x)) \text{ for a.a. } x \in \Omega,\ \forall\, u \in D(\overline{\beta}).$$

Recall that $\overline{\beta}$ is called the canonical extension of $\beta$ to $H$. Obviously, it is maximal monotone (see, e.g., [34], p. 31). Now, problem $(P.1)_\varepsilon$ can be written as the following two-point boundary value problem in $H$

$$\begin{cases} \varepsilon u'' - u' - Ju = -f, \ t \in (0,T), \\ u(0) = u_0, \ u(T) = u_T, \end{cases} \tag{11.4}$$

where $u(t) = u(\cdot, t), \ f(t) = f(\cdot, t), \ t \in (0,T), \ J : D(J) \subset H \to H$,

$$\begin{aligned} D(J) =& H_0^1(\Omega) \bigcap H^2(\Omega) \bigcap D(\overline{\beta}), \\ Ju =& -\Delta u + \overline{\beta}(u) \ \forall \ u \in D(J). \end{aligned}$$

It is well known that under assumption $(h)$ operator $J : D(J) \to H$ is maximal monotone (see, e.g., [34], p. 181). Note that $J$ is also injective. Therefore, if $f \in L^2(\Omega_T)$, $u_0, \ u_T \in D(J)$, and $(h)$ holds then it follows by Theorem 2.0.37 in Chapter 2 that problem (11.4) has a unique solution $u_\varepsilon \in W^{2,2}(0,T;H)$ and $\beta(u_\varepsilon) \in L^2(0,T;H)$.

If, in addition, $f \in W^{1,2}(0,T;H), \ \beta' \in L^\infty(\mathbb{R})$, and denote

$$h_\varepsilon = \varepsilon^{-1}(u_\varepsilon' + \overline{\beta}(u_\varepsilon) - f), \ L_\varepsilon = -\varepsilon^{-1}\Delta,$$

then $h_\varepsilon \in W^{1,2}(0,T;H)$, and problem (11.4) has the form (10.6). Therefore, from Lemma 10.1.1 in Subsection 10.1.2, we derive the following result:

**Theorem 11.1.1.** *Assume that $(h)$ is satisfied, and*

$$f \in L^2(\Omega_T), u_0, \ u_T \in H_0^1(\Omega) \bigcap H^2(\Omega), \ \beta(u_0), \ \beta(u_T) \in L^2(\Omega).$$

*Then, problem $(P.1)_\varepsilon$ has a unique solution $u_\varepsilon \in W^{2,2}(0,T;H)$. If, in addition,*

$$f \in W^{1,2}(0,T;H), \ \beta' \in L^\infty(\mathbb{R}),$$

*then $u_\varepsilon$ belongs to the space $W^{2,2}(0,T;H) \bigcap W^{3,2}_{\text{loc}}(0,T;H)$, with*

$$t^{3/2}(T-t)^{3/2}u''' \in L^2(0,T;H).$$

In the following we deal with problem $P_0$. This can be represented as the Cauchy problem in $H$:

$$\begin{cases} X_0' + JX_0 = f(t), \ t \in (0,T), \\ X_0(0) = u_0, \end{cases} \tag{11.5}$$

where $X_0(t) := X_0(\cdot, t), \ f(t) := f(\cdot, t), \ t \in (0,T)$. As noted above, operator $J$ is maximal monotone. In fact, $J$ is even cyclically monotone. More precisely, since $\beta$

is the derivative of the convex function $j(x) = \int_0^x \beta(y)dy$, operator $J = \partial\psi$, where $\psi : H \to (-\infty, \infty]$,

$$\psi(u) = \begin{cases} \frac{1}{2}\int_\Omega |\nabla u|^2 dx + \int_\Omega j(u)dx, \text{ if } u \in H_0^1(\Omega),\ j(u) \in L^1(\Omega), \\ +\infty, \text{ otherwise,} \end{cases} \tag{11.6}$$

where $|\cdot|$ denotes the norm of $\mathbb{R}^n$. Therefore, by Theorems 2.0.20 and 2.0.24 in Chapter 2, we have

**Theorem 11.1.2.** *Assume that* $(h)$ *holds, and*

$$f \in L^2(\Omega_T),\ u_0 \in H_0^1(\Omega),\ j(u_0) \in L^1(\Omega).$$

*Then, problem* $P_0$ *has a unique solution*

$$X_0 \in W^{1,2}(0,T;L^2(\Omega)),\ X_0(\cdot,t) \in H_0^1(\Omega)\ \ \forall\, t \in [0,T].$$

*If, in addition,*

$$f \in W^{1,1}(0,T;L^2(\Omega)),\ u_0 \in H^2(\Omega)\bigcap H_0^1(\Omega),\ \beta(u_0) \in L^2(\Omega),$$

*then* $X_0$ *belongs to the space* $W^{1,\infty}(0,T;L^2(\Omega))$ *and* $X_0(\cdot,t) \in H_0^1(\Omega)\bigcap H^2(\Omega)$ *for all* $t \in [0,T]$.

By employing what we have proved so far, we can state the following concluding result:

**Corollary 11.1.3.** *Assume that* $(h)$ *is satisfied and*

$$f \in L^2(\Omega_T),\ u_0,\ u_T \in H_0^1(\Omega)\bigcap H^2(\Omega),\ \beta(u_0),\ \beta(u_T) \in L^2(\Omega). \tag{11.7}$$

*Then, problems* $(P.1)_\varepsilon$, $\varepsilon > 0$, *and* $P_0$ *have unique solutions*

$$u_\varepsilon \in W^{2,2}(0,T;L^2(\Omega)),\ X_0 \in W^{1,2}(0,T;L^2(\Omega)).$$

*If, moreover,*

$$f \in W^{1,2}(0,T;L^2(\Omega)),\ \beta' \in L^\infty(\mathbb{R}), \tag{11.8}$$

*then*

$$u_\varepsilon \in W^{2,2}(0,T;L^2(\Omega))\bigcap W^{3,2}_{\text{loc}}(0,T;L^2(\Omega)),\ t^{3/2}(T-t)^{3/2}u''' \in L^2(\Omega_T),$$
$$X_0 \in W^{1,\infty}(0,T;L^2(\Omega)).$$

*Remark* 11.1.4. It should be pointed out that, under assumptions (11.7) of Corollary 11.1.3, $X_0(\cdot,t) \in H_0^1(\Omega)$ $\forall t \in [0,T]$ (see Theorem 11.1.2), thus $i_0(\cdot,\tau) \in H_0^1(\Omega)$ $\forall\, \tau \geq 0$. Therefore, according to $(11.3)_2$, we infer that $R_\varepsilon = 0$ on $\Sigma_T$.

### 11.1.3 Estimates for the remainder

While our expansion (11.1) is well defined under the above assumptions, we can validate it completely by stating and proving the following result:

**Theorem 11.1.5.** *Suppose that* $(h)$ *is satisfied, and* $f \in L^2(\Omega_T)$, $u_0, u_T \in H_0^1(\Omega) \cap H^2(\Omega)$, $\beta' \in L^\infty(\mathbb{R})$. *Then, for every* $\varepsilon > 0$, *the solution of problem* $(P.1)_\varepsilon$ *admits an asymptotic expansion of the form* (11.1) *and the following estimates hold*

$$\|R_\varepsilon\|_{C([0,T];L^2(\Omega))} = O(\varepsilon^{1/4}), \quad \| u_\varepsilon - X_0 \|_{L^2(0,T;H_0^1(\Omega))} = \mathcal{O}(\varepsilon^{1/2}).$$

*In addition,* $u_{\varepsilon t} \to X_{0t}$ *weakly in* $L^2(\Omega_T)$ *as* $\varepsilon \to 0^+$.

*Proof.* First of all, we homogenize the boundary conditions $(11.3)_3$ by the change

$$\begin{aligned}\overline{R}_\varepsilon(x,t) =& R_\varepsilon(x,t) + \alpha_\varepsilon(x,t),\\ \alpha_\varepsilon(x,t) =& \frac{(T-t)}{T} i_0\Big(x, \frac{T}{\varepsilon}\Big), \quad (x,t) \in \Omega_T.\end{aligned}$$

A straightforward computation shows that $\overline{R}_\varepsilon$ satisfies the boundary value problem

$$\begin{cases} \varepsilon(\overline{R}_\varepsilon - \alpha_\varepsilon + X_0)_{tt} - \overline{R}_{\varepsilon t} + \Delta \overline{R}_\varepsilon - \beta(u_\varepsilon) + \beta(X_0) = \overline{h}_\varepsilon \text{ in } \Omega_T, \\ \overline{R}_\varepsilon = 0 \text{ on } \Sigma_T, \\ \overline{R}_\varepsilon(x,0) = \overline{R}_\varepsilon(x,T) = 0, \ x \in \Omega, \end{cases} \tag{11.9}$$

where $\overline{h}_\varepsilon = -\Delta i_0 - \alpha_{\varepsilon t} + \Delta \alpha_\varepsilon$.
From Corollary 11.1.3 and equation (11.1) we infer that

$$R_\varepsilon + X_0 \in W^{2,2}(0,T;H), \ R_\varepsilon \in W^{1,2}(0,T;H).$$

We multiply equation $(11.9)_1$ by $\overline{R}_\varepsilon(t)$ and then integrate the resulting equation over $\Omega_T$. Thus, by applying Green's formula, we obtain

$$\begin{aligned} &\varepsilon \int_{\Omega_T} \left[ \frac{\partial}{\partial t}\Big( \big(\overline{R}_\varepsilon - \alpha_\varepsilon + X_0\big)_t \cdot \overline{R}_\varepsilon \Big) - (\overline{R}_{\varepsilon t})^2 \right] dxdt \\ &- \frac{1}{2} \int_{\Omega_T} \frac{\partial}{\partial t} (\overline{R}_\varepsilon)^2 \, dxdt = \int_{\Omega_T} |\nabla \overline{R}_\varepsilon|^2 dxdt + \int_{\Omega_T} \overline{h}_\varepsilon \overline{R}_\varepsilon \, dxdt \\ &+ \int_{\Omega_T} \overline{R}_\varepsilon (\beta(u_\varepsilon) - \beta(X_0)) \, dxdt + \varepsilon \int_{\Omega_T} X_{0t} \overline{R}_{\varepsilon t} \, dxdt, \end{aligned} \tag{11.10}$$

where $|\cdot|$ denotes the Euclidean norm of $\mathbb{R}^n$.

Since $\beta$ is nondecreasing, it follows that

$$\begin{aligned}\int_{\Omega_T} & |\nabla \overline{R}_\varepsilon|^2 dxdt + \varepsilon \parallel \overline{R}_{\varepsilon t} \parallel^2_{L^2(\Omega_T)} \\ & \leq \parallel \overline{R}_\varepsilon \parallel_{L^2(\Omega_T)} \cdot \parallel \overline{h}_\varepsilon \parallel_{L^2(\Omega_T)} \\ & \quad + \frac{\varepsilon}{2} \parallel \overline{R}_{\varepsilon t} \parallel^2_{L^2(\Omega_T)} + \frac{\varepsilon}{2} \parallel X_{0t} \parallel^2_{L^2(\Omega_T)} \\ & \quad + \parallel \overline{R}_\varepsilon \parallel_{L^2(\Omega_T)} \cdot \parallel \beta(X_0 + i_0 - \alpha_\varepsilon) - \beta(X_0) \parallel_{L^2(\Omega_T)} .\end{aligned} \tag{11.11}$$

On the other hand, we have

$$\begin{aligned} & \parallel i_0 \parallel_{L^2(\Omega_T)} = \mathcal{O}(\varepsilon^{1/2}), \quad \parallel \Delta i_0 \parallel_{L^2(\Omega_T)} = \mathcal{O}(\varepsilon^{1/2}), \\ & \left\| i_0\left(\frac{T}{\varepsilon}\right) \right\| = \mathcal{O}(\varepsilon^j) \ \forall\, j \geq 1, \\ & \parallel \beta(X_0 + i_0 - \alpha_\varepsilon) - \beta(X_0) \parallel_{L^2(\Omega_T)} = \mathcal{O}(\varepsilon^{1/2}), \end{aligned} \tag{11.12}$$

since $\beta' \in L^\infty(\mathbb{R})$.
Estimates (11.11), (11.12) and the Poincaré inequality lead us to

$$\parallel \overline{R}_\varepsilon \parallel_{L^2(\Omega_T)} = \mathcal{O}(\varepsilon^{1/2}), \quad \parallel \overline{R}_{\varepsilon t} \parallel_{L^2(\Omega_T)} = \mathcal{O}(1), \quad \parallel \nabla \overline{R}_\varepsilon \parallel_{L^2(\Omega_T;\mathbb{R}^n)} = \mathcal{O}(\varepsilon^{1/2}),$$

from which we derive the desired estimates. □

*Remark* 11.1.6. Assume that $u_T(x) = X_0(x,T)$, $x \in \Omega$, which implies $i_0 \equiv 0$. Then $\alpha_\varepsilon = \overline{h}_\varepsilon = 0$. If, in addition, $X_0 \in W^{2,2}(0,T;L^2(\Omega))$ (this happens under additional assumptions on $f$ and $u_0$), then, making use of the obvious equality

$$\int_{\Omega_T} X_{0t} \cdot \overline{R}_{\varepsilon t}\, dxdt = - \int_{\Omega_T} X_{0tt} \cdot \overline{R}_\varepsilon\, dxdt,$$

we infer from (11.10) that

$$\parallel R_\varepsilon \parallel_{L^2(\Omega_T)} = \mathcal{O}(\varepsilon), \quad \parallel R_{\varepsilon t} \parallel_{L^2(\Omega_T)} = \mathcal{O}(\sqrt{\varepsilon}).$$

Therefore, in such circumstances, problem $(P.1)_\varepsilon$ is regularly perturbed of order zero, and the following estimates hold

$$\begin{aligned} & \|u_\varepsilon - X_0\|_{C([0,T];L^2(\Omega))} = \mathcal{O}(\varepsilon^{3/4}), \quad \parallel u_{\varepsilon t} - X_{0t} \parallel_{L^2(\Omega_T)} = \mathcal{O}(\varepsilon^{1/2}), \\ & \|u_\varepsilon - X_0\|_{L^2(0,T;H_0^1(\Omega))} = \mathcal{O}(\varepsilon). \end{aligned}$$

*Remark* 11.1.7. If $X_0 \in L^\infty(\Omega_T)$ and $u_T \in L^\infty(\Omega)$ (hence $i_0 \in L^\infty(\Omega \times [0,\infty))$), then it is easily seen that estimation $(11.12)_3$ can be obtained under the following weaker assumption on $\beta$

$(h)'$ $\beta' \in L^\infty_{\mathrm{loc}}(\mathbb{R})$,

and the conclusions of Theorem 11.1.5 hold.

Let us present a particular case in which $(h)'$ is enough. Indeed, one can prove that if $(h)$ holds, and

$$f \in W^{1,1}(0,T;H),\ u_0 \in H^2(\Omega)\bigcap H_0^1(\Omega),\ \beta(u_0) \in L^2(\Omega),$$

then $X_0 \in L^\infty(0,T;H^2(\Omega)\bigcap H_0^1(\Omega))$ (see [9], p. 257).
On the other hand, by the Sobolev-Kondrashov theorem, if $m \cdot p > n$, then $W^{m,p}(\Omega) \subset C(\overline{\Omega})$ compactly. In particular, for $m = p = 2$ and $n \leq 3$, we have $X_0 \in L^\infty(\Omega_T)$, and $u_T \in H^2(\Omega) \Rightarrow u_T \in C(\overline{\Omega}) \subset L^\infty(\Omega)$.

## 11.2 Asymptotic analysis of problem $(P.2)_\varepsilon$

As in the linear case investigated in Section 10.2, the nonlinear problem $(P.2)_\varepsilon$ (formulated in the beginning of this chapter) is regularly perturbed of order zero, but singularly perturbed of any order $\geq 1$ with respect to the norm of $C([0,T];L^2(\Omega))$, with a boundary layer near $\{T\} \times \Omega$. In fact, we shall confine ourselves to finding a first order expansion.

### 11.2.1 A first order asymptotic expansion for the solution of problem $(P.2)_\varepsilon$

Having in mind what we did in the linear case, we suggest a first order asymptotic expansion for the solution of problem $(P.2)_\varepsilon$ in the form

$$u_\varepsilon(x,t) = X_0(x,t) + \varepsilon X_1(x,t) + \varepsilon i_1(x,\tau) + R_\varepsilon(x,t),\ (x,t) \in \Omega_T, \tag{11.13}$$

where: $\tau := (T-t)/\varepsilon$ is the stretched (fast) variable, and the terms of the expansion have the same meaning as in Subsection 10.1.1.

The first regular term of the expansion, $X_0$, should satisfy the reduced problem $P_0$ formulated in the very first part of this chapter, while $X_1$ satisfies the linear problem

$$\begin{cases} X_{1t}(x,t) - \Delta X_1(x,t) + \beta'(X_0(x,t))X_1(x,t) = X_{0tt}(x,t),\ (x,t) \in \Omega_T, \\ X_1(x,t) = 0 \ \text{ for } \ (x,t) \in \Sigma_T, \\ X_1(x,0) = 0,\ x \in \Omega, \end{cases}$$

which will be denoted by $(P.2)_1$.

The first order boundary layer function has the same form as that we derived in Subsection 10.2.1, i.e.,

$$i_1(x,\tau) = (u_T(x) - X_{0t}(x,T))e^{-\tau},\ \tau \geq 0,\ x \in \Omega.$$

Finally, the (first order) remainder $R_\varepsilon$ of expansion (11.13) should satisfy the problem

$$\begin{cases} \varepsilon(R_\varepsilon + \varepsilon X_1)_{tt} - R_{\varepsilon t} + \Delta R_\varepsilon \\ -\beta(u_\varepsilon) + \beta(X_0) + \varepsilon\beta'(X_0)X_1 = -\varepsilon\Delta i_1 \text{ in } \Omega_T, \\ R_\varepsilon = -\varepsilon i_1 \text{ on } \Sigma_T, \\ R_\varepsilon(x,0) = -\varepsilon i_1(x,T/\varepsilon),\ R_{\varepsilon t}(x,T) = -\varepsilon X_{1t}(x,T),\ x \in \Omega. \end{cases} \tag{11.14}$$

### 11.2.2 Existence, uniqueness and regularity of the solutions of problems $(P.2)_\varepsilon$, $P_0$ and $(P.2)_1$

In this subsection we shall use the definitions and notation from Subsection 11.1.2. As far as $(P.2)_\varepsilon$ is concerned, we have the following result:

**Theorem 11.2.1.** *Assume that* $(h)$ *is satisfied, and*

$$f \in L^2(\Omega_T),\ u_0,\ u_T \in H_0^1(\Omega)\bigcap H^2(\Omega),\ \beta(u_0),\ \beta(u_T) \in L^2(\Omega).$$

*Then, problem* $(P.2)_\varepsilon$ *has a unique solution* $u_\varepsilon \in W^{2,2}(0,T;H)$. *If, in addition,*

$$f \in W^{1,2}(0,T;H),\ \beta' \in L^\infty(\mathbb{R}),$$

*then* $u_\varepsilon$ *belongs to the space* $W^{2,2}(0,T;H)\bigcap W^{3,2}_{\text{loc}}(0,T;H)$, *with*

$$t^{3/2}(T-t)^{3/2}u''' \in L^2(0,T;H).$$

The proof of this theorem relies on arguments similar to those from the proof of Theorem 11.1.1.

In the following we shall examine problems $P_0$ and $(P.2)_1$. We anticipate that in establishing estimates for the remainder $R_\varepsilon$ we shall need that $X_1 \in W^{1,2}(0,T;L^2(\Omega))$, hence $X_0 \in W^{2,2}(0,T;L^2(\Omega))$. As we remarked in Subsection 11.1.2, the reduced problem $P_0$ can be written as a Cauchy problem in $H$, namely (11.5), where $J = \partial\psi$, with $\psi$ defined by (11.6). Therefore, one can prove the following result:

**Theorem 11.2.2.** *Assume that* $(h)$ *holds, and*

$$f \in L^2(\Omega_T),\ u_0 \in H_0^1(\Omega),\ j(u_0) \in L^1(\Omega). \tag{11.15}$$

*Then problem* $P_0$ *has a unique solution*

$$X_0 \in W^{1,2}(0,T;L^2(\Omega)),\ X_0(\cdot,t) \in H_0^1(\Omega)\ \forall\, t \in [0,T].$$

*If, in addition,*

$$\begin{aligned} &f \in W^{1,2}(0,T;L^2(\Omega)),\ \beta \in C^1(\mathbb{R}),\ \beta' \in L^\infty(\mathbb{R}),\\ &u_0 \in H^2(\Omega),\ f(\cdot,0) + \Delta u_0 - \beta(u_0) \in H_0^1(\Omega), \end{aligned} \tag{11.16}$$

*then* $X_0$ *belongs to the space* $W^{2,2}(0,T;L^2(\Omega))$, $X_{0t}(\cdot,t) \in H_0^1(\Omega)\ \forall\ t \in [0,T]$, *and problem* $(P.2)_1$ *has a unique solution* $X_1 \in W^{1,2}(0,T;L^2(\Omega))$.

*Proof.* By Theorem 11.1.2 we know that under assumptions $(h)$ and (11.15) problem $P_0$ has a unique solution $X_0 \in W^{1,2}(0,T;L^2(\Omega))$, $X_0(\cdot,t) \in H_0^1(\Omega)\ \forall\, t \in [0,T]$. If, in addition, assumptions (11.16) hold then one can prove by a standard reasoning that $\overline{X}_0 = X_{0t}(\cdot,t)$ satisfies the problem obtained by formal differentiation with respect to $t$ of problem $P_0$:

$$\begin{cases} \overline{X}_0' + L\overline{X}_0 = g \text{ in } (0,T), \\ \overline{X}_0(0) = f(0) - J(u_0), \end{cases}$$

where

$$\begin{aligned} g(t) =& f_t(\cdot,t) - \beta'(X_0(\cdot,t))X_{0t}(\cdot,t),\ t \in (0,T), \\ L : D(L) = H^2(\Omega)&\bigcap H_0^1(\Omega) \to H,\ Lu = -\Delta u\ \forall\ u \in D(L). \end{aligned}$$

Obviously, $g \in L^2(\Omega_T)$ and $L = \partial\psi_0$, where $\psi_0$ is the function defined by (11.6) in which $j \equiv 0$. The last assumption of (11.16) implies that $\overline{X}_0(0) \in D(\psi_0)$. Thus, it follows by Theorem 2.0.24 that $\overline{X}_0 = X_{0t} \in W^{1,2}(0,T;L^2(\Omega)$, $X_{0t}(\cdot,t) \in H_0^1(\Omega)\ \forall\, t \in [0,T]$. Now, we write problem $(P.2)_1$ as the following Cauchy problem in $H$:

$$\begin{cases} X_1' + LX_1 + B(t,X_1(t)) = 0, \text{ in } (0,T), \\ X_1(0) = 0, \end{cases}$$

where $X_1(t) = X_1(\cdot,t)$, $B : [0,T] \times H \to H$,

$$B(t,z) = X_0''(\cdot,t) - \beta'(X_0(\cdot,t))z\ \ \forall\, t \in [0,T],\ z \in H.$$

By Theorem 2.0.28 the above problem has a unique mild solution

$$X_1 \in C([0,T];L^2(\Omega)).$$

Consequently, the function $t \to B(t,X_1(t))$ belongs to $L^2(\Omega_T)$. According to Theorem 2.0.24, the problem

$$\begin{cases} Z'(t) + LZ(t) = -B(t,X_1(t)) \text{ in } (0,T), \\ Z(0) = 0, \end{cases}$$

has a unique strong solution which coincides with $X_1$. Therefore,

$$X_1 \in W^{1,2}(0,T;L^2(\Omega)).$$

□

*Remark* 11.2.3. In the case $n \leq 3$, if we take into account our arguments from Remark 11.1.7, we see that, under assumptions (11.16), the above condition $\beta' \in L^\infty(\mathbb{R})$ can be replaced by a weaker one, namely $\beta' \in L^\infty_{\text{loc}}(\mathbb{R})$, such that the conclusions of Theorem 11.2.2 are preserved.

The following corollary gathers all our results above in an integrated fashion:

**Corollary 11.2.4.** *Assume that* $(h)$ *is satisfied, and*

$$f \in L^2(\Omega_T),\ u_0,\ u_T \in H_0^1(\Omega) \bigcap H^2(\Omega),\ \beta(u_0),\ \beta(u_T) \in L^2(\Omega). \tag{11.17}$$

*Then problems* $(P.2)_\varepsilon$, $\varepsilon > 0$, *and* $P_0$ *have unique solutions*

$$u_\varepsilon \in W^{2,2}(0,T;L^2(\Omega)),\ X_0 \in W^{1,2}(0,T;L^2(\Omega)).$$

*If, in addition,* (11.16) *are fulfilled, then problem* $(P.2)_1$ *has a unique strong solution* $X_1 \in W^{1,2}(0,T;L^2(\Omega))$, *and*

$$u_\varepsilon \in W^{2,2}(0,T;L^2(\Omega)) \bigcap W^{3,2}_{\mathrm{loc}}(0,T;L^2(\Omega)),\ t^{3/2}(T-t)^{3/2}u''' \in L^2(\Omega_T),$$
$$X_0 \in W^{2,2}(0,T;L^2(\Omega)).$$

### 11.2.3 Estimates for the remainder

We conclude this section by proving an estimate for the remainder, which validates completely our first order expansion, as well as other estimates. One of these estimates (see $(11.18)_1$ below) shows that problem $(P.2)_\varepsilon$ is regularly perturbed of order zero with respect to the norm of $C([0,T];L^2(\Omega))$.

**Theorem 11.2.5.** *If* $(h)$ *and* (11.17) *hold, then*

$$\|u_\varepsilon - X_0\|_{C([0,T];L^2(\Omega))} = O(\varepsilon^{1/4}),\ \| u_\varepsilon - X_0 \|_{L^2(0,T;H_0^1(\Omega))} = \mathcal{O}(\varepsilon^{1/2}), \tag{11.18}$$

*and* $u_{\varepsilon t} \to X_{0t}$ *weakly in* $L^2(\Omega_T)$, *where* $u_\varepsilon$ *and* $X_0$ *are the solutions of problems* $(P.2)_\varepsilon$ *and* $P_0$. *If, in addition, assumptions* (11.16) *hold and* $\beta'' \in L^\infty(\mathbb{R})$, *then, for every* $\varepsilon > 0$, $u_\varepsilon$ *admits an asymptotic expansion of the form* (11.13), *and the following estimates hold*

$$\|R_\varepsilon\|_{C([0,T];L^2(\Omega))} = O(\varepsilon^{5/4}),\ \| u_\varepsilon - X_0 - \varepsilon X_1 \|_{L^2(0,T;H_0^1(\Omega))} = \mathcal{O}(\varepsilon^{3/2}). \tag{11.19}$$

*Proof.* Denote $S_\varepsilon = u_\varepsilon - X_0$. By Corollary 11.2.4 we have $S_\varepsilon \in W^{1,2}(0,T;L^2(\Omega))$. Obviously,

$$\begin{cases} \varepsilon u_{\varepsilon tt} - S_{\varepsilon t} + \Delta S_\varepsilon - \beta(u_\varepsilon) + \beta(X_0) = 0 \text{ in } \Omega_T, \\ S_\varepsilon = 0 \text{ on } \Sigma_T, \\ S_\varepsilon(x,0) = 0,\ S_{\varepsilon t}(x,T) = u_T(x) - X_{0t}(x,T),\ x \in \Omega. \end{cases} \tag{11.20}$$

Now, we multiply equation $(11.20)_1$ by $S_\varepsilon(t)$ and then integrate the resulting equation over $\Omega_T$. Thus, by using Green's formula as well as the monotonicity of $\beta$, we obtain

$$\begin{aligned} &\| \nabla S_\varepsilon \|_{L^2(\Omega_T;\mathbb{R}^n)} + \varepsilon \| S_{\varepsilon t} \|^2_{L^2(\Omega_T)} + \frac{1}{2} \| S_\varepsilon(T) \| \\ &\qquad \le \varepsilon\Big( \| S_{\varepsilon t} \|_{L^2(\Omega_T)} \cdot \| X_{0t} \|_{L^2(\Omega_T)} + \| u_T \| \cdot \| S_\varepsilon(T) \| \Big). \end{aligned}$$

This together with the Poincaré inequality leads us to the following estimates

$$\| S_\varepsilon \|_{L^2(\Omega_T)} = \mathcal{O}(\varepsilon^{1/2}), \quad \| S_{\varepsilon t} \|_{L^2(\Omega_T)} = \mathcal{O}(1), \quad \| \nabla S_\varepsilon \|_{L^2(\Omega_T;\mathbb{R}^n)} = \mathcal{O}(\varepsilon^{1/2}),$$

which imply (11.18).

In what follows we suppose that (11.16) hold and that $\beta'' \in L^\infty(\mathbb{R})$. In order to homogenize $(11.14)_3$, we denote

$$\begin{aligned} \overline{R}_\varepsilon(x,t) =& R_\varepsilon(x,t) + \alpha_\varepsilon(x,t), \\ \alpha_\varepsilon(x,t) =& \varepsilon \frac{(T-t)}{T} i_1\Big(x, \frac{T}{\varepsilon}\Big), \quad (x,t) \in \Omega_T. \end{aligned}$$

It is easily seen that $\overline{R}_\varepsilon$ satisfies the following problem

$$\begin{cases} \varepsilon(\overline{R}_\varepsilon - \alpha_\varepsilon + \varepsilon X_1)_{tt} - \overline{R}_{\varepsilon t} + \Delta \overline{R}_\varepsilon \\ -\beta(u_\varepsilon) + \beta(X_0) + \varepsilon\beta'(X_0)X_1 = \overline{h}_\varepsilon \text{ in } \Omega_T, \\ \overline{R}_\varepsilon = 0 \text{ on } \Sigma_T, \\ \overline{R}_\varepsilon(x,0) = 0, \ \overline{R}_{\varepsilon t}(x,T) = -\varepsilon X_{1t}(x,T) + \alpha_{\varepsilon t}(x,T), \ \ x \in \Omega, \end{cases} \tag{11.21}$$

where $\overline{h}_\varepsilon = -\varepsilon \Delta i_1 - \alpha_{\varepsilon t} + \Delta \alpha_\varepsilon$.

Now, we multiply equation $(11.21)_1$ by $\overline{R}_\varepsilon(t)$, then integrate over $\Omega_T$. By a computation similar to that from the first part of the proof, we get

$$\begin{aligned} \| \nabla \overline{R}_\varepsilon \|_{L^2(\Omega_T;\mathbb{R}^n)} &+ \varepsilon \| \overline{R}_{\varepsilon t} \|^2_{L^2(\Omega_T)} + \frac{1}{2} \| \overline{R}_\varepsilon(T) \|^2 \\ &\le \| \overline{R}_\varepsilon \|_{L^2(\Omega_T)} \cdot \| \overline{h}_\varepsilon \|_{L^2(\Omega_T)} \\ &\quad + \varepsilon^2 \| \overline{R}_{\varepsilon t} \|_{L^2(\Omega_T)} \cdot \| X_{1t} \|_{L^2(\Omega_T)} + \varepsilon \| \alpha_{\varepsilon t} \| \cdot \| \overline{R}_\varepsilon(T) \| \\ &\quad + \| \overline{R}_\varepsilon \|_{L^2(\Omega_T)} \cdot \| \beta(u_\varepsilon - \overline{R}_\varepsilon) - \beta(X_0) - \varepsilon\beta'(X_0)X_1 \|_{L^2(\Omega_T)} . \end{aligned} \tag{11.22}$$

Since $\beta', \ \beta'' \in L^\infty(\mathbb{R})$, for every given $j \geq 1$, we have

$$\begin{aligned} &\| \overline{h}_\varepsilon \|_{L^2(\Omega_T)} = \mathcal{O}(\varepsilon^{3/2}), \quad \| \alpha_{\varepsilon t} \|_{L^2(\Omega_T)} = \mathcal{O}(\varepsilon^j), \\ &\| \beta(X_0 + \varepsilon X_1 + \varepsilon i_1 - \alpha_\varepsilon) - \beta(X_0) - \varepsilon \beta'(X_0)X_1 \|_{L^2(\Omega_T)} = \mathcal{O}(\varepsilon^{3/2}). \end{aligned} \tag{11.23}$$

Thus, making use of the Poincaré inequality and (11.22), we find

$$\| \overline{R}_\varepsilon \|_{L^2(\Omega_T)} = \mathcal{O}(\varepsilon^{3/2}), \quad \| \overline{R}_{\varepsilon t} \|_{L^2(\Omega_T)} = \mathcal{O}(\varepsilon), \quad \| \nabla \overline{R}_\varepsilon \|_{L^2(\Omega_T;\mathbb{R}^n)} = \mathcal{O}(\varepsilon^{3/2}),$$

which imply the desired estimates (11.19). □

*Remark* 11.2.6. If it turns out that both $X_0$, $X_1$ belong to $L^\infty(\Omega_T)$, then inequality $(11.23)_3$ holds under the weaker assumption $\beta'' \in L^\infty_{\text{loc}}(\mathbb{R})$. Taking into account the problem satisfied by $X_1$, if $X_{0tt} \in W^{1,2}(0,T;L^2(\Omega))$, we have

$$X_1 \in L^\infty(0,T;H^2(\Omega) \bigcap H^1_0(\Omega)).$$

On the other hand, if the following conditions are added to the set of hypotheses of Theorem 11.2.2

$$f \in W^{2,2}(0,T;L^2(\Omega)),\ f_t(\cdot,0) - \beta'(u_0)(f(\cdot,0) - Ju_0) + \Delta(f(\cdot,0) - Ju_0) \in H_0^1(\Omega),$$

then, indeed,

$$X_0 \in W^{3,2}(0,T;L^2(\Omega)) \bigcap W^{1,\infty}(0,T;H^2(\Omega) \bigcap H_0^1(\Omega)),$$

and therefore, if $n \le 3$, Theorem 11.2.5 remains true if $\beta''$ is assumed to be a function from $L^\infty_{\mathrm{loc}}(\mathbb{R})$ (see also Remark 11.1.7).

## 11.3 Asymptotic analysis of problem $(P.3)_\varepsilon$

In this section we discuss the hyperbolic regularization of problem $P_0$, denoted $(P.3)_\varepsilon$, which comprises the nonlinear equation $(NHE)$, the homogeneous Dirichlet condition $(BC)$ and initial conditions $(IC)_3$, as presented in Chapter 9. Like in the linear case which has been examined in Section 10.3, the present $(P.3)_\varepsilon$ is regularly perturbed of order zero and singularly perturbed of any order $N \ge 1$ with respect to the norm of $C([0,T];H_0^1(\Omega))$, with a boundary layer near the set $\{0\} \times \Omega$.

### 11.3.1 A first order asymptotic expansion

Since problem $(P.3)_\varepsilon$ is nonlinear, we shall restrict our analysis to a first order asymptotic expansion, which is enough to understand the corresponding boundary layer phenomenon. More precisely, based on our experience from the linear case, we seek the solution of problem $(P.3)_\varepsilon$ in the form

$$u_\varepsilon(x,t) = X_0(x,t) + \varepsilon X_1(x,t) + \varepsilon i_1(x,\xi) + R_\varepsilon(x,t),\ (x,t) \in \Omega_T, \tag{11.24}$$

where $\xi = t/\varepsilon$ is the fast variable, and the terms of the expansion have the same meaning as in Subsection 10.3.1. Using the standard matching procedure, we see that $X_0$ should formally satisfy problem $P_0$. In addition, assuming that $X_{0t}(\cdot,0)$ is well defined, we find that $i_1$ satisfies a problem similar to (10.30), i.e.,

$$i_1(x,\xi) = (X_{0t}(x,0) - u_1(x))e^{-\xi},\ x \in \Omega,\ \xi \ge 0.$$

For the first order regular term, we obtain the following problem, denoted $(P.3)_1$:

$$\begin{cases} X_{1t}(x,t) - \Delta X_1(x,t) + \beta'(X_0(x,t))X_1(x,t) = -X_{0tt}(x,t),\ \text{in } \Omega_T, \\ X_1 = 0 \text{ on } \Sigma_T, \\ X_1(x,0) = u_1(x) - X_{0t}(x,0),\ x \in \Omega. \end{cases}$$

Finally, the remainder should satisfy the problem

$$\begin{cases} \varepsilon R_{\varepsilon tt} + R_{\varepsilon t} - \Delta R_\varepsilon + \beta(u_\varepsilon) - \beta(X_0) - \varepsilon\beta'(X_0)X_1 \\ = -\varepsilon^2 X_{1tt} + \varepsilon\Delta i_1 \text{ in } \Omega_T, \\ R_\varepsilon = -\varepsilon i_1 \text{ on } \Sigma_T, \\ R_\varepsilon(x,0) = 0, \ R_{\varepsilon t}(x,0) = -\varepsilon X_{1t}(x,0), \ x \in \Omega. \end{cases} \tag{11.25}$$

### 11.3.2 Existence, uniqueness and regularity of the solutions of problems $(P.3)_\varepsilon$, $P_0$ and $(P.3)_1$

In order to examine problem $(P.3)_\varepsilon$, we assume without any loss of generality that $\varepsilon = 1$. In addition, we shall denote by $u$ the solution of this problem (instead of $u_\varepsilon$). We choose as our framework the real Hilbert space $H_1 = H_0^1(\Omega) \times L^2(\Omega)$, with the scalar product

$$\langle h_1, h_2 \rangle_1 = \int_\Omega \nabla u_1 \cdot \nabla u_2 \, dx + \int_\Omega v_1 v_2 \, dx \ \forall \ h_i = (u_i, v_i) \in H_1, \ i = 1,2,$$

and the corresponding induced norm, denoted $\| \cdot \|_1$.
Define the operator $A : D(A) \subset H_1 \to H_1$,

$$D(A) = \left(H^2(\Omega) \bigcap H_0^1(\Omega)\right) \times L^2(\Omega), \ A(u,v) = (-v, -\Delta u + v) \ \forall \ (u,v) \in H_1.$$

This operator is (linear) maximal monotone, as the Lax-Milgram lemma shows. We also define

$$B : H_1 \to H_1, \ BU = \left(0, \overline{\beta}(u)\right) \ \forall U = (u,v) \in H_1,$$

where $\overline{\beta}(u)$ denotes the composition of functions $\beta$ and $u$. Of course, under assumptions $(i)$, $D(B) = H_1$. With these definitions, we can express problem $(P.3)_\varepsilon$ as the following Cauchy problem in $H_1$:

$$\begin{cases} U'(t) + AU(t) + BU(t) = (0, f(t)), \ t \in (0,T), \\ U(0) = U_0, \end{cases} \tag{11.26}$$

where $U(t) = (u(\cdot,t), u_t(\cdot,t))$, $U_0 = (u_0, u_1)$.

Since $B$ is a Lipschitz perturbation, we can apply Theorem 2.0.20 and Remark 2.0.23 in Chapter 2 to problem (11.26) to derive that this problem has a unique solution $U \in W^{1,\infty}(0,T;H_1)$. Moreover, by a standard device (see, e.g., the proof of Theorem 5.2.3), we obtain

**Theorem 11.3.1.** *Assume that $(i)$ holds and*

$$f \in W^{1,1}(0,T;L^2(\Omega)), \ u_0 \in H^2(\Omega) \bigcap H_0^1(\Omega), \ u_1 \in L^2(\Omega). \tag{11.27}$$

*Then, problem $(P.3)_\varepsilon$ has a unique solution*

$$u_\varepsilon \in C^1([0,T];H_0^1(\Omega)) \bigcap C^2([0,T];L^2(\Omega)) \bigcap C([0,T];H^2(\Omega)).$$

Concerning the solutions $X_0$ and $X_1$ of problems $P_0$ and $(P.3)_1$, we shall need in the next section that $X_1 \in W^{2,2}(0,T;L^2(\Omega)$ and $X_0 \in W^{3,2}(0,T;L^2(\Omega)$. We are going to show that these regularity properties hold under some adequate assumptions on the data. Consider again the Hilbert space $H = L^2(\Omega)$ and the linear self-adjoint operator

$$L : D(L) \subset H \to H,\ D(L) = H^2(\Omega)\bigcap H_0^1(\Omega),\ Lu = -\Delta u\ \ \forall\, u \in D(L).$$

Problem $P_0$ can be expressed as the Cauchy problem in $H$:

$$\begin{cases} X_0'(t) + LX_0(t) + \overline{\beta}(X_0(t)) = f(\cdot,t),\ 0 < t < T, \\ X_0(0) = u_0, \end{cases} \tag{11.28}$$

where $X_0(t) := X_0(\cdot,t)$ and $\overline{\beta}$ is the canonical extension of $\beta$ to $L^2(\Omega)$. On the other hand, problem $(P.3)_1$ reads

$$\begin{cases} X_1'(t) + LX_1(t) + \beta'(X_0(t))X_1(t) = -X_0''(t),\ 0 < t < T, \\ X_1(0) = u_1 - X_0'(0), \end{cases}$$

where $X_1(t) := X_1(\cdot,t),\ t \in [0,T]$.
Now, we are able to state the following result:

**Theorem 11.3.2.** *Assume that assumption (i) holds, and*

$$\begin{aligned} &f \in W^{1,2}(0,T;L^2(\Omega)),\ \beta \in C^1(\mathbb{R}),\ u_0 \in H^2(\Omega)\bigcap H_0^1(\Omega), \\ &f(\cdot,0) + \Delta u_0 - \beta(u_0) =: \zeta \in H_0^1(\Omega). \end{aligned} \tag{11.29}$$

*Then problem $P_0$ has a unique solution*

$$X_0 \in W^{2,2}(0,T;L^2(\Omega)),\ X_0(\cdot,t),\ X_{0t}(\cdot,t)\ \in H_0^1(\Omega)\ \forall t \in [0,T].$$

*If, in addition, $n \le 3$ and*

$$\begin{aligned} &f \in W^{2,2}(0,T;L^2(\Omega)),\ \beta \in C^2(\mathbb{R}), \\ &\zeta \in H^2(\Omega),\ f_t(\cdot,0) - \beta'(u_0)\zeta + \Delta\zeta \in H_0^1(\Omega), \\ &u_1 \in H_0^1(\Omega)\bigcap H^2(\Omega),\ \beta'(u_0)(u_1-\zeta) + \Delta(u_1-\zeta) \in H_0^1(\Omega), \end{aligned} \tag{11.30}$$

*then $X_0$ belongs to the space $W^{3,2}(0,T;L^2(\Omega))$, and problem $(P.3)_1$ has a unique solution $X_1 \in W^{2,2}(0,T;L^2(\Omega))$.*

*Proof.* By Theorem 2.0.20 and Remark 2.0.23 in Chapter 2, we infer that problem (11.28) has a unique strong solution $X_0 \in W^{1,\infty}(0,T;H)$. Moreover, $X_0 \in L^\infty(0,T;H^2(\Omega))$, $X_0(\cdot,t) \in H^2(\Omega)\bigcap H_0^1(\Omega)\ \forall t \in [0,T]$. Since $\beta$ is Lipschitz continuous, it is differentiable almost everywhere and $\beta' \in L^\infty(\mathbb{R})$. Note that $\overline{X}_0 := X_0'$ is the strong solution of the problem

$$\begin{cases} \overline{X}_0'(t) + L\overline{X}_0(t) = -\beta'(X_0(t))X_0'(t) + f'(t), \ 0 < t < T, \\ \overline{X}_0(0) = f(0) - \beta(u_0) - Lu_0 = \zeta, \end{cases} \tag{11.31}$$

since the right-hand side of equation $(11.31)_1$ is a member of $L^2(0,T;H)$ and $\overline{X}_0(0) = \zeta \in D(\psi_0) = H_0^1(\Omega)$ ($\psi_0$ is the function defined by (11.6), where $j$ is the null function), thus $\overline{X}_0 = X_0' \in W^{1,2}(0,T;H)$. In addition, $X_0' \in L^2(0,T;H^2(\Omega))$ and $X_0'(t) \in D(\psi_0) = H_0^1(\Omega)$ $\forall t \in [0,T]$. The first part of the theorem is proved.

In what follows we shall assume in addition that $n \le 3$ and that (11.30) hold. By the Sobolev-Kondrashov theorem, $L^2(0,T;H^2(\Omega)) \subset L^2(0,T;L^\infty(\Omega))$, so the right-hand side of equation $(11.31)_1$ belongs to the space $W^{1,1}(0,T;H)$. Therefore, $X_0' \in C([0,T];H^2(\Omega)) \subset C([0,T];L^\infty(\Omega))$. Thus, $\widetilde{X}_0 := X_0''$ is the strong solution of the Cauchy problem

$$\begin{cases} \widetilde{X}_0'(t) + L\widetilde{X}_0(t) = -\frac{d^2}{dt^2}(\beta(X_0(t))) + f''(t), \ 0 < t < T, \\ \widetilde{X}_0(0) = -\beta'(u_0)\zeta - L(\zeta) + f'(0), \end{cases}$$

and belongs to $W^{1,2}(0,T;H)$.
Analogously, one can prove that problem $(P.3)_1$ has a solution $X_1 \in W^{2,2}(0,T;H)$. □

*Remark* 11.3.3. One can easily formulate sufficient separate conditions on $u_0$, $u_1$, $\beta$, $f$ such that all assumptions (11.30) are fulfilled.

Taking into account the above results, we can state the following

**Corollary 11.3.4.** *Assume that (i) is satisfied, and*

$$\begin{aligned} &f \in W^{1,2}(0,T;L^2(\Omega)), \ \beta \in C^1(\mathbb{R}), \ u_0 \in H_0^1(\Omega) \bigcap H^2(\Omega), \\ &u_1 \in L^2(\Omega), \ f(\cdot,0) + \Delta u_0 - \beta(u_0) \in H_0^1(\Omega). \end{aligned} \tag{11.32}$$

*Then, problems $(P.3)_\varepsilon$, $\varepsilon > 0$, and $P_0$ have unique solutions*

$$u_\varepsilon \in C^2([0,T];L^2(\Omega)) \bigcap C^1([0,T];H_0^1(\Omega)) \bigcap C([0,T];H^2(\Omega)),$$

$$X_0 \in W^{2,2}(0,T;L^2(\Omega)).$$

*If, in addition, $n \le 3$ and (11.30) are fulfilled, then problem $(P.3)_1$ has a unique strong solution $X_1 \in W^{2,2}(0,T;L^2(\Omega))$, and $X_0 \in W^{3,2}(0,T;L^2(\Omega))$.*

### 11.3.3 Estimates for the remainder

We conclude this section by establishing some estimates for the difference $u_\varepsilon - X_0$ and for the first order remainder as well. They validate completely our first order expansion (11.24).

**Theorem 11.3.5.** *If (i) and (11.32) hold, then, the solution $u_\varepsilon$ of problem $(P.3)_\varepsilon$ satisfies*

$$\|u_\varepsilon - X_0\|_{C([0,T];H_0^1(\Omega))} = O(\varepsilon^{1/2}), \quad \| u_{\varepsilon t} - X_{0t} \|_{L^2(\Omega_T)} = \mathcal{O}(\varepsilon^{1/2}). \tag{11.33}$$

*If, in addition, $n \leq 3$, assumptions (11.30) hold and $\beta'' \in L^\infty(\mathbb{R})$, then, for every $\varepsilon > 0$, $u_\varepsilon$ admits an asymptotic expansion of the form (11.24), and the following estimates hold*

$$\|R_\varepsilon\|_{C([0,T];H_0^1(\Omega))} = O(\varepsilon^{3/2}), \quad \| R_{\varepsilon t} \|_{L^2(\Omega_T)} = \mathcal{O}(\varepsilon^{3/2}). \tag{11.34}$$

*Proof.* Denote $S_\varepsilon = u_\varepsilon - X_0$. By Corollary 11.3.4, $S_\varepsilon \in W^{2,2}(0,T;L^2(\Omega))$. Using $P_0$ and $(P.3)_\varepsilon$, we see that

$$\begin{cases} \varepsilon S_{\varepsilon tt} + S_{\varepsilon t} - \Delta S_\varepsilon + \beta(u_\varepsilon) - \beta(X_0) = -\varepsilon X_{0tt} \text{ in } \Omega_T, \\ S_\varepsilon = 0 \text{ on } \Sigma_T, \\ S_\varepsilon(x,0) = 0, \ S_{\varepsilon t}(x,0) = u_1(x) - X_{0t}(x,0), \ x \in \Omega. \end{cases}$$

Now, denote $\overline{S}_\varepsilon = e^{-\gamma_0 t} S_\varepsilon$, where $\gamma_0 > 0$ will be chosen later. A simple computation yields

$$\begin{aligned} &\varepsilon \overline{S}_{\varepsilon tt} + (1 + 2\varepsilon\gamma_0)\overline{S}_{\varepsilon t} + (\gamma_0 + \varepsilon\gamma_0^2)\overline{S}_\varepsilon - \Delta \overline{S}_\varepsilon \\ &\quad + e^{-\gamma_0 t}(\beta(u_\varepsilon) - \beta(X_0)) = \overline{h}_\varepsilon \text{ in } \Omega_T, \end{aligned} \tag{11.35}$$

where $\overline{h}_\varepsilon = -\varepsilon e^{-\gamma_0 t} X_{0tt}$.
Scalar multiplication in $H$ of (11.35) by $\overline{S}_{\varepsilon t}$ and Green's formula lead us to

$$\begin{aligned} &\frac{\varepsilon}{2}\frac{d}{dt} \| \overline{S}_{\varepsilon t}(\cdot,t) \|^2 + \| \overline{S}_{\varepsilon t}(\cdot,t) \|^2 + \frac{\gamma_0}{2}\frac{d}{dt} \| \overline{S}_\varepsilon(\cdot,t) \|^2 \\ &\quad + \langle \nabla \overline{S}_\varepsilon(\cdot,t), \nabla \overline{S}_{\varepsilon t}(\cdot,t) \rangle_{L^2(\Omega;\mathbb{R}^n)} \\ &\quad + e^{-\gamma_0 t} \langle \beta(u_\varepsilon(\cdot,t)) - \beta(X_0(\cdot,t)), \overline{S}_{\varepsilon t}(\cdot,t) \rangle \\ &\leq \langle \overline{h}_\varepsilon(\cdot,t)), \overline{S}_{\varepsilon t}(\cdot,t) \rangle \text{ for a.a. } t \in (0,T), \end{aligned}$$

which yields by integration over $[0,t]$

$$\begin{aligned} &\frac{\varepsilon}{2} \| \overline{S}_{\varepsilon t}(\cdot,t) \|^2 + \int_0^t \| \overline{S}_{\varepsilon t}(\cdot,s) \|^2 \, ds \\ &\quad + \frac{\gamma_0}{2} \| \overline{S}_\varepsilon(\cdot,t) \|^2 + \frac{1}{2} \| \nabla \overline{S}_\varepsilon(\cdot,t) \|^2_{L^2(\Omega;\mathbb{R}^n)} \\ &\leq \frac{1}{2} \int_0^t \| \overline{h}_\varepsilon(\cdot,s) \|^2 \, ds + \frac{1}{2} \int_0^t \| \overline{S}_{\varepsilon t}(\cdot,s) \|^2 \, ds \\ &\quad + \frac{\varepsilon}{2} \| u_1 - X_{0t}(\cdot,0) \|^2 \\ &\quad + \int_0^t e^{-\gamma_0 s} \| \beta(u_\varepsilon(\cdot,s)) - \beta(X_0(\cdot,s)) \| \cdot \| \overline{S}_{\varepsilon t}(\cdot,s) \| \, ds \end{aligned} \tag{11.36}$$

for all $t \in [0,T]$, where we have denoted by $\|\cdot\|$, $\langle\cdot,\cdot\rangle$ the usual norm and scalar product of $L^2(\Omega)$. Since

$$\|\overline{h}_\varepsilon \|_{L^2(\Omega_T)} = \mathcal{O}(\varepsilon),$$

$$\begin{aligned}
\int_0^t & e^{-\gamma_0 s} \| \beta(u_\varepsilon(\cdot,s)) - \beta(X_0(\cdot,s)) \| \cdot \| \overline{S}_{\varepsilon t}(\cdot,s) \| \; ds \\
&\le l \int_0^t \| \overline{S}_\varepsilon(\cdot,s) \| \cdot \| \overline{S}_{\varepsilon t}(\cdot,s) \| \; ds \\
&\le \frac{1}{4} \int_0^t \| \overline{S}_{\varepsilon t}(\cdot,s) \|^2 \, ds + l^2 T \| \overline{S}_\varepsilon \|^2_{C([0,T];L^2(\Omega))}
\end{aligned} \tag{11.37}$$

for all $t \in [0,T]$, we can derive from (11.36) that

$$\begin{aligned}
\frac{1}{4} \| \overline{S}_{\varepsilon t} \|^2_{L^2(\Omega_T)} &+ \frac{\gamma_0}{2} \| \overline{S}_\varepsilon \|^2_{C([0,T];L^2(\Omega)} + \frac{1}{2} \| \nabla \overline{S}_\varepsilon \|^2_{C([0,T];L^2(\Omega;\mathbb{R}^n))} \\
&\le M\varepsilon + l^2 T \| \overline{S}_\varepsilon \|^2_{C([0,T];L^2(\Omega))},
\end{aligned} \tag{11.38}$$

where $M$ is a positive constant, independent of $\varepsilon$. Now, if we choose $\gamma_0 = 2l^2T$, it follows

$$\| \overline{S}_{\varepsilon t} \|_{L^2(\Omega_T)} = \mathcal{O}(\varepsilon^{1/2}), \quad \| \overline{S}_\varepsilon \|_{C([0,T];H_0^1(\Omega))} = \mathcal{O}(\varepsilon^{1/2}),$$

which lead to (11.33).

The proof of the last part of the theorem can be done by similar arguments. □

*Remark* 11.3.6. If $u_1 = f(\cdot,0) + \Delta u_0 - \beta(u_0)$ in $\Omega$, then $S_{\varepsilon t}(\cdot,0) = 0$, so it follows by (11.36) that

$$\begin{aligned}
\|u_\varepsilon - X_0\|_{C([0,T];H_0^1(\Omega))} &= \mathcal{O}(\varepsilon), \; \| u_{\varepsilon t} - X_{0t} \|_{C([0,T];L^2(\Omega))} = \mathcal{O}(\varepsilon^{1/2}), \\
\| u_{\varepsilon t} - X_{0t} \|_{L^2(\Omega_T)} &= \mathcal{O}(\varepsilon).
\end{aligned}$$

*Remark* 11.3.7. If we are interested in a zeroth order approximation only, then obviously less assumptions on the data are needed. It is just a simple exercise to formulate such assumptions. On the other hand, all these results extend to the more general case $\beta = \beta(t,x)$, under adequate conditions.

# Bibliography


[1] R.A. Adams, *Sobolev Spaces*, Academic Press, New York, 1975.

[2] A.R. Aftabizadeh, N.H. Pavel, *Boundary value problems for second order differential equations and a convex problem of Bolza*, Diff. Integral Eqns., 2 (1989), 495–509.

[3] N.U. Ahmed, *Semigroup Theory with Applications to Systems and Control*, Longman Scientific & Technical, New York, 1991.

[4] N. Apreutesei-Dumitriu, G. Moroşanu, *Second Order Differential Equations in Hilbert Spaces*, MatrixRom, Bucharest, 2000 (in Romanian).

[5] H. Attouch, A. Damlamian *Strong solutions for parabolic variational inequalities*, Nonlinear Anal. TMA, 2 (1978), no. 3, 329–353.

[6] L. Barbu, G. Moroşanu, *Asymptotic Analysis of Some Nonlinear Boundary Value Problems*, Editura Academiei, Bucharest, 2000 (in Romanian).

[7] V. Barbu, *Nonlinear Semigroups and Differential Equations in Banach Spaces*, Noordhoff, Leyden, 1976.

[8] V. Barbu, *Nonlinear boundary-value problems for a class of hyperbolic systems*, Rev. Roum. Math. Pures Appl., 22 (1977), 155–168.

[9] V. Barbu, *Analysis and control of nonlinear infinite dimensional systems*, Academic Press, 1993.

[10] H. Brézis, *Opérateurs maximaux monotones et semigroupes de contractions dans les espaces de Hilbert*, North-Holland, Amsterdam, 1973.

[11] H. Brézis, *Analyse fonctionnelle. Théorie et applications*, Masson, Paris, 1983.

[12] R.E. Bruck, *Periodic forcing of solutions of a boundary value problem for a second order differential equations in Hilbert spaces*, J. Math. Anal. Appl. 76(1980), 159–173.

[13] C. Coclici, G. Moroşanu, W.L. Wendland, *One-dimensional singularly perturbed coupled boundary value problems*, Math. Sci. Res. Hot-Line 3 (1999), no. 10, 1–21.

[14] C. Coclici, G. Moroşanu, W.L. Wendland, *On the viscous-viscous and the viscous-inviscid interaction in Computational Fluid Dynamics*, Computing and Visualization in Sci., 2 (1999), 95–105.

[15] K.L. Cooke, D.W. Krumme, *Differential-difference equations and nonlinear initial-boundary value problems for linear hyperbolic partial differential equations*, J. Math. Anal. Appl., 24 (1968), 372–387.

[16] A.K. Datta, *Biological and Bioenvironmental Heat and Mass Transfer*, Marcel Dekker, 2002.

[17] V. Drăgan, A. Halanay, *Perturbaţii singulare. Dezvoltări asimptotice*, Editura Academiei, Bucharest, 1983 (in Romanian).

[18] W. Eckhaus, *Matched Asymptotic Expansions and Singular Perturbations*, North-Holland, 1973.

[19] L.C. Evans, *Partial Differential Equations*, Graduate Studies in Mathematics, vol. 19, AMS, Providence, Rhode Island, 1998.

[20] F. Gastaldi, A. Quarteroni, *On the coupling of hyperbolic and parabolic systems: Analytical and numerical approach*, Appl. Numerical Math., 6 (1989/1990), 3–31.

[21] G.R. Goldstein, J.A. Goldstein, *Regularity for semilinear abstract Cauchy problems*, Theory and applications of nonlinear operators of accretive and monotone type, Lecture Notes in Pure and Appl. Math., 178, Dekker, New York, 1996, 99–105.

[22] J.A. Goldstein, *Semigroups of Linear Operators and Applications*, Oxford University Press, New York, 1985.

[23] V. Hara, *Numerical Approximation of Asymptotic Behaviour of Distributed Parameters Circuits with Transistors*, Doctoral Thesis, Report 42, Department of Mathematics, University of Jyväskylä, Finland, 1988.

[24] V.-M. Hokkanen and G. Moroşanu, *Functional Methods in Differential Equations*, Chapman and Hall/CRC, 2002.

[25] V. Iftimie, *Sur un problème non linéaire pour un système hyperbolique*, Rev. Roum. Math. Pures Appl., 20 (1976), 303–328.

[26] A.M. Ilin, *Matching of Asymptotic Expansions of Solutions of Boundary Value Problems*, American Mathematical Society, Providence, RI, 1992.

[27] E.M. de Jager, *The Theory of Singular Perturbations*, North-Holland, 1996.

[28] T. Kato, *Nonlinear semigroups and evolutions equations*, J. Math. Soc. Japan, 19 (1967), 508–520.

[29] J. Kervorkian, J.D. Cole, *Multiple Scale and Singular Perturbation Methods*, Springer, 1996.

[30] S. Kesavan, *Topics in Functional Analysis and Applications*, John Wiley & Sons, 1989.

[31] J.L. Lions, *Quelques problèmes de la théorie des équations nonlinéaires d'evolution*, Problems in nonlinear analysis, C. I. M. E. Varenna 1970, Ed. Cremonese, 1971, 191–342.

[32] J.L. Lions, *Perturbations singulières dans les problèmes aux limites et en contrôl optimal*, Lecture Notes in Math., vol. 323, Springer, Berlin, 1973.

[33] C.A. Marinov, A. Lehtonen, *Mixed-type circuits with distributed and lumped parameters*, IEEE Trans. Circuits and Systems CAS-36(1989), 1080–1086.

[34] G. Moroşanu, *Nonlinear Evolution Equations and Applications*, D. Reidel, Dordrecht, 1988.

[35] G. Moroşanu, A. Perjan, *The singular limit of telegraph equations*, Comm. Appl. Nonlinear Anal., 3 (1998), no. 1, 101–106.

[36] R.E. O'Malley, *Introduction to Singular Perturbations*, Lecture Notes, Univ. of Edinburgh, 1972.

[37] R.E. O'Malley, *Introduction to Singular Perturbations*, Applied Mathematics and Mechanics, vol. 14, Academic Press, New York-London, 1974.

[38] A. Pazy, *Semigroups of Linear Operators and Applications to Partial Differential Equations*, Springer-Verlag, New York, 1983.

[39] L. Schwartz, *Théorie des distributions à valeurs vectorielles*, Ann. Inst. Fourier, VII, VIII, 1958.

[40] L. Schwartz, *Théorie des distributions*, Herman, Paris, 1967.

[41] I. Straskraba, V. Lovicar, *Qualitative behavior of solutions to* $2 \times 2$ *linear hyperbolic system with nonlinear boundary conditions*, SAACM-Vol.3 (1993), no. 4, 277–290.

[42] V.L. Streeter, E.B. Wylie, *Hydraulic Transients*, McGraw-Hill, New York, 1967.

[43] A.N. Tikhonov, *The dependence of the solutions of differential equations on a small parameter*, Math. Sb., 22 (1948), 193–204 (in Russian).

[44] A.N. Tikhonov, *On systems of differential equations containing parameters*, Math. Sb., 27 (1950), 147–156 (in Russian).

[45] A.N. Tikhonov, *Systems of differential equations containing small parameters multiplying some of derivatives*, Math. Sbornik, 31 (1952), 575–586 (in Russian).

[46] V.A. Trenogin, *Development and Applications of the Lyusternik-Vishik Asymptotic Method*, Russian Math. Surveys, 25 (1970), 119–156.

[47] A.B. Vasilieva, V.F. Butuzov, *Asymptotic Expansions of Solutions of Singularly Perturbed Equations*, Nauka, Moscow, 1973 (in Russian).

[48] A.B. Vasilieva, V.F. Butuzov, L.V. Kalachev, *The Boundary Function Method for Singular Perturbation Problems*, SIAM, Philadelphia, 1995.

[49] M.I. Vishik, *Asymptotic Behaviour of Solutions of Evolutionary Equations*, Cambridge University Press, 1992.

[50] M.I. Vishik, L.A. Lyusternik, *Regular degeneration and boundary layer for linear differential equations with a small parameter multiplying the highest derivatives*, Usp. Mat. Nauk, 12 (1957), 3–122 (in Russian).

[51] A.M. Watts, *A Singular Perturbation Problem with a Turning Point*, Bull. Austral. Math. Soc., 5, 1971, 61–73.

[52] K. Yosida, *Functional Analysis*, Third Ed., Springer-Verlag, Berlin-Heidelberg-New York, 1971.

# Index

# International Series of Numerical Mathematics (ISNM)

Edited by

**Karl-Heinz Hoffmann**, TU München, Germany

**Hans D. Mittelmann**, Arizona State University, Tempe, USA

BIRKHÄUSER

ISNM is open to all aspects of numerical mathematics. Some of the topics of particular interest include free boundary value problems for differential equations, phase transitions, problems of optimal control and optimization, other nonlinear phenomena in analysis, nonlinear partial differential equations, efficient solution methods, bifurcation problems and approximation theory. When possible, the topic of each volume is discussed from three different angles, namely those of mathematical modeling, mathematical analysis, and numerical case studies.

**ISNM 155: Kunisch, K. / Leugering, G. / Sprekels, J. / Tröltsch, F.** (Eds.), Control of Coupled Partial Differential Equations (2007). ISBN 978-3-7643-7720-5

**ISNM 154: Figueiredo, I.N. / Rodrigues, J.F. / Santos, L.** (Eds.), Free Boundary Problems. Theory and Applications (2006). ISBN 978-3-7643-7718-2

**ISNM 153: Roubícek, T.**, Nonlinear Partial Differential Equations with Applications (2005). ISBN 978-3-7643-7293-4

**ISNM 152: Nowak, I.**, Relaxation and Decomposition Methods for Mixed Integer Nonlinear Programming (2005). ISBN 978-3-7643-7238-5

**ISNM 151: Mache, D.H. / Szabados, J. / de Bruin, M.G.** (Eds.), Trends and Applications in Constructive Approximation (2005). ISBN 978-3-7643-7124-1

**ISNM 150: Rohwer, C.**, Nonlinear Smoothing and Multiresolution Analysis (2005). ISBN 978-3-7643-7229-3

**ISNM 149: Voigt, A.** (Ed.), Multiscale Modeling in Epitaxial Growth (2005). ISBN 978-3-7643-7208-8

**ISNM 148: Lagnese, J.E. / Leugering, G.**, Domain Decomposition in Optimal Control of Partial Differential Equations (2004). ISBN 978-3-7643-2194-9

**ISNM 147: Colli, P. / Verdi, C. / Visintin, A.** (Eds.), Free Boundary Problems. Theory and Applications (2003). ISBN 978-3-7643-2193-2

**ISNM 146: Antreich, K. / Bulirsch, R. / Gilg, A. / Rentrop, P.** (Eds.), Modeling, Simulation and Optimization of Integrated Circuits (2003). ISBN 978-3-7643-2192-5

**ISNM 145: Haussmann, W. / Jetter, K. / Reimer, M. / Stöckler, J.** (Eds.), Modern Developments in Multivariate Approximation (2003). ISBN 978-3-7643-2195-6

**ISNM 144: Reimer, M.**, Multivariate Polynomial Approximation (2003). ISBN 978-3-7643-1638-9

**ISNM 143: Desch, W. / Kappel, F. / Kunisch, K.** (Eds.), Control and Estimation of Distributed Parameter Systems (2002). ISBN 978-3-7643-7004-6

**ISNM 142: Buhmann, M.D. / Mache, D.** (Eds.), Advanced Problems in Constructive Approximation (2003). ISBN 978-3-7643-6648-3

**ISNM 141: Freistühler, H. / Warnecke, G.** (Eds.), Hyperbolic Problems: Theory, Numerics, Applications. Volume II (2001). ISBN 978-3-7643-6710-7

**ISNM 140: Freistühler, H. / Warnecke, G.** (Eds.), Hyperbolic Problems: Theory, Numerics, Applications. Volume I (2001). ISBN 978-3-7643-6709-1

**ISNM 140/141:** (Set) ISBN 978-3-7643-6711-4

**ISNM 139: Hoffmann, K.-H. / Lasiecka, I. / Leugering, G. / Sprekels, J.** (Eds.), Optimal Control of Complex Structures (2001). ISBN 978-3-7643-6682-7

**ISNM 138: Hoffmann, K.-H. / Hoppe, R.H.W. / Schulz, V.** (Eds.), Fast Solution of Discretized Optimization Problems (2001). ISBN 978-3-7643-6599-8

**ISNM 137: Haussmann, W. / Jetter, K. / Reimer, M.** (Eds.), Recent Progress in Multivariate Approximation (2001). ISBN 978-3-7643-6505-9

GPSR Compliance

*The European Union's (EU) General Product Safety Regulation (GPSR) is a set of rules that requires consumer products to be safe and our obligations to ensure this.*

*If you have any concerns about our products, you can contact us on ProductSafety@springernature.com*

In case Publisher is established outside the EU, the EU authorized representative is:

Springer Nature Customer Service Center GmbH
Europaplatz 3
69115 Heidelberg, Germany

Batch number: 10578049

Printed by Printforce, the Netherlands